# ADVANCES IN PROTEIN CHEMISTRY AND STRUCTURAL BIOLOGY

Volume 83

Protein Structure and Diseases

# ADVANCES IN PROTEIN CHEMISTRY AND STRUCTURAL BIOLOGY

Protein Structure and Diseases

EDITED BY

ROSSEN DONEV
Institute of Life Science
College of Medicine, Swansea University
Swansea
United Kingdom

MIDDLEBURY COLLEGE LIBRARY

AMSTERDAM • BOSTON • HEIDELBERG • LONDON
NEW YORK • OXFORD • PARIS • SAN DIEGO
SAN FRANCISCO • SINGAPORE • SYDNEY • TOKYO
Academic Press is an imprint of Elsevier

Academic Press is an imprint of Elsevier
The Boulevard, Langford Lane, Kidlington, Oxford OX5 1GB, UK
225 Wyman Street, Waltham, MA 02451, USA
525 B Street, Suite 1900, San Diego, CA 92101-4495, USA

First edition 2011

Copyright © 2011 Elsevier Inc. All rights reserved.

No part of this publication may be reproduced, stored in a retrieval system or transmitted in any form or by any means electronic, mechanical, photocopying, recording or otherwise without the prior written permission of the publisher.

Permissions may be sought directly from Elsevier's Science & Technology Rights Department in Oxford, UK:
phone: (+44) (0) 1865 843830; fax: (+44) (0) 1865 853333;
email: permissions@elsevier.com.

Alternatively you can submit your request online by visiting the Elsevier web site at http://elsevier.com/locate/permissions, and selecting, *Obtaining permission to use Elsevier material.*

Notice
No responsibility is assumed by the publisher for any injury and/or damage to persons or property as a matter of products liability, negligence or otherwise, or from any use or operation of any methods, products, instructions or ideas contained in the material herein. Because of rapid advances in the medical sciences, in particular, independent verification of diagnoses and drug dosages should be made.

ISBN: 978-0-12-381262-9
ISSN: 1876-1623

For information on all Academic Press publications
visit our website at www.elsevierdirect.com

Printed and bound in USA
11  12  13  14     10  9  8  7  6  5  4  3  2  1

Working together to grow
libraries in developing countries

www.elsevier.com | www.bookaid.org | www.sabre.org

ELSEVIER    BOOK AID International    Sabre Foundation

## Contents

### Graphical Representation and Mathematical Characterization of Protein Sequences and Applications to Viral Proteins

Ambarnil Ghosh and Ashesh Nandy

| | |
|---|---|
| I. Introduction | 2 |
| II. Graphical Methods | 9 |
| III. Application to Viral Proteins | 21 |
| IV. Conclusion | 32 |
| References | 32 |

### Structural, Thermodynamic, and Mechanistical Studies in Uroporphyrinogen III Synthase: Molecular Basis of Congenital Erythropoietic Porphyria

Arola Fortian, David Castaño, Esperanza Gonzalez, Ana Laín, Juan M. Falcon-Perez, and Oscar Millet

| | |
|---|---|
| I. Introduction | 44 |
| II. Uroporphyrinogen III Synthase | 47 |
| III. Molecular Basis of CEP | 54 |
| IV. The Hotspot Mutation C73R-UROIIIS | 62 |
| V. Treatment of CEP | 66 |
| VI. Concluding Remarks | 69 |
| References | 71 |

### Role of Fibrin Structure in Thrombosis and Vascular Disease

Amy L. Cilia La Corte, Helen Philippou, and Robert A. S. Ariëns

| | |
|---|---|
| I. Introduction | 76 |
| II. The Coagulation Cascade | 77 |

| III. | Fibrinogen Structure and Function | 81 |
|---|---|---|
| IV. | Factor XIII Structure and Function | 84 |
| V. | Clot Formation and Function | 86 |
| VI. | Fibrinolysis | 91 |
| VII. | Elastic Properties of Fibrin | 94 |
| VIII. | Heterogeneity in Coagulation and Fibrin Structure | 96 |
| IX. | Environmental Factors and Fibrin Structure | 102 |
| X. | Fibrin Density and Thrombosis | 105 |
| XI. | Interactions with Cells and Wound Healing | 107 |
| XII. | Perspectives | 109 |
| | References | 110 |

## Structural, Dynamic, and Functional Aspects of Helix Association in Membranes: A Computational View

ANTON A. POLYANSKY, PAVEL E. VOLYNSKY, AND ROMAN G. EFREMOV

| I. | Introduction | 130 |
|---|---|---|
| II. | Prediction of Structure of HH Dimers | 133 |
| III. | Free Energy of TM Helix–Helix Association | 141 |
| IV. | Combination of Modeling and Experimental Techniques | 146 |
| V. | Limitations and Shortcomings of the Computational Methods | 148 |
| VI. | Comparison with NMR Models: How Reliable Is The Reference? | 150 |
| VII. | Is an Accurate Structure Prediction Always Required? | 151 |
| VIII. | What Makes TM Helices Suitable Pharmacological Targets? | 152 |
| IX. | Modeling of TM Helical Dimers as a First Step Toward 3D Structure of Polytopic MPs | 152 |
| X. | From Structure and Thermodynamics to Function and Design | 153 |
| XI. | Conclusion | 154 |
| | References | 155 |

## Proteins Move! Protein Dynamics and Long-Range Allostery in Cell Signaling

ZIMEI BU AND DAVID J. E. CALLAWAY

| | |
|---|---:|
| I. Introduction | 164 |
| II. NHERF1 Modulates the Macromolecular Assembly, Cell Surface Retention, and Subcellular Localization of Membrane Proteins | 167 |
| III. Signal Transduction by Allosteric Scaffolding Protein Interactions | 171 |
| IV. Structural Basis of Autoinhibition and Long-Range Allostery in NHERF1 | 178 |
| V. Dynamic Propagation of Allosteric Signals by Nanoscale Protein Motion | 186 |
| VI. Summary and Perspective | 213 |
| References | 214 |

## Structural Diversity of Class I MHC-like Molecules and its Implications in Binding Specificities

MD. IMTAIYAZ HASSAN AND FAIZAN AHMAD

| | |
|---|---:|
| I. Introduction | 224 |
| II. Sequence Analysis | 227 |
| III. Structure Analysis | 239 |
| IV. Glycosyalation | 258 |
| V. Conclusion | 261 |
| References | 262 |

| | |
|---|---:|
| AUTHOR INDEX | 271 |
| SUBJECT INDEX | 297 |

# GRAPHICAL REPRESENTATION AND MATHEMATICAL CHARACTERIZATION OF PROTEIN SEQUENCES AND APPLICATIONS TO VIRAL PROTEINS

By AMBARNIL GHOSH* AND ASHESH NANDY[†]

*Physics Department, Jadavpur University, Jadavpur, Kolkata, India
[†]Centre for Interdisciplinary Research and Education, Jodhpur Park, Kolkata, India

| | | |
|---|---|---|
| I. | Introduction | 2 |
| | A. Protein Basics | 2 |
| | B. Drugs and Proteins | 5 |
| | C. Bioinformatics in Protein Studies | 7 |
| II. | Graphical Methods | 9 |
| | A. Graphical Methods for DNA Sequences | 10 |
| | B. Graphical Methods for Protein Sequences | 15 |
| III. | Application to Viral Proteins | 21 |
| | A. Unique Features of Viral Proteins | 21 |
| | B. Two Viral Examples: The Avian and Swine Flu Viruses and the SARS Coronavirus | 23 |
| | C. Graphical Representation Methods in Viral Studies | 25 |
| | D. Results for the Coronavirus and the Flu Viruses | 28 |
| IV. | Conclusion | 32 |
| | References | 32 |

## Abstract

Graphical representation and numerical characterization (GRANCH) of nucleotide and protein sequences is a new field that is showing a lot of promise in analysis of such sequences. While formulation and applications of GRANCH techniques for DNA/RNA sequences started just over a decade ago, analyses of protein sequences by these techniques are of more recent origin. The emphasis is still on developing the underlying technique, but significant results have been achieved in using these methods for protein phylogeny, mass spectral data of proteins and protein serum profiles in parasites, toxicoproteomics, determination of different indices for use in QSAR studies, among others. We briefly mention these in this chapter, with some details on protein phylogeny and viral diseases. In particular, we cover a systematic method developed in GRANCH to determine conserved surface exposed peptide segments in selected viral proteins that can be used for drug and vaccine targeting. The new

GRANCH techniques and applications for DNAs and proteins are covered briefly to provide an overview to this nascent field.

## I. Introduction

### A. Protein Basics

Proteins, the most versatile macromolecules in the living system, primarily constitute complex folded chain of amino acids which are encoded by genes. The information content of the folded complex constitutes a functional unit that plays a crucial role in biological processes. The origin of the word "Protein" is from the Greek "prota" which means "of primary importance." This name was coined by Jöns Jakob Berzelius in 1838 for large organic compounds with a very close similarity in their empirical formulae and of primary importance in animal nutrition, though the evidences were not so prominent at that time. A landmark in protein chemistry came through Frederick Sanger and his colleagues, at the University of Cambridge in 1954 when, after 10 years of hard work, they succeeded in solving the complete primary structure of insulin (Sanger and Tuppy, 1951; Sanger, 1952; Sanger and Thompson, 1953). The very next milestone in protein chemistry was Max Perutz (Perutz and Weisz, 1947; Perutz, 1960; Perutz et al., 1960) and Sir John Cowdery Kendrew (Kendrew et al., 1958, 1960; Kendrew, 1959) solving the 3D structure of hemoglobin and myoglobin. These findings are the basis of modern age of advanced structural protein chemistry research.

Proteins form the building blocks of the structure and function of biological entities. A typical mammalian cell contains as many as 10,000 different proteins having a diverse array of functions (Karp, 2008). The set of proteins expressed in a cell or cell type is called a proteome. Proteins are generally a few hundred amino acids in chain length but can vary in size from a few tens of amino acids to over 34,000 amino acids, for example, the human titins, also known as the largest in protein world (MW = 3–3.7 MDa; Opitz et al., 2003). While a single protein chain can theoretically fold in an unlimited number of ways (Chou and Fasman, 1974b; Fasman, 1989; Feldman and Hogue, 2002), typically a specific amino acid chain folds to a particular structure through a process that is not yet clearly understood (Dill et al., 2007, 2008; Ghosh et al., 2007), but which is the basis for all protein interactions; recent research shows that the folded structure might have conformational changes depending on the environment too (Makowski et al., 2008). Protein structure is often referred to in terms of four aspects: The primary structure consisting of the amino acid chain, the secondary structure which contains regularly repeating structures like alpha helices and beta sheets stabilized by hydrogen bonds, the tertiary structure which is the final folded structure incorporating the various secondary structures, and a quaternary structure where several proteins are bound together to form one protein complex such as are found in the neuraminidase body of an influenza virion (Russell et al., 2006) or the VP7 of a rotavirus particle (Li et al., 2009b). The tertiary and quaternary structures of a large number of proteins have become available through X-ray crystallography and NMR spectroscopy studies and the data are available in Protein Data Bases (PDB) such as World Wide Protein Data Bank (WWPDB; Berman et al., 2007), RCSB Protein Data Bank (RCSB-PDB; Deshpande et al., 2005; Dutta et al., 2007), Protein Data Bank Europe (PDBe; Velankar et al., 2010), Protein Data Bank Japan (PDBj; Nakamura et al., 2002; Kinjo et al., 2010), and Biological Magnetic Resonance Databank (BMRB; Markley et al., 2008). The difficulty of crystallizing proteins has restricted the number of proteins whose structures are sufficiently well known (Chayen, 2004, 2009; Chayen and Saridakis, 2008). However, taking the protein primary structure as the source material for all subsequent structures, structural genomics and protein structure prediction methods theoretically predict protein secondary and tertiary structures based on known structures (Baker and Sali, 2001).

The importance of proteins in biological function have led to wide ranging studies to understand how proteins fold (Dobson, 2004; Dill et al., 2007; Ghosh et al., 2007), interact with other proteins to regulate enzyme activity (Frieden, 1971), oligomerize to form fibrils (Powers and Powers, 2008), aggregate to protein complexes that lead to conformational changes, and enable signaling networks. These interactions are mediated by the chief characteristic of a protein: the ability to bind other molecules specifically and tightly to it. The specificity arises from unique shapes in the tertiary structure of the protein surface (Roach et al., 2005, 2006) where, for example, a depression acts as a binding site or pocket and by the chemical natures of the side chains of the neighboring amino acids. This also results in total inability to bind in cases where changes in the amino acid composition render conformational changes to the binding site (Moscona, 2005). Such changes arising out of mutations in the amino acid chains are among the main factors responsible for development of drug resistance in bacterial and viral diseases (Moscona, 2004). Enzymatic role of proteins helps catalyze metabolic reactions but only a small region of the protein consisting of a few amino acids are active in the catalysis; a noncatalytic example of protein includes the antibodies that are part of the adaptive immune systems and act as a binder to antigens for destruction (MacCallum et al., 1996). Ligand-binding proteins such as hemoglobin bind specific small molecules to transport them to other locations in the body of a multicellular organism (Baldwin and Chothia, 1979). Structural proteins such as actin and tubulin confer stiffness and rigidity to the cytoskeleton (Doherty and McMahon, 2008); other structural proteins such as myosin and kinesin generate mechanical forces and are responsible for the motility of many single cell organisms (Rayment, 1996).

Thus, there are numerous processes, and there are numerous proteins that take part in them. These processes and the functions of the proteins are studied through *in vivo* and *in vitro* analysis. *In vitro* analysis helps understand how a protein functions, *in vivo* analysis often helps in understanding its functional location and related parameters in the living system; however, the specifics of how a protein targets particular organelles or cellular structures are often unclear (Bejarano and Gonzalez, 1999). Site-directed mutagenesis techniques (Ruvkun and Ausubel, 1981) that alter the protein sequence and hence its structure and cellular location/function that help to identify susceptibility to regulation provide guidelines to rational drug design or development of new proteins with novel properties.

Among the simplest of biological entities, and of particular interest for this chapter, is the virus. A virus particle like the influenza or rotavirus contains about 8–11 protein-coding genes in a multiprotein coat that protects the RNA or DNA of the virus and also enables the proteins and genetic materials to enter and leave cells. A great range of variability in amino acid composition is observed for these viral proteins (Reid et al., 2000; Ghosh et al., 2009), specifically the surface situated ones like NA (neuraminidase; Ghosh et al., 2010), HA (hemagglutinin), VP4 (variable protein), VP7 (Gunn et al., 1985), and gp120 (of the HIV) but the functional impact remains the same. Often, a single change in the side chain of a single amino acid is enough for producing a new mutant (Lopez et al., 2005). Viruses use this highly mutable property for escaping the host defense mechanism and they are also frequently found to generate escape mutants against a naturally occurring immunity or artificially designed drug or vaccine (Air and Laver, 1989).

## B. Drugs and Proteins

Proteins are involved by function or malfunction, in diseases of organism. Bacterial, viral, and other pathogens disrupt the normal protein functions and thereby destabilize the infected host organism (Goldsby et al., 2000). While immunological defenses are called into action by the infection, often these are inadequate by themselves and have to be supported by drugs, vaccines, and other therapeutical regimes. Design of drugs and study of their actions have therefore been an important area of research. Drugs can act through formation of drug–DNA complexes (Chaires, 1997, 1998) or protein–drug complexes (Chicault et al., 1981). Major trends of research into drug–DNA relationships have been recently reviewed (Nandy and Basak, 2010). Stated simply, DNA drugs and vaccines are made of plasmids designed to carry a selected gene into cells where it is translated into a protein. In the case of antiviral DNA vaccine, for example, plasmids are created for producing the selected viral protein in the cell and immune systems are expected to act to prevent future infections from the virus (Ulmer et al., 1996a,b; Gurunathan et al., 2000). Advanced techniques such as codon optimization (Deml et al., 2001) are enhancing the protein production from the plasmids and others such as adjuvant incorporation are enhancing the immune response leading to more effective vaccines and therapies, several of which are already available for treatment of specified animals afflicted with the West Nile virus (Kramer et al., 2007), melanoma and fetal loss, while applications for humans for treating HIV, influenza, hepatitis C, and other diseases are under trial (Morrow and Weiner, 2010).

Pharmaceutical proteins effective against a wide range of bacterial infections can be traced to penicillin, and developed into new class of drugs referred to as antibiotics. Conventional production processes for antibiotics are expensive and face many regulatory issues. Vaccines that enhance the body's immune system consist of attenuated viruses but can, in rare cases, harm the host with a full-blown viral occupancy (Ball et al., 1998; Colgrove and Bayer, 2005). Since viruses use the host's cells to replicate, designing safe and effective antiviral drugs is difficult and also makes it difficult to find targets for the drugs that would interfere with the virus without also harming the host organism's cells. But almost all antimicrobials, including antivirals, are subject to drug resistance as the pathogens mutate over time (Gold and Moellering, 1996), becoming less susceptible to the treatment. Small molecules are often used as drugs, but the new technology of recombinant proteins (Geigert, 1989; Dingermann, 2008), commonly produced using bacteria or yeast in a bioreactor, potentially provide greater efficacy and fewer side effects because their action can be more precisely targeted toward the cause of a disease rather than treatment of symptoms, is yet to gain wide acceptance.

Peptide-based drugs operate by stimulating the immune response to the peptide and thereby to the invading pathogen. Peptides play an important role in modulating many physiological processes in our body. Use of peptides as drugs have the benefit that they are small, easily optimized, and can be quickly investigated for therapeutic potential. However, peptide drug screening process (Otvos, 2008), although a well-established approach, is long and arduous resulting in high manufacturing costs, and the fact that they have short half-life, and limited *in vivo* bioavailability hampers their effectiveness; new approaches have been proposed to overcome the difficulty of generating sufficient amount of the required tRNAs (Owens, 2004). The peptides can be naturally derived or chemically synthesized, with the latter method being more prevalent. Novel peptide analogs (Lee et al., 2002) are also being synthesized to create more potent drugs.

In practice, protein and peptide drugs are finding increasing acceptance in therapeutics. A drug's efficiency is related to the degree of its binding with the proteins in blood serum (Meyer and Guttman, 1968; Koch-Weser and Sellers, 1976): The less bound a drug is, the more efficiently it can diffuse through cell membrane. Common drug-binding proteins in plasma are human serum albumin, lipoprotein, glycoprotein, etc. It is the unbound fraction of the drug–protein complex that exhibits therapeutic effect and excessive binding may mitigate against rapid action of the drug. However, the same effect can be used for long-lasting dosage by designing drugs that bind to the protein and act as a reservoir so that the unbound fraction is released slowly.

But degradation of the proteins during storage and drug administration routes remains a challenging problem (Frokjaer and Otzen, 2005). These issues of stability of therapeutic proteins toward aggregation and misfolding in long-term storage as well as means of efficacious delivery that avoid adverse immunogenic side effects are engaging the attention of the pharmaceutical industry (Frokjaer and Otzen, 2005). While invasive routes such as subcutaneous injections are often used, oral delivery faces difficulties in poor permeability across biological membranes due to the hydrophobic nature and large molecular size, susceptibility to enzymatic attack, among others. Formulation strategies for protein therapeutics thus continue to remain a challenging problem.

## C. Bioinformatics in Protein Studies

The complexities of protein function and structure have necessitated the development of computational techniques to analyze available data and help in formulating novel ways to predict structure, function, and interaction of proteins. Especially, in view of the requirements of new approaches to drug development through recombinant proteins, synthesizing new peptides, and investigating drugs–DNA complexes, use of computational methods is now of vital importance.

The increased availability and accessibility of genomic and protein sequence data have opened up new possibilities for the search for target proteins, and the success of protein and peptide therapeutics is revolutionizing the biotech and pharmaceutical market, spurring the creation of next-generation products with reduced immunogenicity (Schellekens, 2002; Tangri et al., 2005), improved safety, and greater effectiveness. The protein engineering market is expected to cross $100 billion in sales in 2010 from about $36 billion 4 years ago. The top-selling therapeutic protein is reported to be Amgen's Aranesp (Locatelli and Vecchio, 2001), a reengineered variant of the company's first-generation product Epogen (recombinant human erythropoietin). A number of such products have been launched by Genetech and others, and nonparenteral delivery systems, alongside parenteral protein and peptide drug delivery systems have also been approved (Packhaeuser et al., 2004). Progress in bioinformatics and computational biology as well as new techniques in protein engineering (recombinant proteins through site-directed mutagenesis and posttranslational modifications) are aiding the development of reengineered, improved, whole antibody, and antibody fragment-based products, reducing immunogenicity by using fully human recombinant antibodies or human antibodies derived from transgenic mice and allowing biosimilar products to be differentiated on the basis of superior characteristics. Screening experiments for appropriate molecules rely critically on bioinformatics support for design of experiments and for interpreting the generated data, for example, to identify interesting differentially expressed genes and to predict the function and structure of putative target proteins (Lengauer and Zimmer, 2000).

Protein characterization and *in silico* protein design and structure analyses form an integral part of these developments. Phylogenetic analyses based on primary sequences have been used to group related proteins and understand their evolutionary history, algorithms have been developed to predict protein secondary structures, and web accessible systems are available to suggest possible folding patterns (Shen and Chou, 2009). A number of epitope prediction tools have been devised with varying degrees of success to aid in drug design (Yang and Yu, 2009); one area of nascent research is concerned with understanding of allosteric conformations that may help or hinder protein interactions (Teague, 2003). In a broader area, computational biology has already proved itself as one of the powerful tools for handling the large genomic databases. The basic applications involve killer tools like sequence alignment, phylogenetic tree drawing, sequence comparison, etc. *In silico* motif search algorithms on primary protein structure can be applied for finding structural information like signal sequence prediction (Menne et al., 2000), cleavage site prediction (Chou, 2001), glycosylation prediction (Blom et al., 2004), posttranslational modifications prediction, etc. Large datasets are frequently found to be utilized in predictions of protein structural levels from primary structure. Software like Modeller (Eswar et al., 2007, 2008), Discovery Studio, etc., can predict 3D structure of proteins from a database of known crystallized proteins. Many theories have been developed in this prediction research but they are often ineffective in case of a completely new protein for comparison with the preexisting database (comparative protein modeling) or a protein without appropriate template. Another very important application of data mining is the use of computational power in handling the proteomics data. In proteomics, proteins are detected by matching a part of it with the whole existing protein database in mass spectrophotometer software (Perkins et al., 1999). The basis of all these data mining and related computational techniques is mathematics and statistics. Different theories like dot-matrix algorithm (Gibbs and McIntyre, 1970), Needleman Wunsch algorithm (Needleman and Wunsch, 1970), Smith Waterman algorithm (Smith and Waterman, 1981; Smith et al., 1985), Hidden Marakov model (Eddy, 1996), Chou-Fasman algorithm (Chou and Fasman, 1974a,b), etc., are widely used.

Some models or algorithms work on interpretation from statistics and probability and others depend on the visual interpretation of genomic data by different techniques (Nandy, 1994; Randic, 2004, 2006; Randic et al., 2005, 2006; Nandy et al., 2007; Basak and Gute, 2008; Gonzalez-Diaz et al., 2008a, 2009).

To aid in protein characterization, ideas of graphical representation and numerical characterization (GRANCH) have been taken up from their success in DNA sequence analysis, but complicated here by the fact that protein sequences are composed of 20 amino acids whereas a DNA sequence is concerned with only the four building blocks of nucleotides. However, while some standard procedures such as dot-matrix plot have been used for a long time, several ingenious schemes have been developed recently that have marked significant success in this nascent field as we show in the next few sections. Coronavirus phylogeny, studies of H5N1 neuraminidase protein mutations and identification of highly conserved peptide stretches on influenza virus and rotavirus proteins that could potentially aid in the development of new drugs and vaccines are some of the significant results of application of these novel techniques. We provide a brief review of these studies in Section III.

## II. Graphical Methods

Graphical methods to display sequences have the advantage of visual indications of trends and inherent features. The familiar dot-matrix type of graphs have been widely used to determine systematics in nucleotide and amino acid sequences. The dots plotted on a 2D grid with the sequence running along the positive $x$- and negative $y$-axes produce a pattern (Gibbs and McIntyre, 1970) that is useful in determining sequence similarity, direct repeats, inverted repeats, etc., and such plots have also been used in RNA secondary structure predictions theories, for example, complementary sequences in a RNA structure in the dot-matrix analysis of nucleotide sequences of potato tuber spindle viroid (Fig. 1). In the case of proteins, one of the more widely used molecular graphs is the hydrophobicity–polarity lattice graph to model structure–activity relationships and folding dynamics in 2D/3D spaces (Jiang and Zhu, 2005; Chikenji et al., 2006). In continuing developments in the field, a new pseudofolding molecular graph or network-type representation has been proposed recently (Fernandez et al., 2008).

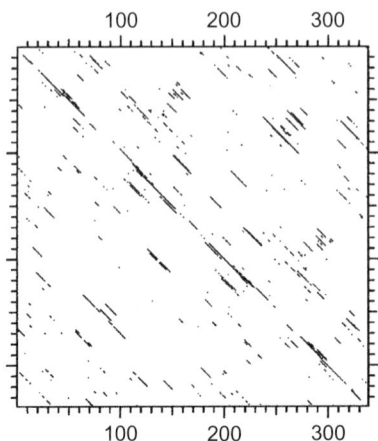

FIG. 1. Dot-matrix graph for RNA secondary structure prediction for Potato Tuber Spindle Viroid RNA sequence. Reproduced with permission from Cold Spring Harbor Laboratory Press, 1 Bungtown Road, Cold Spring Harbor, NY 11724, USA. Web: http://www.bioinformaticsonline.org/. Source: Mount, 2004.

### A. Graphical Methods for DNA Sequences

#### 1. Graphical Representation of DNA Sequences

Much of the recent interest in graphical methods arose from their applications in analysis of DNA and RNA sequences. Representation of the sequence of bases in a DNA or RNA strand using graphical methods was initiated several years ago with a 3D model proposed by Hamori and Ruskin (1983), followed up subsequently by Gates (1986), Nandy (1994), and Leong and Morgenthaler (1995) with 2D representations, while Peng et al. (1992) and Jeffrey (1990) represented sequence data graphically in more abstract forms. The plot of purine–pyrimidines against base numbers devised by Peng et al. (1992) demonstrated the presence of long-range correlations in DNA sequences while Jeffrey's Chaos Game Representation (CGR) method showed visually for the first time the fractal nature embedded in these sequences (Fig. 2), as also the different patterns for mammalian, bacterial, and phage sequences reflecting the inherent differences in their base organization. The utility of the graphical approach have led to many new techniques of GRANCH of DNA and RNA sequences (see review Nandy et al., 2006).

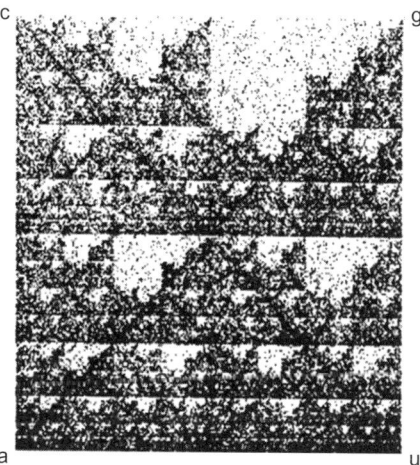

FIG. 2. CGR of human beta-globin (HUMHBB) region of Chromosome 11. Reproduced with permission from Oxford University Press, UK and Copyright Clearance Center (CCC) Web: http://nar.oxfordjournals.org/. Source: Jeffrey, 1990.

The basic approach can be most simply described in the 2D representation where the four cardinal directions are associated with the four bases. Nandy (1994) associated adenine with the negative $x$-axis, cytosine with the $+y$-axis, guanine with $+x$-axis, and the thymine with the $-y$-axis and plotted a sequence starting from the origin and moving, for each base in the sequence, one step at a time in the designated direction depending on the specific base until the entire sequence is plotted. Figure 3 follows the above mentioned direction of graphical representation technique for first 10 nucleotides of neuraminidase RNA (c-DNA) and generates a series of points like a Markov chain that reflects the sequence and distribution of bases in the sequence in the chosen representation. However, this simple approach has the disadvantage of allowing reentry in the random walk path, for example, a sequence like AGAGAG traces only one unit path in the Nandy representation, and several other schemes have been formulated that minimize or eliminate this problem, but with reduced visual appeal (Nandy et al., 2006). Randic and his coworkers, for example, proposed various representations such as "worm" curve (Randic et al., 2003c; Randic, 2004), "four horizontal line" curve (Randic et al., 2003a,b), four-color maps (Randic et al., 2005), "spectrum-like" figures

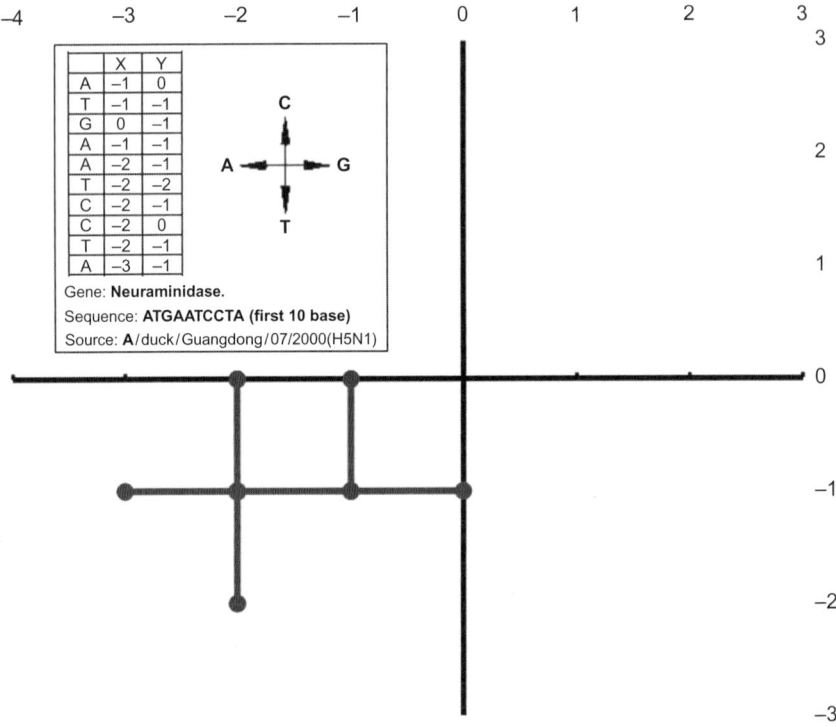

FIG. 3. Graphical representation (according to Nandy, 1994) of first 10 nucleotides of H5N1 neuraminidase RNA (or c-DNA).

(Randic, 2006) among others to reduce or eliminate the degeneracy inherent in the 2D approach. Yau et al. (2003) proposed a 2D graphical representation, where the purines (A, G) and pyrimidines (T, C) are plotted on two quadrants of the Cartesian coordinate system at fixed angles to the $x$-axis; such a system has no degenracy. A sequence is plotted as a progression of points counted along the $x$-axis but rising or falling with the nature of the base thus tracing a pattern that is unique for the particular sequence. Among other proposals, mention may be made of the recent works of Todeschini et al. (2006, 2008) who use partial ordering ideas to compare the first exons of eight beta-globin sequences, and Liu and Wang (2010) who used an 8D representation of DNA sequences for comparison of similarities/dissimilarities of over 40 viral, lipase, phage, and other genes, both of which methods dispense with visual rendering in favor of more rigorous mathematical approaches.

## 2. Numerical Characterization of DNA Sequences

To obtain a quantitative measure of the graphical representations, different techniques have been devised to convert these representations into numbers or vectors that are expected to be characteristic of each sequence. A simple geometrical technique for the 2D graph of a DNA sequence determines the weighted center of mass of a plot ($\mu_x$, $\mu_y$) and a graph radius ($g_R$), and therefrom the distance ($\Delta g_R$) of two sequences, using Euclidean measures (Raychaudhury and Nandy, 1999):

$$\mu_x = \frac{\sum x_i}{N}, \quad \mu_y = \frac{\sum y_i}{N} \quad \text{and} \quad g_R = \sqrt{\mu_x^2 + \mu_y^2} \qquad (1)$$

$$\Delta g_R = \sqrt{(\mu_{1x} - \mu_{2x})^2 + (\mu_{1y} - \mu_{2y})^2} \qquad (2)$$

where the ($x_i$, $y_i$) represent the coordinates of each point on the plot, $N$ is the total number of the bases in the segment and the $\mu_1$ and $\mu_2$ refer to two different DNA sequences. The $g_R$ here represents a base distribution index that is critically dependent upon the position of each base in the sequence and together with the $\mu_x$, $\mu_y$ form a set of biodescriptors for the sequence. The $g_R$ and the $\Delta g_R$ have been found to be very sensitive measures of the sequence composition and distribution (Nandy and Nandy, 2003). The difference index, $\Delta g_R$, provides a quantitative comparison between the sequences: the smaller the $\Delta g_R$, the more similar are the underlying DNA sequences and the higher the $\Delta g_R$, the more dissimilar are the sequences.

A matrix method of determining numerical indexes for DNA sequences was proposed by Randic et al. (2000) in a 3D graphical representation in which the position of every base of a sequence was related to all other bases through a Euclidean and graph–theoretic distance. The ratios of these distances, $D_E/D_G$, formed the elements of a DD matrix. Since matrices are well-known objects with well-defined properties, the leading eigenvalues of a DD matrix are considered to be characteristic, or invariants, of the matrix and, by association, to be descriptors of the DNA sequence itself. The authors calculated the leading eigenvalues of the first exon sequence of the *beta-globin* gene of eight species and determined the similarity/dissimilarity between the various sequences. This was followed by successive proposals for different graphical representations that similarly used matrix methods to determine invariants to characterize each sequence and form vectors of such invariants to estimate the degrees of similarities and dissimilarities between members of a family of DNA sequences (Nandy et al., 2006). The works of Randic, Todischini, and Wang referred to earlier use these techniques of DD matrices or Hasse matrices to compute the distances between species from their gene sequences.

## 3. Applications

The new GRANCH techniques gave a rich view of the complexities of DNA sequences. Among the first applications of these techniques to human diseases, Liao et al. (2006) showed that mathematical techniques can be used to analyze the underlying DNA/RNA sequences by studying the severe acute respiratory syndrome (SARS) coronavirus, and, separately, that GRANCH techniques could do away with multiple alignment requirements to study gene families. Larionov et al. (2008) broadened the usage by showing that plots of human and mouse chromosomal sequences in a graphical representation were able to reveal long-range palindromes. The 3D and 2D graphical representations visually highlighted the base preferences along a DNA sequence (Hamori and Ruskin, 1983; Nandy, 1994), while 2D representations showed long runs of duplications of a motif as simple runs on the graphs (Nandy and Nandy, 1995). Gates had remarked on large-scale complex repeats that were revealed by 2D graphs (Gates, 1986); Nandy showed that conserved genes have shapes on the 2D maps that are similar across species (Nandy, 1994), a visual rendition no doubt of homology. Viewing a number of maps of the H5N1 neuraminidase gene revealed a conserved region (Nandy et al., 2007), and numerical characterization of the maps, in the whole RNA sequence and in segments, has allowed reconstruction of the wide dissemination and possible recombination of segments of the gene not reported heretofore (Ghosh et al., 2009). In a novel application, using a variation of the 2D graphical representation, Wiesner and Wiesnerova (2010) studied multiallelic marker loci from *Begonia × tuberhybrida*. They found significant correlation of graph invariants to genetic diversity of the marker loci and suggested that DNA walk representation may predict allele-rich loci solely from their primary sequences, which improves current design of new DNA germplasm identificators. Recently, Nandy has shown from inspection of conserved gene representations on 2D maps (Nandy, 2009) that effects of point mutations in gene sequences over evolutionary time scales indicate a polynomial relationship between the intrapurine intrapyrimidine differences on each strand of a DNA sequence.

## B. Graphical Methods for Protein Sequences

### 1. Graphical Representation of Protein Sequences

The experience with GRANCH techniques for DNA and RNA sequences led to many proposals for GRANCH methods for protein sequences, although complicated by the necessity of accommodating 20 residues for proteins compared with four bases for the nucleotide sequences. One of the earliest attempts is the dot-matrix plot for protein sequences, but other techniques were also developed. Among the pioneer works for representing the chemical information in a protein graphically is the representation of protein bonds through the Ramachandran plot (Ramachandran et al., 1963; Ramachandran and Sasisekharan, 1968) and, for protein primary sequences, the Hydropathy plot (Engelman et al., 1986). The former can extract the secondary structural information from protein's bond angles, while the latter draws the graph from thermodynamical and chemical properties of amino acids. The DNA graphical representation methodology led Randic to propose a Magic Circle representation (Randic et al., 2006), where the total protein sequence is represented in a unit circle and the graph starts from the center following the sequence by moving half way toward the corresponding amino acids which are positioned equally spaced on the circumference. The result of the complete execution of the protein sequence within the circle produces a typical graph for a particular protein (Fig. 4), except for large protein sequences which are often found to have lesser visual benefits. Li et al. used a reduction model of abstracting the protein sequence in a five-letter code (Wang and Wang, 1999; Li et al., 2008) each representing a specific group of amino acids and generated a 2D-graph by plotting the reduced sequence on the $x$-axis and all five group representatives horizontally at equal intervals along the $y$-axis resulting in a zig-zag like graphical

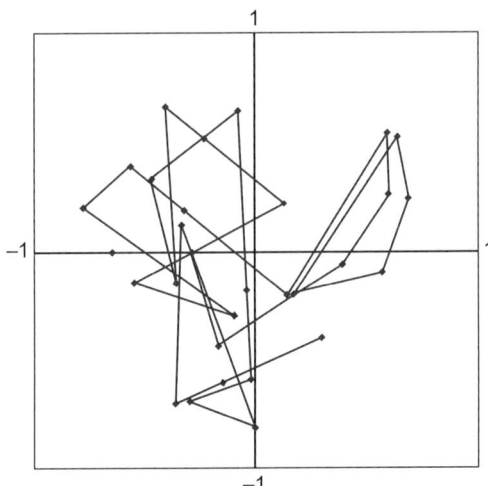

Fig. 4. Magic Circle representation of protein sequence WFFESRNDPAND-PIILWLNGGPGCSSFTGL. Reproduced with permission from Elsevier provided by Copyright Clearance Center (CCC). Web: http://www.sciencedirect.com/science/journal/00092614. Source: Randic et al., 2006.

representation of the sequence (Fig. 5). 2D graphical representations based on nucleotide triplet codons (Bai and Wang, 2006) have been proposed for sequence comparison and start–stop sign of a coding region. Liao et al. (2006) used this approach to study 24 coronavirus genomic sequences which have ~29,000 bases each. They classified the 20 amino acids of a protein sequence into four separate groups according to the chemistry of their R groups: amino acids A, V, F, P, M, I, L belong to the hydrophobic chemical group; amino acids D, E, K, R belong to charged chemical group; amino acids S, T, Y, H, C, N, Q, W belong to polar chemical group; the unique G amino acid belongs to glycine chemical group. Starting with the nucleotide sequence, this enabled them to construct three 2D graphs (one for each reading frame) for each gene sequence and compute a distance matrix between the 24 coronaviruses from which they could generate a phylogenetic tree relating all the sequences without the need for any multiple alignments. Gonzalez-Diaz and his coworkers have used 2D lattice graphs for proteins (Aguero-Chapin et al., 2006; Gonzalez-Diaz et al., 2008a), constructed in a similar way to the DNA representations of Nandy and adapted to proteins

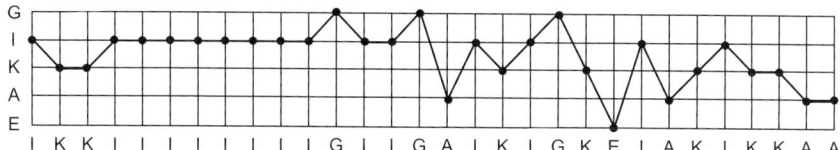

FIG. 5. Zig-zag curve generated from the 2D graphical representation of five-letter coded amino acids IKKIIIIIIIIGIIGAIKIGKEIAKIKKAA. Reproduced with permission from BMB reports, Web: http://www.bmbreports.org. Source: Li et al., 2008.

according to a proposed protocol (Estrada, 2002), and extended to other graph representations such as spiral and star networks (Aguero-Chapin et al., 2008a,b; Dea-Ayuela et al., 2008; Vilar et al., 2008; Munteanu et al., 2009); for example, for a star network, starting from the beginning of the sequence, the amino acids are placed in the corresponding branches transforming the protein sequence into modified branch connectivity graph from which connectivity indices (CIs) can be derived. Other authors have proposed higher dimensional representations such as the 3D model of Bai and Wang (2006) who embedded a dodecahedron in 3D space where each corner represented one of the 20 amino acids and thus generated a walk for the protein sequence, and the 20D representation of Novic and Randic (2008). The present authors have proposed an alternate 20D representation in Euclidean space (Nandy et al., 2009), where each amino acid is assigned to one axis in the 20D space and the sequence plotted using algorithms similar to the random walk model for 2D graphical representation of DNA sequences. This procedure generates a graph in the abstract 20D space from which consequences can be calculated to characterize proteins and quantify similarities and dissimilarities.

## 2. Numerical Characterization of Protein Sequences

Analogously to the GRANCH techniques of DNA sequences, to obtain quantitative measures for protein sequences, Randic, Li, Humberto Gonzales-Diaz, and several other authors have extended the methodologies of numerical descriptors for DNA sequences and of topological indices (TIs) used in QSAR studies to analysis of protein sequences, viral surfaces, and RNA secondary structures (Estrada and Uriarte, 2001; Randic et al., 2004, 2008; Bai et al., 2005; Gonzalez-Diaz et al., 2007b; Li et al., 2009a) leading to more general biological applications. González-Díaz and collaborators have done extensive work on extension of these representations to the study of protein sequences (Aguero-Chapin et al., 2009) and applied to mass spectral data of proteins and protein serum profiles in parasites (Gonzalez-Diaz et al., 2008b), toxicoproteomics, and diagnosis of cancer patients (Cruz-Monteagudo et al., 2008; Gonzalez-Diaz et al., 2008a). Their group has used mathematical biodescriptors derived from toxicoproteomics maps in conjunction with chemodescriptors of toxic molecules to predict their toxicity (Hawkins et al., 2006). Integrated QSARs (Basak et al., 2006) developed using chemodescriptors for ligands and biodescriptors of a molecular entity, for example, connect structural information of drug molecules, DNA and RNA sequences, or RNA secondary and protein tertiary structures and may be used to predict parameters for new entities (Gonzalez-Diaz et al., 2008a). It has been found that using different type of numerical indices derived from the protein 2D molecular graphics to perform QSAR studies is simpler than having to work with the protein 3D structures (Gonzalez-Diaz et al., 2009; Vilar and González-Díaz, 2010). These indices describe graph/network topology, connectivity, or branching, often referred to as the graph TIs or network CIs used to determine structure–function relationships in cellular biochemistry (Chou and Cai, 2003), and have been applied in theoretical biology and bioinformatics of small-size molecules, macromolecules, proteome mass spectra, and protein interaction networks (Aguero-Chapin et al., 2006; Gonzalez-Diaz et al., 2007a, 2008a). Basak et al. (2011) have in a pathbreaking work using a new differential QSAR approach for study of dihydrofolate reductases (DHFR) from multiple strains of *Plasmodium falciparum* shown that DHFR from the wild strain is substantially different from four mutant strains of their study and remark that the protocols indicated in the paper can be used for the development of drugs to combat drug-resistant pathogens arising continuously in nature due to mutations. Bai and Wang (2005) proposed to numerically characterize protein sequences through the nucleotide triplet codons by using a 2D graphical representation system similar to that of Yau et al. (2003) to generate protein descriptors for the *Homo sapiens* X-linked nuclear protein (ATRX). An intuitively simpler indexation scheme based on the 20D graphical representation of protein sequences proposed by the present authors (Nandy et al., 2009) and described below has been found useful in generating phylogenetic relationships between sequences without necessity of multiple alignments and for determining conserved surface exposed stretches on viral proteins that could be useful in drug and vaccine designs (Ghosh et al., 2010).

*3. Protein Similarities/Dissimilarities and Phylogeny by Graphical Methods*

Numerical characterization of sequences have also been targeted at the challenging problem of determining evolutionary relationships in protein families; for example, the multiplicity of voltage-gated sodium channel proteins from one for the bacteria (e.g., *Bacillus halodurans*) to 10 in humans, the development of the globin genes, the growth of differences in the highly conserved histones. Popular software like PHYLIP (Retief, 2000), MEGA (Tamura et al., 2007; Kumar et al., 2008), etc., are available for phylogenetic analysis, based generally on complex multiple sequence alignment (MSA) algorithms. Graphical methods like k-tuple, dot-matrix method (Gibbs and McIntyre, 1970), etc., are found as an integral part of MSA algorithm, and other graphical methods assess the extent of similarity/dissimilarity between protein sequence and serve as inputs to the software packages to generate the phylogenetic trees.

Bai and Wang (2006) derived the phylogenetic relationships for selected proteins using their 3D graphical representation where the amino acids are plotted on the corners of a dodecahedron. From the curve of the protein sequence obtained as a walk within this 3D space (Fig. 6), they derive a quotient matrix similar to the DD matrix discussed earlier for the DNA plots (Li and Wang, 2005), from which they can calculate the distance matrix between a set of protein sequences. The application of a similar procedure to a set of nine nerve genes from various organisms led to the generation of phylogenetic trees (Fig. 7). While usual methods of generating such trees are difficult due to the varying lengths of the sequences, the matrix method with leading eigenvalues do not have such problems and generates fairly acceptable relationships, although some of the details show, as the authors point out, that the method requires further refinement. The method is also useful in that it allows visible inspection of protein sequence characteristics and thus is good for comparative study of proteins too.

Li et al. (2009a) proposed a 3D graphical representation of protein sequences where the amino acids were classified into five separate groups based on their interactions. Thus, in terms of the one-letter code of amino acids, Group 1 consisted of the amino acids C, M, F, I, L, V, W, Y; Group 2

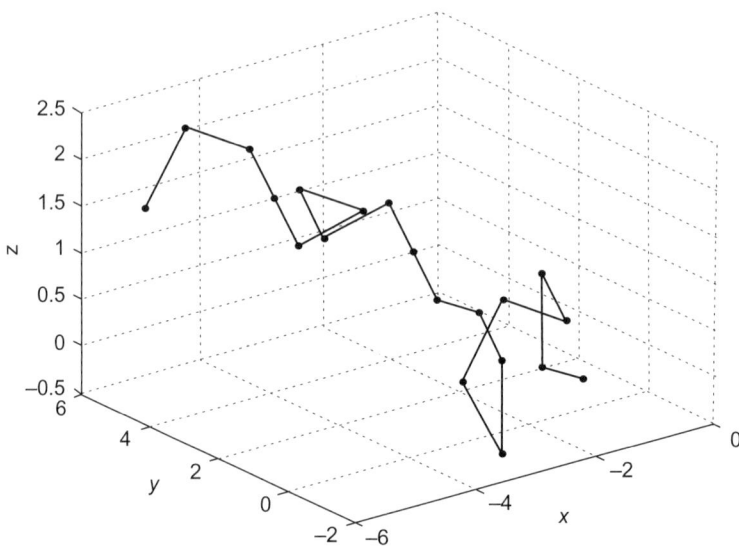

FIG. 6. 3D curve of protein sequence MGAPFVWALGLLMLQMLLFV, proposed by Bai and Wang, 2006. Reproduced (or published after acceptance) with permission from Adenine Press, 2066 Central Ave, New York, USA (Web: http://www.jbsdonline.com). Source: Bai and Wang, 2006.

of A, T, H; Group 3 of G, P; Group 4 of D, E; and the last Group 5 of S, N, Q, R, K. Representing each group by one amino acid, a protein sequence can be reduced to a sequence of five letters only, which can then be used to generate a random walk in a 3D Euclidean space where the steps are in designated cardinal directions (Fig. 8). Taking a cue from the work by Gonzalez-Diaz et al. (2005), they use the charge information of the amino acids and the number of amino acids at each node of the walk to define four charge coupling numbers for each sequence from which, after some combinatorics, they generate a 60-component vector for each sequence. Applying this technique to beta-globin protein sequences from 15 species, they were able to quantitatively assess the similarities and dissimilarities between the proteins from comparison of the sequence vectors. This also led to generation of a distance matrix which, though not explicitly shown by the authors, can be used to draw the phylogenetic tree for this protein family. The results obtained by the authors' prescription are analogous to established data.

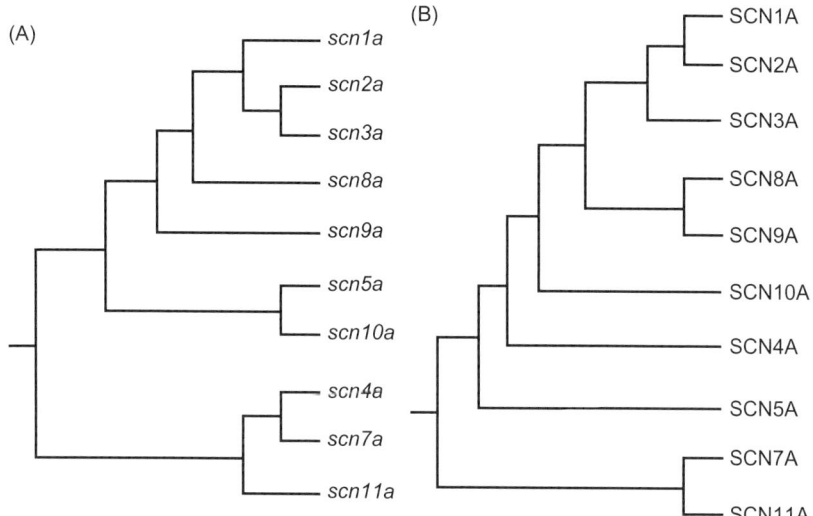

FIG. 7. Phylogenetic trees generated from $\Delta p_R$-matrix of nine-rat (*scn1a* to *scn11a*) voltage-gated sodium channel isoforms (A) and nine-human (SCN1A to SCN11A) voltage-gated sodium channel isoforms (B). Here the $\Delta p_R$-matrix is generated from the 20D algorithm proposed by Nandy et al. Reproduced with permission from IOS Press BV, Nieuwe Hemweg 6B, 1013 BG Amsterdam, The Netherlands. Web: www.bioinfo.de/isb/. Source: Nandy et al., 2009.

Thus, GRANCH methods are seen to be useful techniques to represent protein characteristics that can be easily computed while avoiding complications arising out of the need for multiple alignments (Altschul, 1989; Gotoh, 1993) and other modeling assumptions. The usefulness of these approaches and the reasonable agreement that we observe with standard results provide a good basis to investigate new phenomena such as viral issues which are the subjects of the next section.

## III. Application to Viral Proteins

### A. Unique Features of Viral Proteins

Viruses, the smallest biological entities, possess distinct groups of proteins holding a number of unique properties like high adaptability, high mutation rate, high structural flexibility, loose packing of the core, high

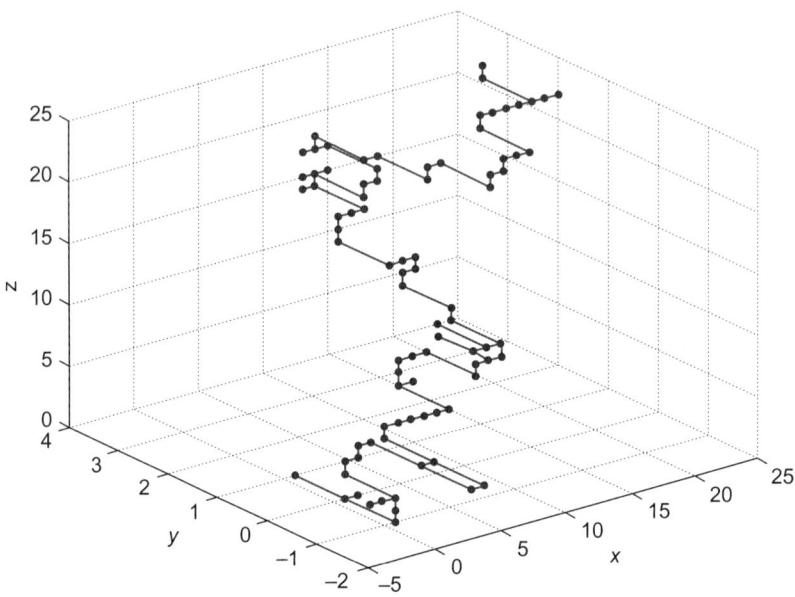

FIG. 8. 3D graph of the five-letter sequence of first 31 residues of Gorilla beta-protein IIAIAGEEKKAIAAIIGKIKIEEIGGEAIGK; each node may contain more than one amino acid. Reproduced with permission from Elsevier provided by Copyright Clearance Center (CCC). Web: http://www.sciencedirect.com/science/journal/03784371. Source: Li et al., 2009a.

proportion of disordered segments, among others (Koonin et al., 2009; Tokuriki et al., 2009; Kristensen et al., 2010). At the genome level, a very specific example of viral uniqueness resides in the existence of virus hallmark genes (Koonin et al., 2006), which play a central role in viral replication and structure, and are shared by a broad variety of viruses. In contrast to thermostable proteins like the heat-shock protein *Thermotoga maritime* (Tokuriki et al., 2009) which have specialized characteristics like high contact density, highly stable sequence composition, and highly compact structural scaffold, viral proteins necessarily have to be more complex to retain their functional characteristics in spite of the high variability. Another remarkable feature of viruses is the diversity in their genetic cycle. Altogether the variety of genetic strategies, genomic complexity, and global ecology of viral evolution lead to the formation of an infective long existing noncellular life form.

Currently prevention and treatment of viral diseases such as influenza rely on inactivated vaccines and antiviral drugs. Impact of mutational changes in amino acid residues on the stability, activity, and sensitivity of the target protein is a widely studied topic in antiviral drug design and for adequate remedy. The general causes like high mutability, altered specificity, environmental adaptability, etc., that are involved in generation of antiviral-resistant variety of the strains have been the main target of the major researches. Several investigations have focused upon phylogenetic relationships in viral evolution and transmission (Vijaykrishna et al., 2010) and reassortment (Lam et al., 2008; Owoade et al., 2008) and some other researchers are trying to correlate the evolution of genomic influenza varieties that affect humans and those that infect other life forms, for example, avian populations with a view that characterization of the causative proteins and determination or isolation of the conserved parts are useful approaches to combating viral diseases. Application of GRANCH techniques to viral proteins indicates one path to achieve this goal.

### B. Two Viral Examples: The Avian and Swine Flu Viruses and the SARS Coronavirus

The smallest unit that makes a particular protein identifiable is an eight to nine amino acid long peptide segment. This fundamental unit is frequently used in wet lab and dry lab researches involving protein mass spectrometry data analysis, sequence alignment and phylogenetic algorithms, protein database handling software (Perkins et al., 1999), structure-based drug design, etc. Comparison of a large group of protein sequences often involves comparing the basic units of the proteins (single amino acids to complex structural levels like peptides, secondary structures, domains, etc.) and their organization. GRANCH provides novel ways for identifying sequences or peptides by generating an identifier (Nandy et al., 2009) with the aim to uniquely prescribe a protein and its compositional information. GRANCH techniques for protein sequences have emerged recently with promising applications to studies of coronavirus and the avian and swine flu viruses. We briefly describe the characteristics of the viruses, cover the GRANCH methods used in these studies, and state the significant results.

## 1. Avian and Swine Flus

The H5N1 avian flu erupted in Hong Kong in 1997 (Hatta et al., 2001) and got carried by migratory birds from its place of origin in South Central China to the rest of Asia and to Europe and Africa. The existence of the virus gene pool in China and continuous mutations among the virus strains have led to continued rapid spread worldwide by different carriers with sudden conflagrations erupting at different locations, among aquatic birds, poultry, and farm animals, and also infecting humans resulting in over 300 deaths out of 505 confirmed cases (from World Health Organization; updated August 31, 2010). The H5N1 virus, like all other influenza viruses, is an enveloped virus with an eight-segment single-stranded RNA in the core and two surface proteins on the envelope, the hemagglutinin and the neuraminidase, which are responsible for the glycosilation necessary for cell entry and exit. Although the number of fatalities in humans from this virus appears small, the rapid mutations that can occur in the RNA genome, and the possibility of whole gene or gene fragments shuffling between avian and mammalian hosts (Wu et al., 2008), are considered to carry the potential to cause a pandemic challenge. Since the inhibitors of this influenza virus, principally oseltamivir and zanamivir, act on the neuraminidase component of the H5N1 protein, continuous monitoring of the mutational changes in this gene assumes significance.

The H1N1 swine flu outbreak of 2009, often referred to as Mexican flu or just swine flu, though less severe pathogenically than the H5N1 avian flu, infected humans and spread worldwide rapidly enough to lead the WHO to declare it as a pandemic. The genomic structure closely parallels the H5N1 genome except for the important difference in the hemagglutinin subtype and the virus has responded well to the osletamivir therapy, implying again the importance of the neuraminidase in the control and remedy of these forms of influenza (Moscona, 2005). However, an escape route (Moscona, 2004, 2009) from this standard treatment through genetic mutations remains highly probable and provides ample impetus for continued research into development of alternate therapeutic strategies.

## 2. SARS Coronavirus

The SARS erupted on the world stage in 2003 (Gorbalenya et al., 2004) from its origins in South East Asia and was established to have been caused by a novel form of the coronavirus. Coronaviruses also are enveloped viruses with a single-stranded multisegment RNA genome, but ranging in size from 16 to 31 kb (Lai, 1990). The virus primarily infects the upper respiratory and gastrointestinal tracts of mammals and birds, but the human SARS coronavirus also affects the lower respiratory tract. Experimental studies are complicated by the fact that the human coronaviruses are difficult to grow in the laboratory. While earlier only two coronaviruses were known, the HcoV-229E and HcoV-OC43 (Gorbalenya et al., 2004), after the SARS epidemic three more coronaviruses were identified by 2005, the SARS-CoV, the NL63 (van der Hoek et al., 2004), and HKU1 leading to interest in the evolutionary history of this virus.

### C. Graphical Representation Methods in Viral Studies

#### 1. The 2D Method of Liao et al. (2006)

As mentioned earlier (Section II), Liao et al. (2006) constructed 2D graphs with the four R-groups of the amino acids at predetermined angles on either side of the $x$-axis. For each nucleotide sequence, they constructed three separate graphs for the three reading frames of the gene sequence. For each graph, they defined a geometric center of mass $x_0$, $y_0$ and a covariance matrix CM as

$$x_0 = \sum \frac{x_i}{N}, \quad y_0 = \sum \frac{y_i}{N} \qquad (3)$$

$$CM_{xx} = \sum \frac{(x_i - x_0)(x_i - x_0)}{N} \qquad (4)$$

$$CM_{xy} = CM_{yx} = \sum \frac{(x_i - x_0)(y_i - y_0)}{N} \qquad (5)$$

$$CM_{yy} = \sum \frac{(y_i - y_0)(y_i - y_0)}{N} \qquad (6)$$

where the summations are over the subscript $i$ which runs from 1 to $N$, the length of the sequence. The covariance matrix CM is a $2 \times 2$ square matrix with a leading eigenvalue $\lambda$. Thus, for the three graphs of each sequence there will be a set of three geometric centers of masses and three leading eigenvalues. From these eigenvalues, they defined a distance measure for two sequences $i$ and $j$ as

$$d_{ij} = \sqrt{\left\{(\lambda_{i1} - \lambda_{j1})^2 + (\lambda_{i2} - \lambda_{j2})^2 + (\lambda_{i3} - \lambda_{j3})^2\right\}} \tag{7}$$

which can be used for studies of evolutionary relationships between species without having to make any evolutionary model assumptions or multiple alignments of the sequence.

2. *The 2D Method of Li et al. (2008)*

Another 2D graphical method has been described by Li et al. (2008) where they ascribe a 60-component vector to each of the proteins and construct a distance matrix

$$d_{ij} = \sqrt{\sum (x_{ir} - x_{jr})^2} \tag{8}$$

where $i$ and $j$ refer to two different sequences and $r = 1,2,3,\ldots, 60$. This structure allows them to generate a phylogenetic tree in similar fashion to Liao et al.

3. *The 20D Method of Nandy et al.*

Taking a cue from the graphical representations of DNA sequences, Nandy et al. (2009) proposed an abstract 20D Cartesian coordinate system to generate a protein sequence walk by plotting one point for each amino acid in the sequence along a designated axis for that acid as shown in Table I; the choice of association is equivalent for all residues and can be arbitrarily assigned but once assigned will be fixed for the duration of the computation.

The walk as per the sequence will result in a series of points in the abstract 20D space generating a curve, each point on the walk being specified by 20 coordinate values. For example, for a protein sequence like MVHLTPEEKS the coordinate of the end point will then be (0,0,0,2,0,0,1,0,1,1,1,0,1,0,0,1,1,1,0,0) and the exercise can likewise be performed for any protein sequence.

Unlike some of the 2D graphical representation of DNA and protein sequences, there are no degeneracies or path retracements (Nandy et al., 2009) in this representation and all amino acids are represented on equal footing. While the disadvantage of this method is clear that the graph cannot be visualized, numerical characterization of the sequences

TABLE I
Assignment of Axis to Individual Amino Acids

| Axis No. | Amino acid | Three-letter code | Single-letter code |
|---|---|---|---|
| 1 | Alanine | Ala | A |
| 2 | Cysteine | Cys | C |
| 3 | Aspartic acid | Asp | D |
| 4 | Glutamic acid | Glu | E |
| 5 | Phenylalanine | Phe | F |
| 6 | Glycine | Gly | G |
| 7 | Histidine | His | H |
| 8 | Isoleucine | Ile | I |
| 9 | Lysine | Lys | K |
| 10 | Leucine | Leu | L |
| 11 | Methionine | Met | M |
| 12 | Asparagine | Asn | N |
| 13 | Proline | Pro | P |
| 14 | Glutamine | Gln | Q |
| 15 | Arginine | Arg | R |
| 16 | Serine | Ser | S |
| 17 | Threonine | Thr | T |
| 18 | Valine | Val | V |
| 19 | Tryptophan | Trp | W |
| 20 | Tyrosine | Tyr | Y |

can be easily computed as described below and used for comparison between sequences irrespective of sequence lengths (Nandy et al., 2009).

The quantification procedure in this representation characterizes a sequence by a weighted center of mass approach first used for DNA sequences with the CM coordinates given by

$$\mu_1 = \sum \frac{x_1}{N}, \mu_2 = \sum \frac{x_2}{N}, \mu_3 = \sum \frac{x_3}{N}, \ldots, \mu_{20} = \sum \frac{x_{20}}{N} \quad (9)$$

here the $x_i$'s are the coordinate values of each point on the abstract curve and $N$, a normalization factor for the $\mu_i$'s, is the number of amino acids in the protein chain. Using these weighted averages, the procedure defines a protein graph vector $p_R$ ($\mu_1, \mu_2, \ldots, \mu_{20}$) and a protein graph radius

$$p_R = \sqrt{(\mu_1^2 + \mu_2^2 + \cdots + \mu_n^2)} \quad (10)$$

Again, the distance between two sequences $i$ and $j$ can be defined as

$$\Delta p_R = \sqrt{\left\{\left(\mu_{i_1} - \mu_{j_1}\right)^2 + \left(\mu_{i_2} - \mu_{j_2}\right)^2 + \cdots + \left(\mu_{i_{20}} - \mu_{j_{20}}\right)^2\right\}} \quad (11)$$

where the sum is taken over all 20 coordinates. Obtaining a distance matrix from comparison of a family of sequences can enable generating a phylogenetic tree to study evolutionary relationships, again, as in all GRANCH methods, without having to introduce multiple alignments or any other model dependencies. Nandy et al. (2009) has successfully applied this algorithm in tree construction for human globin variants and between voltage-gated sodium channel isoforms.

It is to be noted that this numerical characterization method refers strictly to the identities of the amino acids and is transparent to their chemical properties, that is, no distinctions are made between residues that are mutationally conservative or nonconservative, between polar and nonpolar residues, between basic and acidic residues, etc., and all residues are treated at par. As in the case of the $g_R$ for the DNA sequences, the $p_R$ values also are found to be sensitive to changes in the amino acid sequences (Ghosh et al., 2009, 2010), and equal values of the $p_R$ imply exact duplication of the amino acid composition and distribution along the sequences.

### D. Results for the Coronavirus and the Flu Viruses

#### 1. Phylogenetic Studies

For the 24 coronavirus genomes selected for their study, Liao et al. constructed a 24 × 24 distance matrix (Liao et al., 2006) from which they were able to generate a phylogenetic tree of the whole genome of the virus for different species using their 2D GRANCH technique. MSA, the popular phylogenetic tree generation algorithm, does not work properly for the whole genome, and the evolutionary model used may produce a wrong interpretation (Liao et al., 2006). Here, the phylogenetic tree (Fig. 9) obtained by their method clearly defined the evolutionary relationships between the whole genomes of 24 different species of the coronaviruses.

Liu and Wang (2010) using an L-tuple-based DNA representation constructed a set of $L \times L$ matrices whose mathematical characterization led to

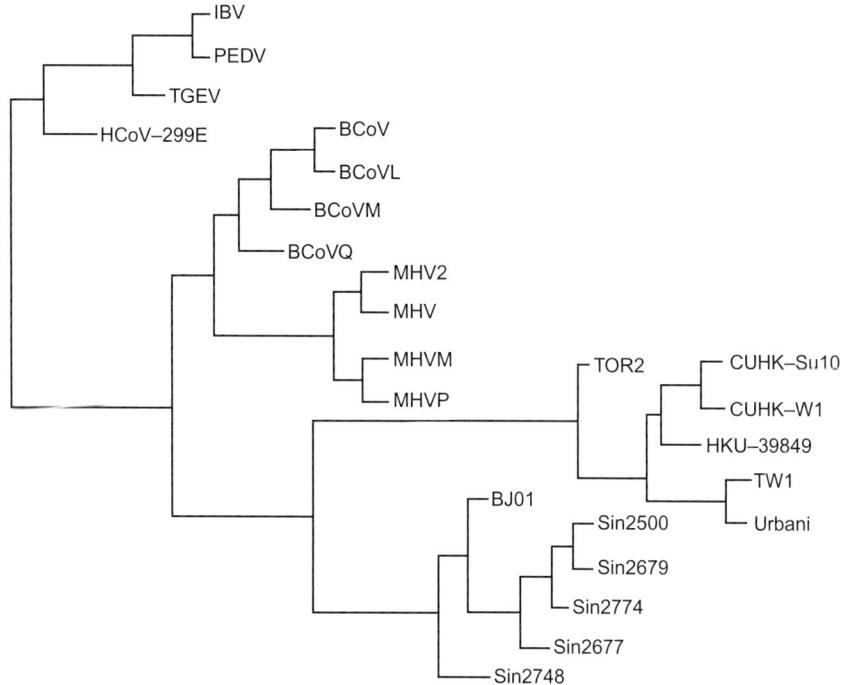

FIG. 9. The phylogenetic tree of the whole genome of 24 Coronavirus species. Reproduced with permission from Elsevier provided by Copyright Clearance Center (CCC). Web: http://www.sciencedirect.com/science/journal/00092614. Source: Liao et al., 2006.

characterization of the DNA sequences. Obtaining the distance matrices between a set of eight H5N1 avian flu genomes, they were able to generate a phylogenetic tree where the evolutionary relationships between the various strains of the virus were clearly identified.

2. *Similarity/Dissimilarity Analyses*

Nandy et al. (2007) and Ghosh et al. (2009) used the 2D GRANCH techniques for DNA sequences and the 20D methods for the protein sequences for analyses of global characteristics of over 680 H5N1 neuraminidase sequences to determine any systematic and exceptional behavior that may have arisen from mutational changes.

They found from detailed comparison of the $g_R$ and $p_R$ values that, at the protein level, only about 62% unique strains are observed, whereas for the nucleotide sequences the percentage of unique strains is considerably high at 80%, implying that about 22% (percentage of synonymous sequences to uniques) of the purportedly new strains of the neuraminidase gene have synonymous mutations (Ghosh et al., 2009). Considering the neuraminidase's segmented structure of transmembrane, stalk, and body regions, it was found that the body region appears proportionately less stable than the transmembrane or the stalk regions (Ghosh et al., 2009). In contrast, a 50-base segment at the 5′-end of the gene is found highly stable, mutations there being observed in less than 4.5% of the sequences at the RNA level and about 1.9% at the protein level raising the possibility of investigating this region as potentially useful for designing novel neuraminidase inhibitors.

The duplicate sequences identified by the $p_R$ analysis showed sequence duplication across species and distributed over substantial distances in space and time (Ghosh et al., 2009). While localized or cosynchronous distributions can be expected to occur due to rapid dissemination of specific strains through viral shedding as one mechanism, the appearance of identical strains in geographically widely separated locations several thousand kilometers apart, or after a lapse of 2 years or more, is puzzling since viral genes are known to mutate rapidly in replication. The authors hypothesized that this may arise out of viral shedding in aquatic and nonaquatic habitats that are subsequently spread across wide regions by the migratory or local birds who themselves might not be infected but act merely as carrier agents. The $p_R$ analysis from this technique also showed for the first time that recombinations between segments in the neuraminidase gene may have been taking place. Thus, sequence similarities and dissimilarities analysis done comparatively easily through the numerical descriptors can reveal many interesting features of viral spread and mutational changes.

### 3. Conserved and surface exposed peptide stretch identification

In nature, viruses are found to carry a great quantity of sequence variation in both the RNA and the protein level. These variations in viral sequences (Phillips et al., 1991; Chen and Deng, 2009) generally come from spontaneous mutation, adaptive forces, various mutagenic effects, sequence recombination, etc. Such mutations are observed more in the parts in contact with the environment and therefore readily develop resistance to drugs. Conserved region in such parts of the protein, when determined, can be used for many purposes like structure-based drug design, viral proteins activity determination, vaccine design, etc. Ghosh et al. have applied the methods of GRANCH to determine just such regions in the H5N1 avian flu neuraminidase protein (Ghosh et al., 2010).

Using the 20D similarity/dissimilarity technique (Ghosh et al., 2009) through comparisons of $p_R$ values, all the proteins in the dataset were scanned by a window size of 6–14 amino acids and the $p_R$ values compared to find regions of least variability. This variability profile is then compared with a solvent accessibility profile to determine regions of low variability and high solvent accessibility implying that these identified regions would be accessible to drugs and vaccines and also offer target sites over many cycles of mutations. A view of the 3D structure also ensures that high surface exposed regions are actually selected. The authors determined six such regions on the neuraminidase protein (Fig. 10), of which the most promising appears to be the 50-base (16 amino acid) stretch at the 5′-end of the gene mentioned earlier. A special feature of this 16 amino acid long

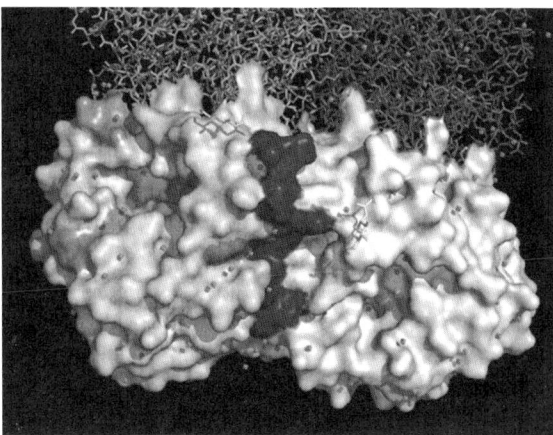

FIG. 10. Conserved surface exposed regions are shown in different colors in the cyan colored monomer of neuraminidase (other monomers are colored in magenta, green, and yellow). Here the six conserved regions are shown in six different colors. The conserved C-terminal portion is shown in blue. (See color plate 1).

peptide is its location on the dimeric interface (indicated by blue color in Fig. 10) of the quaternary structure of the neuraminidase protein, implying that any disruption of this stretch could interfere with the stability of the protein itself.

Nandy (2010) has reported on a similar work done on the rotavirus in association with several others. There, seven such distinct conserved surface exposed regions have been identified with this procedure. The most promising four regions have tested positive by epitope prediction servers (Peters et al., 2005; Vita et al., 2010) and reportedly hold promise for peptide drug and vaccine development.

## IV. Conclusion

Thus, the GRANCH techniques for protein sequences are turning out to be quite useful novel method for analysis of proteins. Extension of these techniques to applications of different measurements as espoused by Gonzalez-Diaz et al. and others are opening up new methods to visualize and analyze experimental data and provide new insights. Applications by various authors to viral issues have generated new model independent ways to establish evolutionary relationships. In particular, the GRANCH techniques have provided for the first time a systematic method to determine conserved surface exposed peptide stretches on viral proteins that could be potentially very useful for drug and vaccine development.

### References

Aguero-Chapin, G., Antunes, A., Ubeira, F. M., Chou, K. C., Gonzalez-Diaz, H. (2008). Comparative study of topological indices of macro/supramolecular RNA complex networks. *J. Chem. Inf. Model.* **48**, 2265–2277.

Aguero-Chapin, G., Gonzalez-Diaz, H., de la Riva, G., Rodriguez, E., Sanchez-Rodriguez, A., Podda, G., et al. (2008). MMM-QSAR recognition of ribonucleases without alignment: comparison with an HMM model and isolation from Schizosaccharomyces pombe, prediction, and experimental assay of a new sequence. *J. Chem. Inf. Model.* **48**, 434–448.

Aguero-Chapin, G., Gonzalez-Diaz, H., Molina, R., Varona-Santos, J., Uriarte, E., Gonzalez-Diaz, Y. (2006). Novel 2D maps and coupling numbers for protein sequences. The first QSAR study of polygalacturonases; isolation and prediction of a novel sequence from *Psidium guajava* L. *FEBS Lett.* **580**, 723–730.

Aguero-Chapin, G., Varona-Santos, J., de la Riva, G. A., Antunes, A., Gonzalez-Vlla, T., Uriarte, E., et al. (2009). Alignment-free prediction of polygalacturonases with pseudofolding topological indices: experimental isolation from *Coffea arabica* and prediction of a new sequence. *J. Proteome Res.* **8**, 2122–2128.

Air, G. M., Laver, W. G. (1989). The neuraminidase of influenza virus. *Proteins* **6**, 341–356.

Altschul, S. F. (1989). Gap costs for multiple sequence alignment. *J. Theor. Biol.* **138**, 297–309.

Bai, F., Wang, T. (2005). A 2-D graphical representation of protein sequences based on nucleotide triplet codons. *Chem. Phys. Lett.* **413**, 458–462.

Bai, F., Wang, T. (2006). On graphical and numerical representation of protein sequences. *J. Biomol. Struct. Dyn.* **23**, 537–546.

Bai, F., Zhu, W., Wang, T. (2005). Analysis of similarity between RNA secondary structures. *Chem. Phys. Lett.* **408**, 258–263.

Baker, D., Sali, A. (2001). Protein structure prediction and structural genomics. *Science* **294**, 93–96.

Baldwin, J., Chothia, C. (1979). Haemoglobin: the structural changes related to ligand binding and its allosteric mechanism. *J. Mol. Biol.* **129**, 175–220.

Ball, L. K., Evans, G., Bostrom, A. (1998). Risky business: challenges in vaccine risk communication. *Pediatrics* **101**, 453–458.

Basak, S. C., Gute, B. D. (2008). Mathematical biodescriptors of proteomics maps: background and applications. *Curr. Opin. Drug Discov. Devel.* **11**, 320–326.

Basak, S. C., Mills, D., Gute, B. D., Natarajan, R. (2006). Predicting pharmacological and toxicological activity of heterocyclic compounds using QSAR and molecular modeling. *In* QSAR and Molecular Modeling Studies of Heterocyclic Drugs I, Gupta, S. P. (Ed.), pp. 39–80. Springer-Verlag, Berlin-Heidelberg-New York.

Basak, S. C., Mills, D., Hawkins, D. M. (2011). Characterization of dihydrofolate reductases from multiple strains of *Plasmodium falciparum* using mathematical descriptors of their inhibitors. *Chem. Biodivers* **8**, 440–453.

Bejarano, L. A., Gonzalez, C. (1999). Motif trap: a rapid method to clone motifs that can target proteins to defined subcellular localisations. *J. Cell Sci.* **112**(Pt 23), 4207–4211.

Berman, H., Henrick, K., Nakamura, H., Markley, J. L. (2007). The worldwide Protein Data Bank (wwPDB): ensuring a single, uniform archive of PDB data. *Nucleic Acids Res.* **35**, D301–D303.

Blom, N., Sicheritz-Ponten, T., Gupta, R., Gammeltoft, S., Brunak, S. (2004). Prediction of post-translational glycosylation and phosphorylation of proteins from the amino acid sequence. *Proteomics* **4**, 1633–1649.

Chaires, J. B. (1997). Energetics of drug-DNA interactions. *Biopolymers* **44**, 201–215.

Chaires, J. B. (1998). Drug–DNA interactions. *Curr. Opin. Struct. Biol.* **8**, 314–320.

Chayen, N. E. (2004). Turning protein crystallisation from an art into a science. *Curr. Opin. Struct. Biol.* **14**, 577–583.

Chayen, N. E. (2009). High-throughput protein crystallization. *Adv. Protein. Chem. Struct. Biol.* **77**, 1–22.

Chayen, N. E., Saridakis, E. (2008). Protein crystallization: from purified protein to diffraction-quality crystal. *Nat Methods* **5**, 147–153.

Chen, J., Deng, Y. M. (2009). Influenza virus antigenic variation, host antibody production and new approach to control epidemics. *Virol. J.* **6**, 30.

Chicault, M., Luu Duc, C., Boucherle, A. (1981). Drug protein interactions. *Arzneimittelforschung* **31**, 1015–1020.

Chikenji, G., Fujitsuka, Y., Takada, S. (2006). Shaping up the protein folding funnel by local interaction: lesson from a structure prediction study. *Proc. Natl. Acad. Sci. USA* **103**, 3141–3146.

Chou, K. C. (2001). Prediction of protein signal sequences and their cleavage sites. *Proteins* **42**, 136–139.

Chou, K. C., Cai, Y. D. (2003). Prediction and classification of protein subcellular location-sequence-order effect and pseudo amino acid composition. *J. Cell. Biochem.* **90**, 1250–1260.

Chou, P. Y., Fasman, G. D. (1974a). Conformational parameters for amino acids in helical, beta-sheet, and random coil regions calculated from proteins. *Biochemistry* **13**, 211–222.

Chou, P. Y., Fasman, G. D. (1974b). Prediction of protein conformation. *Biochemistry* **13**, 222–245.

Colgrove, J., Bayer, R. (2005). Could it happen here? Vaccine risk controversies and the specter of derailment. *Health Aff. (Millwood)* **24**, 729–739.

Cruz-Monteagudo, M., Gonzalez-Diaz, H., Borges, F., Dominguez, E. R., Cordeiro, M. N. (2008). 3D-MEDNEs: an alternative "in silico" technique for chemical research in toxicology. 2. quantitative proteome-toxicity relationships (QPTR) based on mass spectrum spiral entropy. *Chem. Res. Toxicol.* **21**, 619–632.

Dea-Ayuela, M. A., Perez-Castillo, Y., Meneses-Marcel, A., Ubeira, F. M., Bolas-Fernandez, F., Chou, K. C., et al. (2008). HP-Lattice QSAR for dynein proteins: experimental proteomics (2D-electrophoresis, mass spectrometry) and theoretic study of a *Leishmania infantum* sequence. *Bioorg. Med. Chem.* **16**, 7770–7776.

Deml, L., Bojak, A., Steck, S., Graf, M., Wild, J., Schirmbeck, R., et al. (2001). Multiple effects of codon usage optimization on expression and immunogenicity of DNA candidate vaccines encoding the human immunodeficiency virus type 1 Gag protein. *J. Virol.* **75**, 10991–11001.

Deshpande, N., Addess, K. J., Bluhm, W. F., Merino-Ott, J. C., Townsend-Merino, W., Zhang, Q., et al. (2005). The RCSB Protein Data Bank: a redesigned query system and relational database based on the mmCIF schema. *Nucleic Acids Res.* **33**, D233–D237.

Dill, K. A., Ozkan, S. B., Shell, M. S., Weikl, T. R. (2008). The protein folding problem. *Annu. Rev. Biophys.* **37**, 289–316.

Dill, K. A., Ozkan, S. B., Weikl, T. R., Chodera, J. D., Voelz, V. A. (2007). The protein folding problem: when will it be solved? *Curr. Opin. Struct. Biol.* **17**, 342–346.

Dingermann, T. (2008). Recombinant therapeutic proteins: production platforms and challenges. *Biotechnol. J.* **3**, 90–97.

Dobson, C. M. (2004). Principles of protein folding, misfolding and aggregation. *Semin. Cell Dev. Biol.* **15**, 3–16.

Doherty, G. J., McMahon, H. T. (2008). Mediation, modulation, and consequences of membrane-cytoskeleton interactions. *Annu. Rev. Biophys.* **37**, 65–95.

Dutta, S., Berman., H. M., Bluhm, W. F. (2007). Using the tools and resources of the RCSB protein data bank. *Curr. Protoc. Bioinformatics* **20**, 1.9.1–1.9.24.

Eddy, S. R. (1996). Hidden Markov models. *Curr. Opin. Struct. Biol.* **6**, 361–365.

Engelman, D. M., Steitz, T. A., Goldman, A. (1986). Identifying nonpolar transbilayer helices in amino acid sequences of membrane proteins. *Annu. Rev. Biophys. Biophys. Chem.* **15**, 321–353.

Estrada, E. (2002). Characterization of the folding degree of proteins. *Bioinformatics* **18**, 697–704.

Estrada, E., Uriarte, E. (2001). Recent advances on the role of topological indices in drug discovery research. *Curr. Med. Chem.* **8**, 1573–1588.

Eswar, N., Eramian, D., Webb, B., Shen, M. Y., Sali, A. (2008). Protein structure modeling with MODELLER. *Methods Mol. Biol.* **426**, 145–159.

Eswar, N., Webb, B., Marti-Renom, M. A., Madhusudhan, M. S., Eramian, D., Shen, M.-y., Pieper, U., Sali, A. (2007). Comparative protein structure modeling using MODELLER. *Curr. Protoc. Protein Sci.* **50**, 2.9.1–2.9.31.

Fasman, G. D. (1989). Protein conformational prediction. *Trends Biochem. Sci.* **14**, 295–299.

Feldman, H. J., Hogue, C. W. (2002). Probabilistic sampling of protein conformations: new hope for brute force? *Proteins* **46**, 8–23.

Fernandez, M., Caballero, J., Fernandez, L., Abreu, J. I., Acosta, G. (2008). Classification of conformational stability of protein mutants from 3D pseudo-folding graph representation of protein sequences using support vector machines. *Proteins* **70**, 167–175.

Frieden, C. (1971). Protein-protein interaction and enzymatic activity. *Annu. Rev. Biochem.* **40**, 653–696.

Frokjaer, S., Otzen, D. E. (2005). Protein drug stability: a formulation challenge. *Nat. Rev. Drug Discov.* **4**, 298–306.

Gates, M. A. (1986). A simple way to look at DNA. *J. Theor. Biol.* **119**, 319–328.

Geigert, J. (1989). Overview of the stability and handling of recombinant protein drugs. *J. Parenter. Sci. Technol.* **43**, 220–224.

Ghosh, A., Nandy, A., Nandy, P. (2010). Computational analysis and determination of a highly conserved surface exposed segment in H5N1 avian flu and H1N1 swine flu neuraminidase. *BMC Struct. Biol.* **10**, 6.

Ghosh, A., Nandy, A., Nandy, P., Gute, B. D., Basak, S. C. (2009). Computational study of dispersion and extent of mutated and duplicated sequences of the H5N1 influenza neuraminidase over the period 1997–2008. *J. Chem. Inf. Model.* **49**, 2627–2638.

Ghosh, K., Ozkan, S. B., Dill, K. A. (2007). The ultimate speed limit to protein folding is conformational searching. *J. Am. Chem. Soc.* **129**, 11920–11927.

Gibbs, A. J., McIntyre, G. A. (1970). The diagram, a method for comparing sequences. Its use with amino acid and nucleotide sequences. *Eur. J. Biochem.* **16**, 1–11.

Gold, H. S., Moellering, R. C., Jr. (1996). Antimicrobial-drug resistance. *N. Engl. J. Med.* **335**, 1445–1453.

Goldsby, R. A., Kindt, T. J., Osborne, B. A. (2000). Overview of Immune System. Kuby Immunology. W.H. Freeman and Company, United State of America, pp. 3.
Gonzalez-Diaz, H., Gonzalez-Diaz, Y., Santana, L., Ubeira, F. M., Uriarte, E. (2008). Proteomics, networks and connectivity indices. *Proteomics* **8**, 750–778.
Gonzalez-Diaz, H., Molina, R., Uriarte, E. (2005). Recognition of stable protein mutants with 3D stochastic average electrostatic potentials. *FEBS Lett.* **579**, 4297–4301.
Gonzalez-Diaz, H., Perez-Montoto, L. G., Duardo-Sanchez, A., Paniagua, E., Vazquez-Prieto, S., Vilas, R., et al. (2009). Generalized lattice graphs for 2D-visualization of biological information. *J. Theor. Biol.* **261**, 136–147.
Gonzalez-Diaz, H., Prado-Prado, F., Ubeira, F. M. (2008). Predicting antimicrobial drugs and targets with the MARCH-INSIDE approach. *Curr. Top. Med. Chem.* **8**, 1676–1690.
Gonzalez-Diaz, H., Saiz-Urra, L., Molina, R., Gonzalez-Diaz, Y., Sanchez-Gonzalez, A. (2007). Computational chemistry approach to protein kinase recognition using 3D stochastic van der Waals spectral moments. *J. Comput. Chem.* **28**, 1042–1048.
Gonzalez-Diaz, H., Vilar, S., Santana, L., Uriarte, E. (2007). Medicinal chemistry and bioinformatics–current trends in drugs discovery with networks topological indices. *Curr. Top. Med. Chem.* **7**, 1015–1029.
Gorbalenya, A. E., Snijder, E. J., Spaan, W. J. (2004). Severe acute respiratory syndrome coronavirus phylogeny: toward consensus. *J. Virol.* **78**, 7863–7866.
Gotoh, O. (1993). Optimal alignment between groups of sequences and its application to multiple sequence alignment. *Comput. Appl. Biosci.* **9**, 361–370.
Gunn, P. R., Sato, F., Powell, K. F., Bellamy, A. R., Napier, J. R., Harding, D. R., et al. (1985). Rotavirus neutralizing protein VP7: antigenic determinants investigated by sequence analysis and peptide synthesis. *J. Virol.* **54**, 791–797.
Gurunathan, S., Klinman, D. M., Seder, R. A. (2000). DNA vaccines: immunology, application, and optimization. *Annu. Rev. Immunol.* **18**, 927–974.
Hamori, E., Ruskin, J. (1983). H curves, a novel method of representation of nucleotide series especially suited for long DNA sequences. *J. Biol. Chem.* **258**, 1318–1327.
Hatta, M., Gao, P., Halfmann, P., Kawaoka, Y. (2001). Molecular basis for high virulence of Hong Kong H5N1 influenza A viruses. *Science* **293**, 1840–1842.
Hawkins, D. M., Basak, S. C., Kraker, J., Geiss, K. T., Witzmann, F. A. (2006). Combining chemodescriptors and biodescriptors in quantitative structure-activity relationship modeling. *J. Chem. Inf. Model.* **46**, 9–16.
Jeffrey, H. J. (1990). Chaos game representation of gene structure. *Nucleic Acids Res.* **18**, 2163–2170.
Jiang, M., Zhu, B. (2005). Protein folding on the hexagonal lattice in the HP model. *J. Bioinform. Comput. Biol.* **3**, 19–34.
Karp, G. (2008). Biological molecules. Cell and Molecular Biology. John Wiley & Sons (Asia) Pte Ltd, Asia, p. 31.
Kendrew, J. C. (1959). Structure and function in myoglobin and other proteins. *Fed. Proc.* **18**, 740–751.
Kendrew, J. C., Bodo, G., Dintzis, H. M., Parrish, R. G., Wyckoff, H., Phillips, D. C. (1958). A three-dimensional model of the myoglobin molecule obtained by x-ray analysis. *Nature* **181**, 662–666.

Kendrew, J. C., Dickerson, R. E., Strandberg, B. E., Hart, R. G., Davies, D. R., Phillips, D. C., et al. (1960). Structure of myoglobin: A three-dimensional Fourier synthesis at 2 A. resolution. *Nature* **185**, 422–427.

Kinjo, A. R., Yamashita, R., Nakamura, H. (2010). PDBj Mine: design and implementation of relational database interface for Protein Data Bank Japan. *Database (Oxford) Database published online August 25, 2010)* **2010**, baq021. http://database.oxfordjournals.org/cgi/crossref-forward-links/2010/0/baq021.

Koch-Weser, J., Sellers, E. M. (1976). Binding of drugs to serum albumin (first of two parts). *N. Engl. J. Med.* **294**, 311–316.

Koonin, E. V., Senkevich, T. G., Dolja, V. V. (2006). The ancient Virus World and evolution of cells. *Biol. Direct* **1**, 29.

Koonin, E. V., Wolf, Y. I., Nagasaki, K., Dolja, V. V. (2009). The complexity of the virus world. *Nat. Rev. Microbiol.* **7**, 250.

Kramer, L. D., Li, J., Shi, P. Y. (2007). West Nile virus. *Lancet Neurol.* **6**, 171–181.

Kristensen, D. M., Mushegian, A. R., Dolja, V. V., Koonin, E. V. (2010). New dimensions of the virus world discovered through metagenomics. *Trends Microbiol.* **18**, 11–19.

Kumar, S., Nei, M., Dudley, J., Tamura, K. (2008). MEGA: a biologist-centric software for evolutionary analysis of DNA and protein sequences. *Brief Bioinform.* **9**, 299–306.

Lai, M. M. (1990). Coronavirus: organization, replication and expression of genome. *Annu. Rev. Microbiol.* **44**, 303–333.

Lam, T. T., Hon, C. C., Pybus, O. G., Kosakovsky Pond, S. L., Wong, R. T., Yip, C. W., et al. (2008). Evolutionary and transmission dynamics of reassortant H5N1 influenza virus in Indonesia. *PLoS Pathog.* **4**, e1000130.

Larionov, S., Loskutov, A., Ryadchenko, E. (2008). Chromosome evolution with naked eye: palindromic context of the life origin. *Chaos* **18**, 013105.

Lee, D. G., Kim, P. I., Park, Y., Woo, E. R., Choi, J. S., Choi, C. H., et al. (2002). Design of novel peptide analogs with potent fungicidal activity, based on PMAP-23 antimicrobial peptide isolated from porcine myeloid. *Biochem. Biophys. Res. Commun.* **293**, 231–238.

Lengauer, T., Zimmer, R. (2000). Protein structure prediction methods for drug design. *Brief. Bioinform.* **1**, 275–288.

Leong, P. M., Morgenthaler, S. (1995). Random walk and gap plots of DNA sequences. *Comput. Appl. Biosci.* **11**, 503–507.

Li, Z., Baker, M. L., Jiang, W., Estes, M. K., Prasad, B. V. (2009). Rotavirus architecture at subnanometer resolution. *J. Virol.* **83**, 1754–1766.

Li, C., Wang, J. (2005). New invariant of DNA sequences. *J. Chem. Inf. Model.* **45**, 115–120.

Li, C., Xing, L., Wang, X. (2008). 2-D graphical representation of protein sequences and its application to coronavirus phylogeny. *BMB Rep.* **41**, 217–222.

Li, C., Yu, X., Yang, L., Zheng, X., Wang, Z. (2009). 3-D maps and coupling numbers for protein sequences. *Physica A* **388**, 1967–1972.

Liao, B., Liu, Y., Li, R., Zhu, W. (2006). Coronavirus phylogeny based on triplets of nucleic acids bases. *Chem. Phys. Lett.* **421**, 313–318.

Liu, Y. Z., Wang, T. M. (2010). Vector representations and related matrices of DNA primary sequence based on L-tuple. *Math. Biosci.* **227**, 147–152.

Locatelli, F., Vecchio, L. D. (2001). Darbepoetin alfa Amgen. *Curr. Opin. Investig. Drugs* **2**, 1097–1104.

Lopez, J. A., Maldonado, A. J., Gerder, M., Abanero, J., Murgich, J., Pujol, F. H., et al. (2005). Characterization of neuraminidase-resistant mutants derived from rotavirus porcine strain OSU. *J. Virol.* **79**, 10369–10375.

MacCallum, R. M., Martin, A. C., Thornton, J. M. (1996). Antibody-antigen interactions: contact analysis and binding site topography. *J. Mol. Biol.* **262**, 732–745.

Makowski, L., Rodi, D. J., Mandava, S., Minh, D. D., Gore, D. B., Fischetti, R. F. (2008). Molecular crowding inhibits intramolecular breathing motions in proteins. *J. Mol. Biol.* **375**, 529–546.

Markley, J. L., Ulrich, E. L., Berman, H. M., Henrick, K., Nakamura, H., Akutsu, H. (2008). BioMagResBank (BMRB) as a partner in the Worldwide Protein Data Bank (wwPDB): new policies affecting biomolecular NMR depositions. *J. Biomol. NMR* **40**, 153–155.

Menne, K. M., Hermjakob, H., Apweiler, R. (2000). A comparison of signal sequence prediction methods using a test set of signal peptides. *Bioinformatics* **16**, 741–742.

Meyer, M. C., Guttman, D. E. (1968). The binding of drugs by plasma proteins. *J. Pharm. Sci.* **57**, 895–918.

Morrow, M. P., Weiner, D. B. (2010). DNA drugs come of age. *Sci. Am.* **303**, 49–53.

Moscona, A. (2004). Oseltamivir-resistant influenza? *Lancet* **364**, 733–734.

Moscona, A. (2005). Oseltamivir resistance–disabling our influenza defenses. *N. Engl. J. Med.* **353**, 2633–2636.

Moscona, A. (2009). Global transmission of oseltamivir-resistant influenza. *N. Engl. J. Med.* **360**, 953–956.

Mount, D. (2004). Bioinformatics, sequence and genome analysis. 2nd edn. Cold Spring Harbor Press, Cold Spring Harbor, NY, pp. 337.

Munteanu, C. R., Magalhaes, A. L., Uriarte, E., Gonzalez-Diaz, H. (2009). Multi-target QPDR classification model for human breast and colon cancer-related proteins using star graph topological indices. *J. Theor. Biol.* **257**, 303–311.

Nakamura, H., Ito, N., Kusunoki, M. (2002). Development of PDBj: Advanced database for protein structures. *Tanpakushitsu Kakusan Koso* **47**, 1097–1101.

Nandy, A. (1994). A new graphical representation and analysis of DNA sequence structure: I. methodology and application to globin genes. *Curr. Sci.* **66**, 309–314.

Nandy, A. (2009). Empirical relationship between intra-purine and intra-pyrimidine differences in conserved gene sequences. *PLoS ONE* **4**, e6829.

Nandy, A. (2010). Towards stable vaccines: Contributions from DNA and protein numerical characterization studies. 50th Anniversary Celebration with Mathematical Chemistry, Universidad de Pamplona, Pamplona, Colombia.

Nandy, A., Basak, S. C. (2010). New approaches to drug-DNA interactions based on graphical representation and numerical characterization of DNA sequences. *Curr. Comput. Aided Drug Des.* **6**, 283–289.

Nandy, A., Basak, S. C., Gute, B. D. (2007). Graphical representation and numerical characterization of H5N1 avian flu neuraminidase gene sequence. *J. Chem. Inf. Model.* **47**, 945–951.

Nandy, A., Ghosh, A., Nandy, P. (2009). Numerical characterization of protein sequences and application to voltage-gated sodium channel alpha subunit phylogeny. *In Silico Biol.* **9**, 77–87.

Nandy, A., Harle, M., Basak, S. C. (2006). Mathematical descriptors of DNA sequences: development and applications. *ARKIVOC* **9**, 211–238.

Nandy, A., Nandy, P. (1995). Graphical analysis of DNA sequence structure: II. Relative abundances of nucleotides in DNAs, gene evolution and duplication. *Curr. Sci.* **68**, 75–85.

Nandy, A., Nandy, P. (2003). On the uniqueness of quantitative DNA difference descriptors in 2D graphical representation models. *Chem. Phys. Lett.* **368**, 102–107.

Needleman, S. B., Wunsch, C. D. (1970). A general method applicable to the search for similarities in the amino acid sequence of two proteins. *J. Mol. Biol.* **48**, 443–453.

Novic, M., Randic, M. (2008). Representation of proteins as walks in 20-D space. *SAR QSAR Environ. Res.* **19**, 317–337.

Opitz, C. A., Kulke, M., Leake, M. C., Neagoe, C., Hinssen, H., Hajjar, R. J., et al. (2003). Damped elastic recoil of the titin spring in myofibrils of human myocardium. *Proc. Natl. Acad. Sci. USA* **100**, 12688–12693.

Otvos, L., Jr. (2008). Peptide-based drug design: here and now. *Methods Mol. Biol.* **494**, 1–8.

Owens, J. (2004). Building blocks for peptide drugs. *Nat. Rev. Drug Discov.* **3**, 476.

Owoade, A. A., Gerloff, N. A., Ducatez, M. F., Taiwo, J. O., Kremer, J. R., Muller, C. P. (2008). Replacement of sublineages of avian influenza (H5N1) by reassortments, sub-Saharan Africa. *Emerg. Infect. Dis.* **14**, 1731–1735.

Packhaeuser, C. B., Schnieders, J., Oster, C. G., Kissel, T. (2004). In situ forming parenteral drug delivery systems: an overview. *Eur. J. Pharm. Biopharm.* **58**, 445–455.

Peng, C. K., Buldyrev, S. V., Goldberger, A. L., Havlin, S., Sciortino, F., Simons, M., et al. (1992). Long-range correlations in nucleotide sequences. *Nature* **356**, 168–170.

Perkins, D. N., Pappin, D. J., Creasy, D. M., Cottrell, J. S. (1999). Probability-based protein identification by searching sequence databases using mass spectrometry data. *Electrophoresis* **20**, 3551–3567.

Perutz, M. F. (1960). Structure of hemoglobin. *Brookhaven Symp. Biol.* **13**, 165–183.

Perutz, M. F., Rossmann, M. G., Cullis, A. F., Muirhead, H., Will, G., North, A. C. (1960). Structure of haemoglobin: a three-dimensional Fourier synthesis at 5.5-A. resolution, obtained by X-ray analysis. *Nature* **185**, 416–422.

Perutz, M. F., Weisz, O. (1947). Crystal structure of human carboxyhaemoglobin. *Nature* **160**, 786.

Peters, B., Sidney, J., Bourne, P., Bui, H. H., Buus, S., Doh, G., et al. (2005). The design and implementation of the immune epitope database and analysis resource. *Immunogenetics* **57**, 326–336.

Phillips, R. E., Rowland-Jones, S., Nixon, D. F., Gotch, F. M., Edwards, J. P., Ogunlesi, A. O., et al. (1991). Human immunodeficiency virus genetic variation that can escape cytotoxic T cell recognition. *Nature* **354**, 453–459.

Powers, E. T., Powers, D. L. (2008). Mechanisms of protein fibril formation: nucleated polymerization with competing off-pathway aggregation. *Biophys. J.* **94**, 379–391.

Ramachandran, G. N., Ramakrishnan, C., Sasisekharan, V. (1963). Stereochemistry of polypeptide chain configurations. *J. Mol. Biol.* **7**, 95–99.

Ramachandran, G. N., Sasisekharan, V. (1968). Conformation of polypeptides and proteins. *Adv. Protein Chem.* **23**, 283–438.

Randic, M. (2004). Graphical representations of DNA as 2-D map. *Chem. Phys. Lett.* **386**, 468.

Randic, M. (2006). Spectrum-like graphical representation of DNA based on codons. *Acta Chim. Slov.* **53**, 477–485.

Randic, M., Butina, D., Zupan, J. (2006). Novel 2-D graphical representation of proteins. *Chem. Phys. Lett.* **419**, 528–532.

Randic, M., Lers, N., Plavsic, D., Basak, S. C., Balaban, A. T. (2005). Four-color map representation of DNA or RNA sequences and their numerical characterization. *Chem. Phys. Lett.* **407**, 205–208.

Randic, M., Novic, M., Vracko, M. (2008). On novel representation of proteins based on amino acid adjacency matrix. *SAR QSAR Environ. Res.* **19**, 339–349.

Randic, M., Vracko, M., Lers, N., Plavsic, D. (2003a). Analysis of similarity/dissimilarity of DNA sequences based on novel 2-D graphical representation. *Chem. Phys. Lett.* **371**, 202.

Randic, M., Vracko, M., Lers, N., Plavsic, D. (2003b). Novel 2-D graphical representation of DNA sequences and their numerical characterization. *Chem. Phys. Lett.* **368**, 1.

Randic, M., Vracko, M., Nandy, A., Basak, S. C. (2000). On 3-D graphical representation of DNA primary sequences and their numerical characterization. *J. Chem. Inf. Comput. Sci.* **40**, 1235–1244.

Randic, M., Vracko, M., Zupan, J., Novic, M. (2003c). Compact 2-D graphical representation of DNA. *Chem. Phys. Lett.* **373**, 558.

Randic, M., Zupan, J., Balaban, A. T. (2004). Unique graphical representation of protein sequences based on nucleotide triplet codons. *Chem. Phys. Lett.* **397**, 247–252.

Raychaudhury, C., Nandy, A. (1999). Indexing scheme and similarity measures for macromolecular sequences. *J. Chem. Inf. Comput. Sci.* **39**, 243–247.

Rayment, I. (1996). Kinesin and myosin: molecular motors with similar engines. *Structure* **4**, 501–504.

Reid, A. H., Fanning, T. G., Janczewski, T. A., Taubenberger, J. K. (2000). Characterization of the 1918 "Spanish" influenza virus neuraminidase gene. *Proc. Natl. Acad. Sci. USA* **97**, 6785–6790.

Retief, J. D. (2000). Phylogenetic analysis using PHYLIP. *Methods Mol. Biol.* **132**, 243–258.

Roach, P., Farrar, D., Perry, C. C. (2005). Interpretation of protein adsorption: surface-induced conformational changes. *J. Am. Chem. Soc.* **127**, 8168–8173.

Roach, P., Farrar, D., Perry, C. C. (2006). Surface tailoring for controlled protein adsorption: effect of topography at the nanometer scale and chemistry. *J. Am. Chem. Soc.* **128**, 3939–3945.
Russell, R. J., Haire, L. F., Stevens, D. J., Collins, P. J., Lin, Y. P., Blackburn, G. M., et al. (2006). The structure of H5N1 avian influenza neuraminidase suggests new opportunities for drug design. *Nature* **443**, 45–49.
Ruvkun, G. B., Ausubel, F. M. (1981). A general method for site-directed mutagenesis in prokaryotes. *Nature* **289**, 85–88.
Sanger, F. (1952). The arrangement of amino acids in proteins. *Adv. Protein Chem.* **7**, 1–67.
Sanger, F., Thompson, E. O. (1953). The amino-acid sequence in the glycyl chain of insulin. I. The identification of lower peptides from partial hydrolysates. *Biochem. J.* **53**, 353–366.
Sanger, F., Tuppy, H. (1951). The amino-acid sequence in the phenylalanyl chain of insulin. I. The identification of lower peptides from partial hydrolysates. *Biochem. J.* **49**, 463–481.
Schellekens, H. (2002). Bioequivalence and the immunogenicity of biopharmaceuticals. *Nat. Rev. Drug Discov.* **1**, 457–462.
Shen, H. B., Chou, K. C. (2009). Predicting protein fold pattern with functional domain and sequential evolution information. *J. Theor. Biol.* **256**, 441–446.
Smith, T. F., Waterman, M. S. (1981). Identification of common molecular subsequences. *J. Mol. Biol.* **147**, 195–197.
Smith, T. F., Waterman, M. S., Burks, C. (1985). The statistical distribution of nucleic acid similarities. *Nucleic Acids Res.* **13**, 645–656.
Tamura, K., Dudley, J., Nei, M., Kumar, S. (2007). MEGA4: Molecular Evolutionary Genetics Analysis (MEGA) software version 4.0. *Mol. Biol. Evol.* **24**, 1596–1599.
Tangri, S., Mothe, B. R., Eisenbraun, J., Sidney, J., Southwood, S., Briggs, K., et al. (2005). Rationally engineered therapeutic proteins with reduced immunogenicity. *J. Immunol.* **174**, 3187–3196.
Teague, S. J. (2003). Implications of protein flexibility for drug discovery. *Nat. Rev. Drug Discov.* **2**, 527–541.
Todeschini, R., Ballabio, D., Consonni, V., Mauri, A. (2008). A new similarity/diversity measure for the characterization of DNA sequences. *Croat. Chem. Acta* **81**, 657–664.
Todeschini, R., Consonni, V., Mauri, A., Ballabio, D. (2006). Characterization of DNA primary sequences by a new similarity/diversity measure based on the partial ordering. *J. Chem. Inf. Model.* **46**, 1905–1911.
Tokuriki, N., Oldfield, C. J., Uversky, V. N., Berezovsky, I. N., Tawfik, D. S. (2009). Do viral proteins possess unique biophysical features? *Trends Biochem. Sci.* **34**, 53–59.
Ulmer, J. B., Deck, R. R., Yawman, A., Friedman, A., Dewitt, C., Martinez, D., et al. (1996). DNA vaccines for bacteria and viruses. *Adv. Exp. Med. Biol.* **397**, 49–53.
Ulmer, J. B., Sadoff, J. C., Liu, M. A. (1996). DNA vaccines. *Curr. Opin. Immunol.* **8**, 531–536.
van der Hoek, L., Pyrc, K., Jebbink, M. F., Vermeulen-Oost, W., Berkhout, R. J., Wolthers, K. C., et al. (2004). Identification of a new human coronavirus. *Nat. Med.* **10**, 368–373.

Velankar, S., Best, C., Beuth, B., Boutselakis, C. H., Cobley, N., Sousa Da Silva, A. W., et al. (2010). PDBe: Protein Data Bank in Europe. *Nucleic Acids Res.* **38**, D308–D317.

Vijaykrishna, D., Poon, L. L., Zhu, H. C., Ma, S. K., Li, O. T., Cheung, C. L., et al. (2010). Reassortment of pandemic H1N1/2009 influenza A virus in swine. *Science* **328**, 1529.

Vilar, S., González-Díaz, H. (2010). QSPR models for human Rhinovirus surface networks. *In* Topological Indices for Medicinal Chemistry, Biology, Parasitology, Neurological and Social Networks, (González-Díaz, H. and Munteanu, C. R. Eds.), pp. 145–161.

Vilar, S., Gonzalez-Diaz, H., Santana, L., Uriarte, E. (2008). QSAR model for alignment-free prediction of human breast cancer biomarkers based on electrostatic potentials of protein pseudofolding HP-lattice networks. *J. Comput. Chem.* **29**, 2613–2622.

Vita, R., Zarebski, L., Greenbaum, J. A., Emami, H., Hoof, I., Salimi, N., et al. (2010). The immune epitope database 2.0. *Nucleic Acids Res.* **38**, D854–D862.

Wang, J., Wang, W. (1999). A computational approach to simplifying the protein folding alphabet. *Nat. Struct. Biol.* **6**, 1033–1038.

Wiesner, I., Wiesnerova, D. (2010). 2D random walk representation of Begonia × tuberhybrida multiallelic loci used for germplasm identification. *Biologia Plantarum* **54**, 353–356.

Wu, W. L., Chen, Y., Wang, P., Song, W., Lau, S. Y., Rayner, J. M., et al. (2008). Antigenic profile of avian H5N1 viruses in Asia from 2002 to 2007. *J. Virol.* **82**, 1798–1807.

Yang, X., Yu, X. (2009). An introduction to epitope prediction methods and software. *Rev. Med. Virol.* **19**, 77–96.

Yau, S. S., Wang, J., Niknejad, A., Lu, C., Jin, N., Ho, Y. K. (2003). DNA sequence representation without degeneracy. *Nucleic Acids Res.* **31**, 3078–3080.

# STRUCTURAL, THERMODYNAMIC, AND MECHANISTICAL STUDIES IN UROPORPHYRINOGEN III SYNTHASE: MOLECULAR BASIS OF CONGENITAL ERYTHROPOIETIC PORPHYRIA

By AROLA FORTIAN, DAVID CASTAÑO, ESPERANZA GONZALEZ, ANA LAÍN, JUAN M. FALCON-PEREZ, AND OSCAR MILLET

CIC bioGUNE, Bizkaia Technology Park, Derio, Spain

| | | |
|---|---|---|
| I. | Introduction | 44 |
| II. | Uroporphyrinogen III Synthase | 47 |
| | A. The Catalytic Mechanism of UROIIIS | 47 |
| | B. Interdomain Flexibility is Coupled to the Catalytic Reaction | 51 |
| | C. Thermodynamic Stability of UROIIIS | 52 |
| III. | Molecular Basis of CEP | 54 |
| | A. CEP-Producing Mutations in the *URO-Synthase* Gene | 54 |
| | B. Effect of the Pathogenic Mutations on the Catalytic Activity of UROIIIS | 57 |
| | C. UROIIIS Destabilization Induced by the Pathogenic Mutants | 59 |
| IV. | The Hotspot Mutation C73R-UROIIIS | 62 |
| | A. Structural Characterization of C73R *In Vitro* | 62 |
| | B. Expression of C73R *In Vivo* | 64 |
| V. | Treatment of CEP | 66 |
| | A. Marrow Suppression, BMT, and Gene Therapy | 67 |
| | B. Molecular Chaperones and Inhibitors of the Proteasome | 68 |
| VI. | Concluding Remarks | 69 |
| | References | 71 |

## Abbreviations

| | |
|---|---|
| BMT | bone marrow transplantation |
| CEP | congenital erythropoietic porphyria |
| CFTR | cystic fibrosis transmembrane conductance regulator |
| Gfp | green fluorescent protein |
| HMB | hydroxymethylbilane |
| PBG | porphobilinogen |
| PBGD | porphobilinogen deaminase |
| qRT-PCR | quantitative real-time polymerase chain reaction |

rESA    relative specific activity
SDS     sodium dodecyl sulfate
UROIIIS uroporphyrinogen III synthase

## Abstract

Congenital erythropoietic porphyria (CEP) is a rare autosomal disease ultimately related to deleterious mutations in uroporphyrinogen III synthase (UROIIIS), the fourth enzyme of the biosynthetic route of the heme group. UROIIIS catalyzes the cyclization of the linear tetrapyrrol hydroxymethylbilane (HMB), inverting the configuration in one of the aromatic rings. In the absence of the enzyme (or when ill-functioning), HMB spontaneously degrades to the by-product uroporphyrinogen I, which cannot lead to the heme group and accumulates in the body, producing some of the symptoms observed in CEP patients. In the present chapter, clinical, biochemical, and biophysical information has been compiled to provide an integrative view on the molecular basis of CEP. The high-resolution structure of UROIIIS sheds light on the enzyme reaction mechanism while thermodynamic analysis revealed that the protein is thermolabile. Pathogenic missense mutations are found throughout the primary sequence of the enzyme. All but one of these is rarely found in patients, whereas C73R is responsible for more than one-third of the reported cases. Most of the mutant proteins (C73R included) retain partial catalytic activity but the mutations often reduce the enzyme's stability. The stabilization of the protein *in vivo* is discussed in the context of a new line of intervention to complement existing treatments such as bone marrow transplantation and gene therapy.

## I. Introduction

All natural tetrapyrroles, including hemes, sirohaems, chlorophylls, and vitamin B12, share porphobilinogen (PBG) as a common precursor (Battersby, 2000). The heme group is essential for the survival of the vast majority of living organisms and, in animals, the biosynthetic route for this prosthetic group starts from glycine, whereas in plants an alternative route from glutamic acid has also been described (Leeper, 1985). After the formation of 5-aminolevulinic acid, the subsequent steps leading to the formation of uroporphyrinogen III are conserved and are required for all organisms (see Fig. 1). Further, decarboxylation and metal chelation leads to the formation

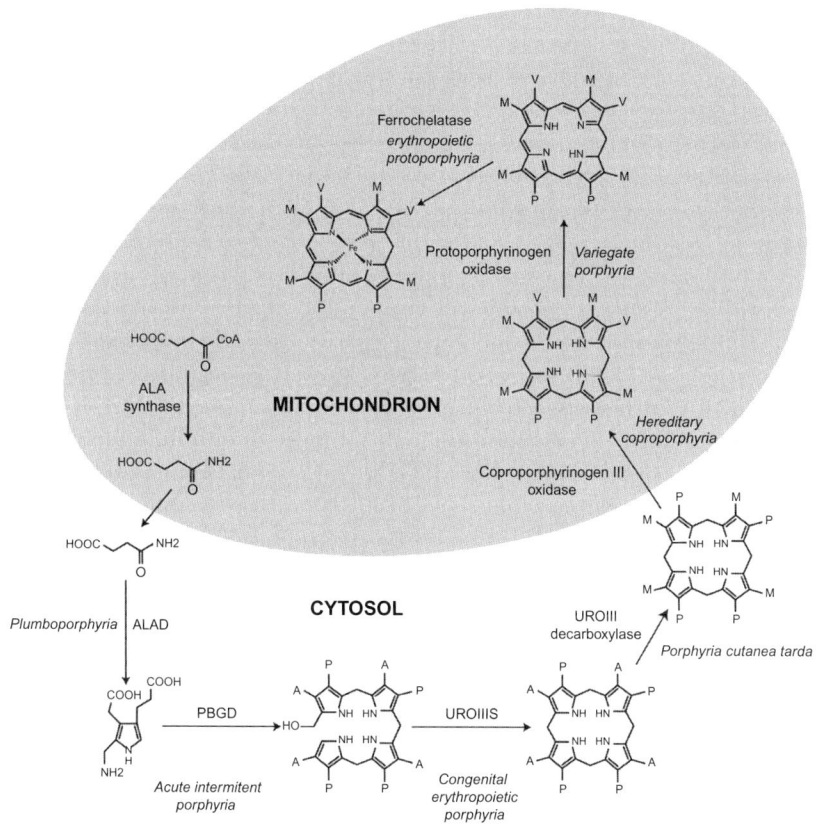

FIG. 1. The human heme group biosynthetic pathway is localized between the mitochondria and the cytoplasm. A deficiency in any of the eight enzymes that belong to this route results in a form of porphyria. Porphyrias are classified according to the enzyme that is ill-functioning.

of the heme group in the cell. Heme group-derivatized cofactors can bind metals and are found associated with proteins active in processes such as electron transfer (cytochrome), oxygen binding (hemoglobin), and oxygen metabolism (hydrolases, catalases, and peroxidases; Dailey, 1997).

Enzymes belonging to this biosynthetic route have been object of intense mechanistic and structural research with high-resolution structures available for all of the involved players. In the PBG synthase mechanism, a monovalent ion (typically a sodium cation) is required to enhance the production of PBG (Frere et al., 2002). This metabolite is the substrate for porphobilinogen deaminase (PBGD), which catalyzes the condensation of four units of PBG to yield the linear tetrapyrrol hydroxymethylbilane (HMB). Interestingly, there is only one active site indicating that a domain rearrangement has to occur during the reaction pathway to accommodate the intermediates. Tetrapyrrol HMB is unstable and is immediately cycled by the following enzyme in the pathway, uroporphyrinogen III synthase (UROIIIS) to produce uroporphyrinogen III.

Just one mutation in the primary sequence of any of the involved enzymes is enough to produce a congenital disease, generically called porphyria, although the name of the pathology and the severity of the disease are largely dependent on the specific enzyme affected (see Fig. 1 for the disease nomenclature). Depending on whether the excess production of porphyrin precursors occurs primarily in the liver or in the erythron, porphyrias are also classified as being hepatic or erythropoietic (Anderson et al., 2001). Congenital erythropoietic porphyria (CEP) is an autosomal recessive rare disease that belongs to the latter group and is the consequence of a deficient (but not absent) activity in the fourth enzyme of the heme biosynthetic route (UROIIIS). CEP is panethnic and the originating mutations are generally widespread throughout the sequence of the enzyme. The mutations show a low degree of repetition among patients. One striking exception is C73R, present in more than one-third of all reported cases. Decreased levels of uroporphyrinogen III constitute the *enzymatic defect* and are accompanied by the *metabolic defect*: the accumulation of the substrate (HMB) which is quickly nonenzymatically converted to uroporphyrinogen I. This by-product (all as isomer I porphyrins) cannot be properly metabolized and generates coproporphyrinogen I and coproporphyria I, which accumulates in large amounts in the body.

The detection of abnormal levels of by-products and intermediate metabolites are the usual way for diagnosing CEP, later confirmed with a genetic analysis of the patient and the progenitors. Patients suffering from this disease show very heterogeneous clinical manifestations, including skin photosensitization, hemolytic anemia, and splenomegaly, and they have red-colored urine. Regular treatment of this disease is largely restricted to preventive practices and to palliative care of the existing symptoms. CEP patients should avoid unprotected exposure to sunlight or ultraviolet light. Erythropoiesis is often reduced by blood transfusions which can be sporadic or chronic depending on the severity of the phenotype. Bone marrow transplantation (BMT) has been proven to be the only therapy capable of curing the disease, but with uneven degrees of success. The capability of this surgery for alleviating the severe manifestations of this disease has provided the rationale for hematopoietic stem cell therapy, which is currently under investigation.

In diseases like CEP, it is diagnostically and therapeutically essential to understand the multiple mechanisms that relate the specific mutants with the pathology. CEP is related to UROIIIS malfunction and this chapter will first describe the current knowledge on the structural basis for the catalytic mechanisms and the folding thermodynamics of this enzyme. We will then turn to the molecular basis of the pathology based on the structural and thermodynamic analyses of the mutants. The special attention that the hotspot mutation C73R has received from the scientific community will also be described in detail. Finally, based on the accumulated data, all possible lines of therapeutic intervention will be discussed.

## II. Uroporphyrinogen III Synthase

UROIIIS (EC 4.2.1.75), the central enzyme in the heme biosynthetic route, accelerates the production of uroporphyrinogen III (an energetically unfavorable chemical reaction) at the same time as it suppresses the spontaneous reaction pathway to yield uroporphyrinogen I (Fig. 2). The current knowledge on the structure and molecular biophysics of this intriguing enzyme are discussed in this chapter.

### A. The Catalytic Mechanism of UROIIIS

The enzyme UROIIIS catalyzes the cyclization of PBG and inverts the D ring to render uroporphyrinogen III. The synthase is a very efficient enzyme, with a high-turnover rate ($2240 \, \text{min}^{-1}$) and a low $K_m$ value (0.15 µM; Cunha et al., 2007). The abiotic condensation of the precursor yields the I isomer (Fig. 2). In the consensus mechanism, the enzymatic reaction starts with the removal of the hydroxyl group from C20 in HMB to yield a reactive azafulvenium cation (Fig. 3). A subsequent electrophilic attachment at C16 yields a spirocyclic pyrroleine intermediate, which may break in the opposite direction to form a new azafulvenium cation. Rotation of the D ring is required to bring C19 close to the reactive carbocation C15, allowing the formation of the macrocycle. In favor of

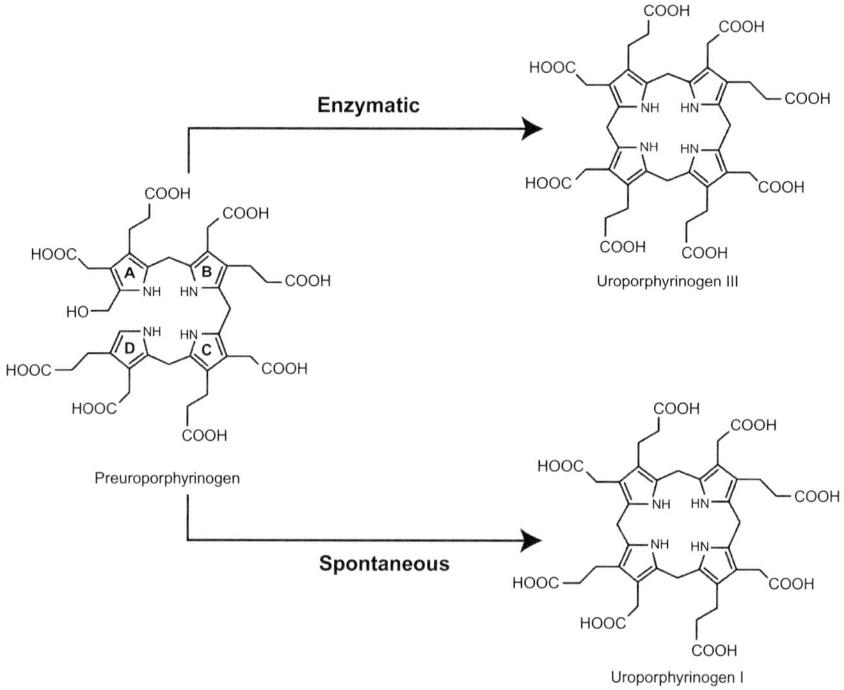

FIG. 2. Uroporphyrinogen III synthase catalyzes the cyclization of the linear tetrapyrrol preuroporphyrinogen with inversion in the configuration of the ring D to produce uroporphyrinogen III. In the absence of enzyme (or when it is ill-functioning), the substrate degrades spontaneously to uroporphyrinogen I.

this mechanism, a plethora of biochemical data has been obtained using radiolabeled substrates and a spirolactamic inhibitor which acts as a potent inhibitor of the enzyme (Petersen et al., 1998). The electrophilic attack on C19 is thermodynamically more favorable (Tietze and Geissler, 1993) and leads to the nonenzymatic product uroporphyrinogen I. In principle, the spirocyclic intermediate could also be obtained after the electrophilic attack on C19 by a 1,5-sigma tropic shift, but density function theory (DAFT) calculations have shown that this route is a kinetic dead-end (Silva and Ramos, 2008). Thus, the substrate must bind the enzyme in a conformation that prevents C19 from reacting with C20.

The three-dimensional structure of human UROIIIS ($\approx 28$ kDa) revealed that this protein consists of a bilobed structure with two domains

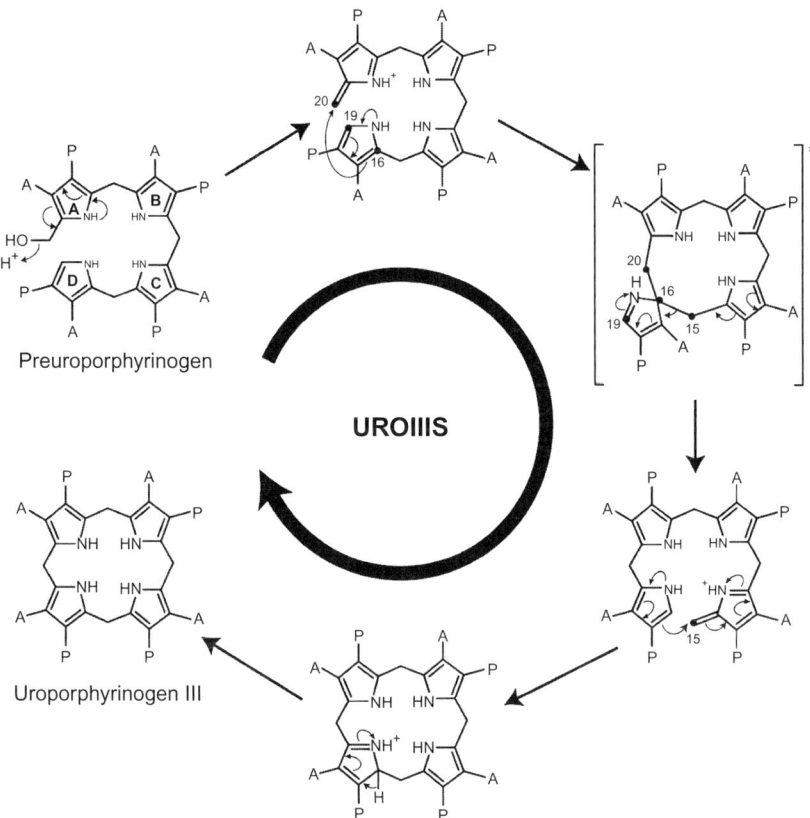

Fig. 3. Enzymatic conversion of preuroporphyrinogen into uroporphyrinogen III. The enzyme creates an azafulvene intermediate that evolves to the product via a transition state that has a spiranic topology for the ring D.

of a similar fold of a β-sheet surrounded by α-helices (Mathews et al., 2001; Fig. 4A). Both domains are separated in the human protein by a two stranded β-sheet, but this linker is much more flexible in UROIIIS from *Thermus thermophilus*. The available structure of the complex between the product and the thermophilic protein (Schubert et al., 2008) revealed that the active site is located in the cleft between the two domains, in full agreement with an NMR titration of UROIIIS with some weak inhibitors (Cunha et al., 2007). In the complex, the tetrapyrrole is completely surrounded by the enzyme even though most of the interactions are

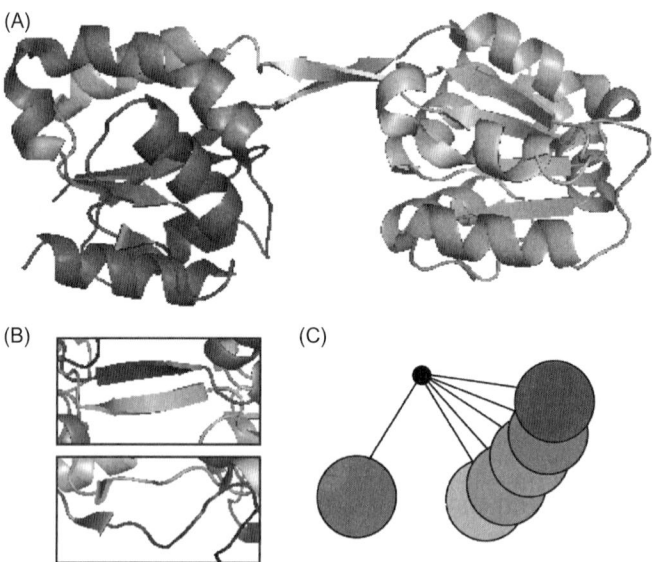

FIG. 4. (A) Ribbon representation of human uroporphyrinogen III synthase (1jr2). Both domains share a β-sheet/α-helix topology. (B) Detail of the hinge region for the human (top) and *Thermus thermophilus* (bottom) where the different degree of order can be observed. (C) Cartoon representing the interdomain reorientations found in the different *T. thermophilus* UROIIIS structures available. The closure angle experiences changes up to 90°.

mediated by ordered water molecules. Direct interactions between the product and the enzyme are between carboxylate side chains and amide groups of the protein. Remarkably, rings A and B form tight interactions with the enzyme whereas carboxylates belonging to the D ring extend into solvents, having no interactions with the enzyme. The lack of nonpolar interactions and the nonspecific coordination explains the very low sequence homology in UROIIIS (for instance, it only shares 14% sequence identity with the *T. thermophilus* enzyme). However, the few conserved residues are mostly located in the active site of the molecule, probably to confer substrate specificity.

A model for the catalytic mechanism of the enzyme has emerged based on the structural data available (Schubert et al., 2008): the A ring seems to be responsible for the placement of the substrate in the active site and eventually for the enzyme's conformational rearrangement (*vide infra*).

Upon binding, Tyr 168 might be properly positioned at the hydrogen bond with the hydroxyl adjacent to C20 in HMB, contributing to the loss of a water molecule from the substrate to generate the azafulvene intermediate. The Y168 residue is one of the few residues conserved among species and it shows a significant decrease in activity upon mutation to alanine in the *Anacystis nidulans* homologue protein (Roessner et al., 2002). Once the azafulvene intermediate is formed, UROIIIS forces a conformation in the tetrapyrrol that separates C19 from C20 to avoid the electrophilic attack that would lead to the formation of the I isomer. The high degree of flexibility conferred to the D ring may be required for the flip that this ring has to undergo prior to the formation of the III isomer product.

### B. Interdomain Flexibility is Coupled to the Catalytic Reaction

The set of high-resolution structures available reveal that the interdomain orientation is very variable (Schubert et al., 2008), most likely due to thermal motion between the two domains caused by the shallow energy landscape of the native state of UROIIIS. The analysis of amide transversal relaxation rates (unpublished data) was also incompatible with a static view of the protein. Figure 4B shows an overlap of the different structures, aligned by the N-domain to show the wide range of rotations that can reach 90° with respect to the closure axis placed on the hinge region. According to the structural model, the A ring creates contact with the two domains upon binding. Thus, the enzyme likely undergoes domain closure upon substrate binding following a "Venus flytrap mechanism," as in bacterial periplasmic-binding proteins (Millet et al., 2003; Castano and Millet, 2010). Nevertheless, the catalytic efficiency of the enzyme ($k_{cat}/K_m$) is in the order of the diffusion limit indicating that such motion has to be fast because it does not interfere with the binding of the substrate to the enzyme.

It has long been speculated that UROIIIS could form a metabolon with PBGD, the previous enzyme in the biosynthetic route. HMB is very unstable and such complexes would minimize the formation of uroporphyrinogen I. Nuclear magnetic resonance studies performed in our laboratory and by others (Cunha et al., 2007) have shown that this complex is not produced under the *in vitro* conditions of the NMR sample. However, this experiment has been performed with *apo* UROIIIS and the accessible interdomain motions could have prevented the formation of such complexes. Thus, the formation of a ternary complex between HMB, UROIIIS, and PBGD cannot be completely ruled out.

### C. Thermodynamic Stability of UROIIIS

From the first purification attempts of UROIIIS, it became clear that this enzyme is thermolabile (Bogorad, 1958). Despite this early observation, a detailed thermodynamic characterization has only recently been published (Fortian et al., 2009). Circular dichroism is sensitive to the secondary structure of the protein (Tadeo et al., 2009) and it can be used to determine the temperature when half of the protein is unfolded, also named the melting temperature ($T_m$), the most common parameter for defining protein thermal stability (Becktel and Schellman, 1987). In UROIIIS, the $T_m$ values are apparent since they depend on the scanning rate, suggesting the presence of an irreversible denaturation process (Miles et al., 1995; Plaza del Pino et al., 2000). UROIIIS is indeed metastable and its native concentration decays over time. Such decay can be observed by the time dependence of the CD ellipticity and is characterized by an apparent kinetic rate ($k_{app}$). Not surprisingly, the loss of catalytic activity parallels the exponential decay of the ellipticity, indicating that enzyme misfunction over time is due to protein unfolding. However, at physiological temperature and *in vitro*, UROIIIS has a half-life time of 61.1 h, a long time for the enzyme to exert its function in the cell.

The simplest model that explains this dynamic behavior of the protein is the three-state model, used to describe the so-called kinetically stable proteins (Baker and Agard, 1994):

$$N \underset{k_{UN}}{\overset{k_{NU}}{\rightleftharpoons}} U \overset{k_{UF}}{\rightarrow} F$$

where $k_{NU}$, $k_{UN}$, and $k_{UF}$ are the kinetic rates of the process, N is the native state and U the unfolded state that spontaneously converts into the more stable (often inactive) final state F. In this model, the only certainty is that F is thermodynamically more stable than N (Fig. 5). The unfolded state is usually lowly populated without any restriction in stability and striking examples have been reported where U is more stable than N (Jaswal et al., 2002). However, the folded conformation is temporarily stabilized due to its unfolding kinetics (kinetic stability; Rodriguez-Larrea et al., 2006).

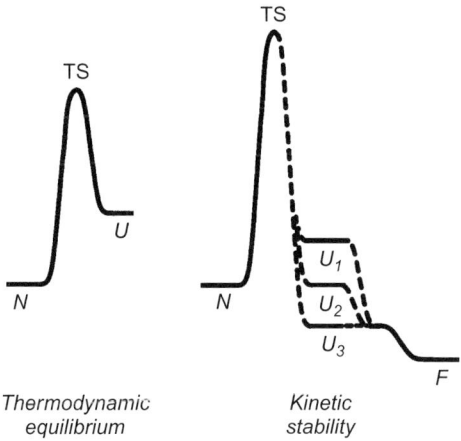

FIG. 5. Thermodynamic versus kinetic stability. In the thermodynamic equilibrium, the native state is always more stable that the unfolded one. When the kinetic stability regime applies, the folded state is always less stable than another conformation of the protein (generically called F). The N state only exists for a certain time and due to the large energy barrier that separates it from the transition state. Under these circumstances, the unfolded state can have less, the same or more stability than the native form of the protein.

In the case of UROIIIS, dynamic light-scattering measurements revealed that the F state is an aggregate form of UROIIIS (Fortian et al., 2009). Nuclear magnetic resonance experiments monitoring of the irreversible unfolding process could not detect signals for the unfolded state, indicating that it does not accumulate because $k_{UF}$ is much larger than $k_{UN}$. Under these conditions the apparent constant $k_{app}$ is equivalent to the unfolding rate ($k_{app} \approx k_{NU}$; Duy and Fitter, 2005). At physiological temperature, $k_{NU}$ is equal to $1.6 \pm 0.2 \times 10^{-6}\,\text{s}^{-1}$ for wild-type URIIIIS. The temperature dependence of the unfolding rate constant can be used to extract the enthalpic contribution to the activation energy $E_a$, assuming the Arrhenius model:

$$k_{NU} = A \times e^{-E_a/RT} \quad (1)$$

where $R$ is the universal gas constant and $A$ is the frequency factor that includes the entropic contribution. The activation energy for wild-type UROIIIS, derived from the slope of the linear plot of $\ln(k_{NU})$ versus $1/T$ is

101.5 kcal mol$^{-1}$, and $E_a$ is the enthalpy difference between the active folded conformation (N state) and the highest energy conformation in the path toward unfolding (Fig. 5). For wild-type UROIIIS, this barrier is large, falling in the range of values of proteins that experience kinetic stability (Duy and Fitter, 2005).

In summary, UROIIIS is an unstable protein that remains folded due to kinetic stability. Although this intrinsic instability does not compromise the catalytic role of the enzyme, we will show in the following chapters that this mechanism is crucial for understanding the effect that many pathogenic mutations have on UROIIIS activity.

### III. Molecular Basis of CEP

Genome analysis of CEP patients is the first step in understanding the molecular mechanisms of this disease and its relationship with the phenotype. In the case of this rare disease, such analysis has been limited by the extremely low number of reported case studies. However, such a low number of cases have enabled a thorough characterization of each and every one of the missense mutations. Advances in the knowledge of the molecular pathology of CEP are described in this chapter.

#### A. CEP-Producing Mutations in the URO-Synthase Gene

A decrease in UROIIIS activity produced by mutations is ultimately responsible for human CEP (Desnick and Astrin, 2002). Genotyping of more than 110 CEP patients identified different types of alterations at the DNA level. The normal URO-synthase gene is constituted by 10 exons and 10 introns. In four pathogenic samples, DNA insertions in the coding regions of the sequence were observed (Boulechfar et al., 1992), with two of them of more than 400 bp (Shady et al., 2002). Deletions and rearrangements are also rare, with only three cases reported (Shady et al., 2002). Defects in the promoter region can also lead to CEP, with four cases reported in the literature (Desnick and Astrin, 2002; Stenson et al., 2008). Alternative splicing can also be responsible for CEP (Bishop et al., 2010). Nonetheless, single point mutations in the coding region are, by far, the most common source of pathogenesis with 24 missense mutations and one nonsense mutation observed. This set of mutants was

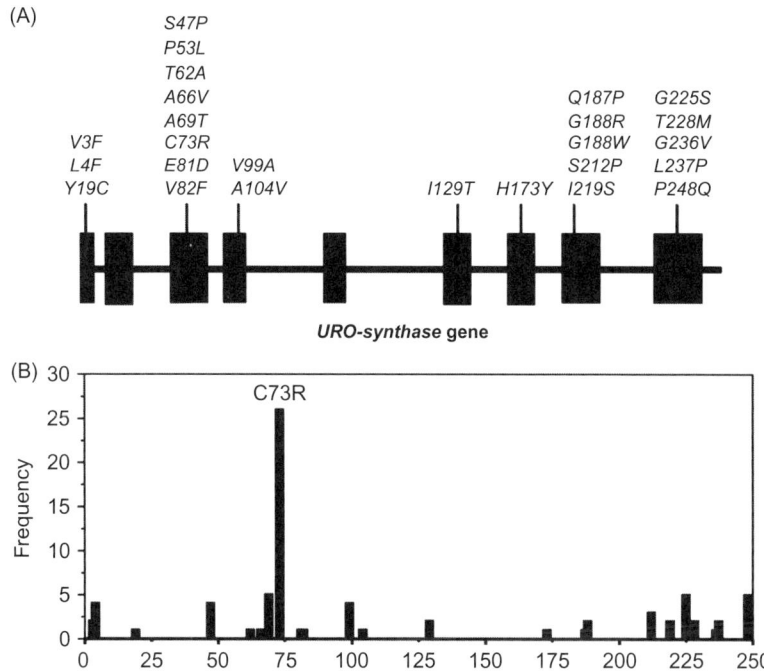

FIG. 6. (A) The CEP reported mutations are distributed throughout seven out of the 10 exons in the *URO-synthase* gene. (B) The average frequency is 2.3 patients per mutation except for C73R that is present in at least one-third of all CEP patients.

found throughout the sequence (Fig. 6A), with only two mutants originating at the same codon (G188R and G188W). Missense mutants affect the two domains of the protein, with the substrate-binding region not showing a special concentration of mutations. All but one mutation (V82F) occurs in amino acid residues that are also conserved in the murine UROIIIS polypeptide sequence.

The distribution of mutations among patients is quite shallow with a mean value of two patients per mutation (see Fig. 6B). A striking exception is C73R, which was present in about one-third of the total reported cases. This hotspot mutation is panethnic, with patients belonging to very diverse racial and demographic groups (Stenson et al., 2003). The reasons why C73R is so frequent remain unclear: C73 does not belong to a CpG dinucleotide region and haplotype analysis of three unrelated families indicated that C73R was due to independent mutational events (Frank et al., 1998). Most of the other mutations are much more traceable. An example is S47P, which is only found in patients belonging to the same Palestinian family (Ged et al., 2004).

The recessive characteristic of CEP implies that both progenitors carry a mutation in the enzyme. Figure 7 shows a chart with all the mutations that have been found color coded depending on the severity of the disease (according to the classification established in the literature; Warner et al., 1992; Fontanellas et al., 1996). Mild patients (yellow) mainly suffer

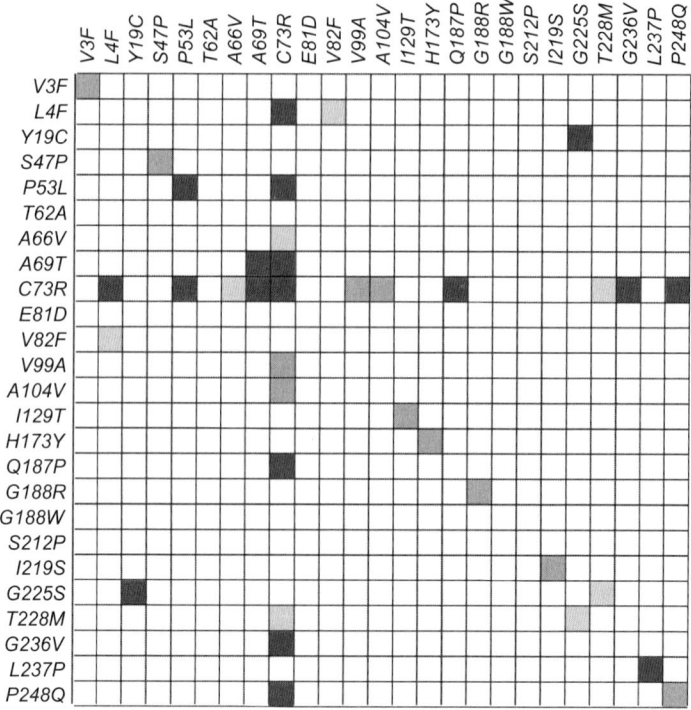

FIG. 7. Genotype/phenotype analysis of the studied cases of congenital erythropoietic porphyria. Mild, moderate, and severe phenotypes are colored in yellow, orange, and red, respectively. Patients with different phenotypes but sharing genotype are represented by triangles. (See color plate 1).

cutaneous symptoms and can often live a normal life in adulthood. Moderate patients (orange) have mild anemia and/or weak splenomegaly with skin lesions. Severe patients (red) are transfusion-dependent and sometimes show reduced life expectancy. Two conclusions can be extracted from inspection of Fig. 7. First, most of the patients inherited the same mutation from both progenitors (the colored squares fall in the diagonal). This coincidence can be attributed to a local spread of the mutation within demographic groups. Second, C73R emerges as the most aggressive mutation and the vast majority of patients carrying this mutation are considered "severe." Thus, C73R is not only the most frequent but also the most pathogenic mutation and it has received special attention in the literature.

## B. Effect of the Pathogenic Mutations on the Catalytic Activity of UROIIIS

In a recent study, wild-type UROIIIS and 25 point mutants were expressed in *Escherichia coli* and isolated with a final purity beyond 97% as determined by SDS gel electrophoresis (Fortian et al., 2009). The yields of pure protein for the different mutants, relative to wild-type UROIIIS, are listed in Table I. All proteins but one showed a reduction in the amount of the molecule obtained per liter of culture as compared to wild-type UROIIIS, with yields spanning from 1.2% to 55%, while in the case of P53L no expression was detected. The decrease in the yield can be as a result of codon alteration at the DNA level or it can be produced by changes in the physicochemical properties of the protein, such as a decrease in the resistance to proteolysis or a reduction in the chelating properties of the mutated enzyme.

The specific activity of UROIIIS can be measured by using a method that uses this enzyme in tandem with PBGD to generate preuroporphyrinogen *in situ* from PBG (Shoolingin-Jordan and Leadbeater, 1997). In this assay, uroporphyrinogen III formed by the synthase is subsequently chemically oxidized to uroporphyrin III and separated from the I isomer using high-pressure liquid chromatography. Figure 8 shows chromatograms for some representative mutants where uroporphyrin III and uroporphyrin I were eluted at 6.5 and 5.3 min, respectively. The relative specific activities (rESA) for the full mutant set are also listed in Table I as compared to wild-type UROIIIS. The UROIIIS mutant set presents a

TABLE I
Specific Activity Values for the UROIIIS Mutants Related to CEP

| Mutant | Purification yield[a] (%) | rESA[b] (%) | Purification × rESA (%) |
|---|---|---|---|
| V3F | 18.7 | 19.3±2.8 | 3.6 |
| L4F | 3.2 | 20.2±1.2 | 0.6 |
| Y19C | 16.2 | 13.1±2.2 | 2.1 |
| S47P | 1.7 | 100.0±1.1 | 1.7 |
| P53L | 0.0 | n.a. | n.a. |
| T62A | 27.7 | 1.2±0.1 | 0.3 |
| A66V | 54.8 | 95.6±0.2 | 52.3 |
| A69T | 13.6 | 24.4±1.3 | 3.3 |
| C73R | 4.4 | 14.5±1.2 | 0.6 |
| E81D | 39.4 | 100.0±3.0 | 39.4 |
| V82F | 26.3 | 93.8±1.8 | 24.6 |
| V99A | 9.4 | 88.2±5.1 | 8.2 |
| A104V | 23.0 | 60.6±3.7 | 13.9 |
| I129T | 1.9 | 20.0±4.0 | 0.4 |
| H173Y | 4.0 | 72.6±1.4 | 2.9 |
| Q187P | 1.2 | 15.0±1.9 | 0.2 |
| G188R | 2.2 | 41.4±7.6 | 0.9 |
| G188W | 2.3 | 31.6±4.3 | 0.7 |
| S212P | 8.3 | 20.0±4.1 | 1.6 |
| I219S | 3.0 | 85.0±5.0 | 2.6 |
| G225S | 26.6 | 32.4±10.2 | 8.6 |
| T228M | 54.8 | 97.5±2.5 | 53.4 |
| G236V | 5.8 | 34.0±3.9 | 2.0 |
| L237P | 2.8 | 57.9±2.2 | 1.6 |
| P248Q | 2.5 | 29.2±3.2 | 0.7 |

[a]Recombinant protein purification yield, referred to wild-type UROIIIS.
[b]Referred to the wild-type UROIIIS specific activity. Errors have been estimated from duplicate measurements.

large variability in the rESA, with values ranging between 1.2% and 100% with an average value of 50.3%. These high values are in good agreement with the recessive characteristic of the disease: a person carrying a single mutated allele would still retain between 50.6% and 100% of the total specific activity. Ten out of 25 mutants (V3F, L4F, Y19C, T62A, A69T, C73R, I129T, Q187P, S212P, and P248Q) showed a significant drop in specific activity, with values below 30%. Most of these residues were identified as putative substrate binders in an NMR chemical-shift perturbation experiment (Cunha et al., 2007) and they probably belong to the

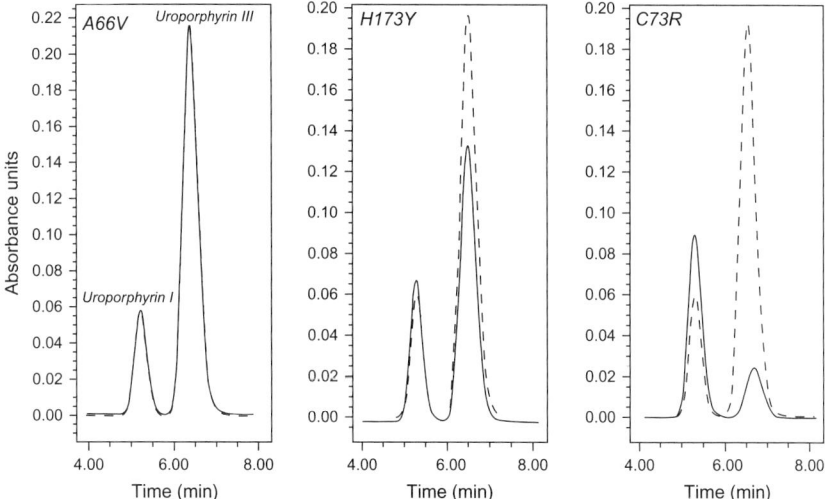

FIG. 8. HPLC separation of uroporphyrin I (retention time: 5.3 min) and uroporphyrin III (retention time: 6.5 min) for three representative mutants. Solid and dashed lines correspond to the mutant and the wild-type chromatograms, respectively.

active site or have some catalytic role. On the opposite side, a second subset of seven mutants (S47P, A66V, E81D, V82F, V99A, I219S, and T228M) showed a relative specific activity of 85% or more, indicating that a different mechanism is responsible for the CEP phenotype.

The effects of mutants producing CEP are often reported in the literature as "expressed enzymatic activity," (Desnick and Astrin, 2002) where the activity is directly measured in cell lysates instead of purified protein. Actually, there is a good agreement between the expressed enzymatic-activity values and the purification yield times and the rESA, also reported in Table I. Remarkably, the very low expressed activity values (often close to zero) were mainly due to a drop in the expression (and purification) yield, instead of alterations in the specific activity of the mutant.

### C. UROIIIS Destabilization Induced by the Pathogenic Mutants

The large rESA values obtained for most of the mutants indicate that another protein property is affected upon mutation. Since UROIIIS is a thermolabile enzyme, the thermodynamic and kinetic stability of

CEP-associated mutations has also been explored (Fortian et al., 2009). For every pathogenic missense mutation, experimental data was analyzed using a similar protocol to the one described in Section II.C for wild-type UROIIIS. Specifically, the unfolding rate constants were measured at different temperatures and used to obtain the half-life of denaturation at physiological temperature and the activation energy barrier between the folded and transition states. For the mutant dataset, the half-lives of the folded state are very diverse at physiological temperature, ranging from 0.16 to 339 h. Changes in the unfolding rate relative to the wild type are shown in Fig. 9A. Whereas the majority of the mutations only cause small perturbations in the equilibrium between folded and unfolded conformations, a subset of them drastically increases the unfolding rate.

Figure 9B shows a ribbon diagram of wild-type human UROIIIS with the mutated positions highlighted and color coded to display the $E_a$ values obtained upon mutation (Fortian et al., 2009). Mutants of residues belonging to the third α-helix of the N-domain result in a significant increase in the unfolding rate at 37 °C, accompanied by a large decrease in the activation energy barrier. Remarkably, C73 belongs to this helix. The C73R mutation causes the largest increase in $k_{NU}$, with an unfolding rate (at 37 °C) 380 times faster than wild-type UROIIIS, although the $E_a$ for C73R parallels activation energies from other destabilizing mutants (within experimental error). A second cluster with large destabilizing effects upon mutation can be identified in the region between S212 and I219.

It may be illustrative to combine activity and stability data to establish a qualitative genotype/phenotype relationship. However, such correlations have to be taken with extreme caution since the phenotype exhibited by the patient is very often multifactorial. A mutant subset (S47P, E81D, V82F, V99A, A104V, and T228M) retains very high synthase activity and shows no decrease in stability. The S47P mutant is the CEP-associated mutant closest to the wild type, according to our analysis. It is normally found in mild CEP patients, with the only reported case of a homozygotic mutation in UROIIIS being found in a healthy individual (Ged et al., 2004). The E81D and V82F mutants have been reported to produce impaired splicing at exon 4 to generate a truncated version of the protein in the cell (Desnick and Astrin, 2002). All of the other mutants from this group have always been found accompanying mutants related to more severe symptoms (V99A/C73R, A104V/C73R, T228M/C73R, T228M/

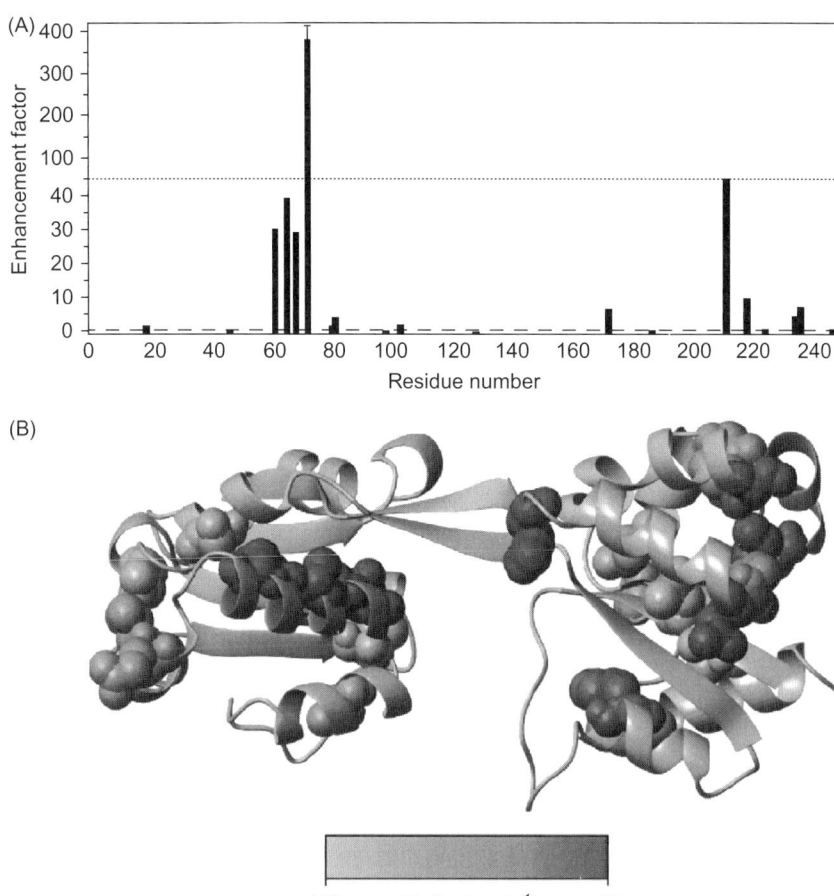

FIG. 9. (A) Enhancement factor in the unfolding rate $(k_{NU}(\text{Mut})/k_{NU}(\text{WT}))$ calculated at 37 °C for each of the mutants. There is a change in the ordinate scale, indicated by the dotted line in the center of the panel. (B) Ribbon representation of wild-type human UROIIIS. Residues that result in congenital erythropoietic porphyria upon mutation study are highlighted by CPK representation. Color code reflects the obtained activation energy value upon mutation according to the legend displayed in the figure.

G225S; Warner et al., 1992; Xu et al., 1995; Fontanellas et al., 1996), suggesting that these mutations may just be gene polymorphisms that are probably also present in healthy individuals. A second subset (V3F, L4F, Y19C, T62A, A69T, C73R, I129T, Q187P, S212P, and P248Q) includes all the mutations with lowered enzymatic activity but with a kinetic stability similar to the wild type, characteristic of residues affecting the catalytic site of the molecule. Finally, a third subset (C73R, Y19C, L237P) that shows decreased enzyme activity and reduced kinetic stability also develops into severe phenotypes.

In summary, most of the mutations drastically reduce the kinetic stability of the protein while retaining much of the catalytic activity. Similar to the two reported defects (*genetic* and *metabolic*), the main molecular mechanism which links the mutation with CEP should be named the *stability defect*.

## IV. The Hotspot Mutation C73R-UROIIIS

The C73R mutation is a very frequent mutation among CEP patients which also results in a very aggressive form of the disease. Therefore, this mutant phenotype has been the object of intense research efforts, up to the point that a murine knock-in model of disease carrying this mutation is now available (Bishop et al., 2006). This chapter is dedicated to the accumulated knowledge on C73R UROIIIS.

### A. *Structural Characterization of C73R In Vitro*

What are the structural changes of UROIIIS introduced with the C73R mutation? The far-UV region of the CD spectrum of wild type and C73R UROIIIS are significantly similar, consistent with the idea that both proteins share the same fold (Fig. 10A). The lower intensity at 222 nm found in C73R spectra is indicative of the mutant having a lower helical content (as compared to the wild type). Tryptophan fluorescence is also an adequate technique for monitoring the local structural changes induced upon mutation since UROIIIS has two tryptophan residues, both at the N-domain, and W83 is only 10 Å away from C73. The emission spectrum for wild-type UROIIIS (Fig. 10B, black line) shows a maximum at 330 nm, indicating that both tryptophans are buried in the solvent, in agreement with the high-resolution structural data: W83 (W91) only exposes 4.6% (0.1%) of the area to the solvent. The fluorescence spectrum of C73R UROIIIS presents decreased intensity (due to a drop in the quantum yield) and a shift of the maximum to 338 nm, clear indications of a more polar environment for the tryptophan residues.

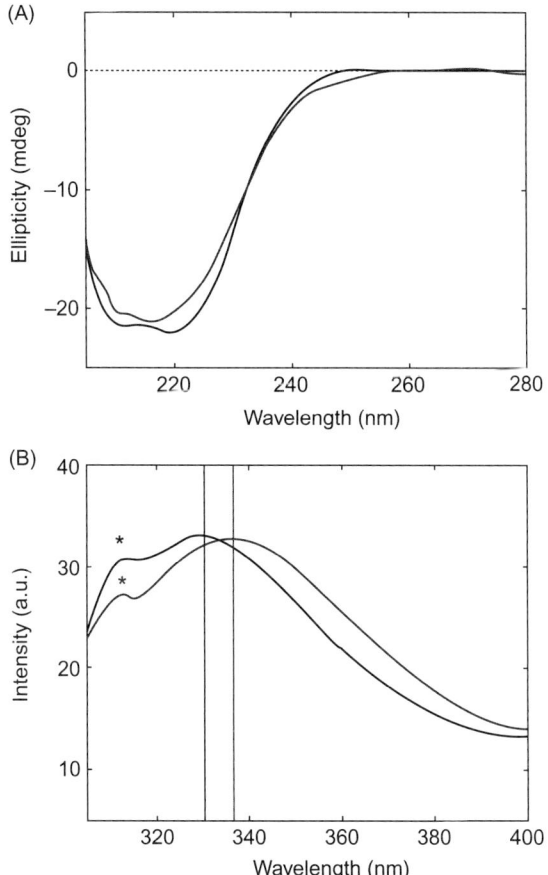

Fig. 10. (A) Far-UV region of the circular dichroism (CD) spectra for wild type (black line) or C73R (blue line) UROIIIS. (B) Tryptophan emission fluorescence spectra for wild type (black line) or C73R (blue line) UROIIIS. The position of the maximum in each spectrum is highlighted by the vertical lines. The peaks labeled with an asterisk correspond to the Raman spectrum of water. (See color plate 2).

A structural model of the mutation (Fig. 11) suggests that the introduction of arginine should create steric impediments with nearby residues (in particular, A69 and L43). Consistently, the mutant structure is locally more relaxed than wild-type UROIIIS to accommodate the bulkier side chain. In the model, Arg73 is buried within the hydrophobic core of the

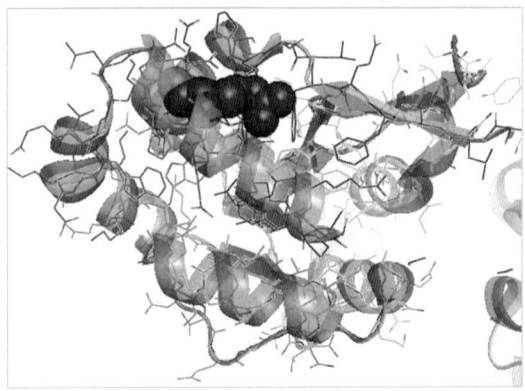

Fig. 11. Overlay of wild-type UROIIIS (1jr2, light brown) and the modeled structure for C73R (dark brown). The side chains for C73 and R73 are represented by red and blue spheres, respectively. (See color plate 3).

domain and a significant energetic penalty due to steric impediment is expected. Such a destabilizing contribution provides a plausible explanation for the observed decrease of the catalytic activity over time in C73R UROIIIS (10-fold compared to the wild type at physiological temperature). As proof of this principle, C73R UROIIIS was reengineered with the aim of releasing the local steric impediment. In particular, two mutants were designed to reduce the side chains in the hydrophobic core: A69G and L43V. The single mutants had a negligible effect on the stability and on the catalytic activity of UROIIIS. However, the double mutants (C73R/A69G and C73R/L43V) were unsuccessful in recovering the activity over time. In addition, the model also failed to predict changes in the local environment of W83 and, therefore, a bigger conformational change is expected than the one shown in Fig. 11.

## B. *Expression of C73R* In Vivo

The protein expression levels of wild type and C73R UROIIIS have been determined *in vivo* to investigate whether or not they paralleled the loss of activity observed upon the mutation *in vitro*. Several tagged versions of wild type and mutant proteins in two different cell types were analyzed (Fortian et al., submitted for publication): a eukaryotic cell line, which was a human fibroblast-derived cell line (named M1) and a murine

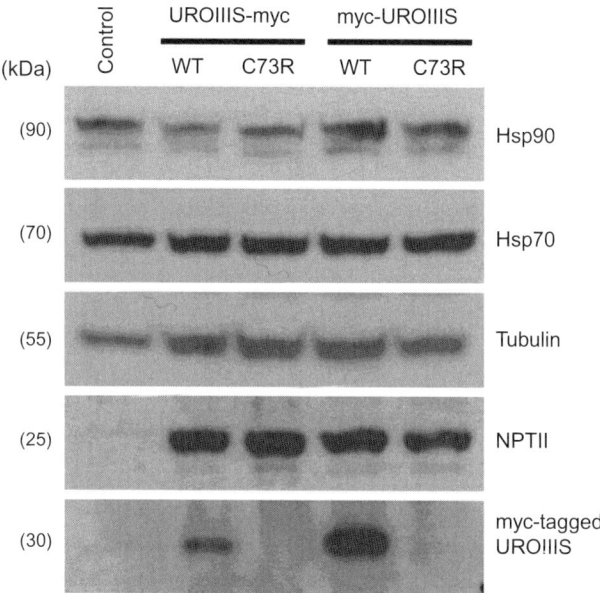

FIG. 12. Western blot analysis of unmodified M1 cell line (control) or stably expressing WT or C73R versions of myc-tagged UROIIIS proteins. Hsp70, Hsp90, and tubulin proteins serve as protein loading control. Specific antibodies against neomycin phosphotransferase II (NPTII) and myc-epitope were used to detect proteins from transfected vectors. Molecular weights of detected proteins are indicated.

hepatocyte-derived cell line (MLP29). In this study, M1 cell lines stably expressing N-terminal or C-terminal myc-tagged versions of wild type and C73R mutant proteins were generated and the levels of ectopically expressed proteins were analyzed by Western blotting. A clear specific band was observed in stable cell lines expressing wild-type UROIIIS (Fig. 12). For the mutant, no specific bands were detected in stable cell lines expressing the C73R mutant protein. This result could not be attributed to the additional stress induced by ectopic expression of the different versions of UROIIIS proteins because no significant differences were detected when two endogenous stress-inducible chaperones (Hsp70 and Hsp90) were analyzed (Duncan, 2005). The mammalian vector pCR3.1 carrying the gene coding for neomycin phosphotransferase II (NPTII) was also detected at similar levels in all of the stable cell lines, indicating that all of them contained functional vectors.

Quantitative real-time PCR (qRT-PCR) with the stable cell lines was used to reject the notion that the lack of mutant protein expression was not due to a failure at the transcriptional level. In the qRT-PCR assay, similar expression levels of NPTII were observed in stable cell lines transfected with the plasmids coding for the C-terminal myc-tagged UROIIIS (wild type or C73R), whereas the expression levels of C73R UROIIIS-myc were twofold higher than wild-type UROIIIS-myc. In the myc-UROIIIS construct, a reduction in the transcript levels was observed upon mutation, but this was unrelated to the mutation since the transcript levels of NPTII and myc-UROIIIS were similar. Altogether, the results obtained from the Western blot analysis and from the qRT-PCR assay indicate that a defect in the transcription of the myc-tagged UROIIIS mutant versions cannot be the origin of the lack of expression generated by the C73R mutation, supporting the idea that the protein mutant is unstable and quickly degraded in the cell, in agreement with the results obtained by the structural and thermodynamic analysis of the mutant.

To investigate the intracellular degradation of C73R UROIIIS, MLP29 cells stably expressing C73R fused to green fluorescent protein (gfp) were incubated in the presence of proteosome (MG132) inhibitors (Fortian et al., submitted for publication) which rendered the accumulation of C73R along with p21 and ubiquitinated proteins. The protein accumulation induced by MG132 was independent of the tag position, as indicated by flow cytometry analysis. This treatment also caused an increase in the level of WT proteins. However, the intracellular localization of WT and C73R were different: a cytosolic staining was observed for WT UROIIIS while a partial aggregated pattern was detected for C73R UROIIIS. A fraction of these cells were disrupted and tested for enzyme activity. Wild type and C73R UROIIIS were both catalytically active and, remarkably, the mutant retained 50% of the activity when normalized with respect to wild type. Altogether, these results demonstrate the implication of the proteosome degradation pathway in the intracellular clearance of the wild type and the misfolded C73R mutant.

## V. Treatment of CEP

The ultimate goal of biomedical research in the field of rare diseases is the development of efficient therapies that would lead to the cure of patients. CEP is still considered a chronic disease and the nonpalliative types of care available are all costly and experimental. Nevertheless, the acquired knowledge on the biochemistry and biophysics of UROIIIS may pave the way for new lines of therapeutic intervention.

## A. Marrow Suppression, BMT, and Gene Therapy

As stated earlier, the CEP phenotype implies a deficit in heme production combined with an increase in nonproductive porphyrin derivatives. Depending on the severity of the symptoms, patients may become blood transfusion-dependent. Chronic transfusions once or twice per month can reduce or suppress erythropoiesis, decreasing the production of porphyrin in bone marrow (Piomelli ct al., 1986). In this context, treatment with hydroxyurea to reduce the bone marrow porphyrin has been successfully explored (Guarini et al., 1994). Heme therapy can also be very helpful in the treatment of porphyrias, but its administration to CEP patients has not been extensively studied (Watson et al., 1974).

BMT has been proven to be effective in curing CEP (Harada et al., 2001; Shaw et al., 2001). At least, seven patients underwent BMT and in six cases it was successful with different levels of reduction in the intracellular porphyrin concentration and reduced skin photosensitivity. In fact, patients who underwent bone marrow transplants and who were subsequently exposed to sunlight without protection did not blister nor did their urinary porphyrin levels increase (Shaw et al., 2001). The sources of stem cells included bone marrow and umbilical cord from siblings as well as unrelated human leukocyte antigen-matched marrow. Thus, BMT clearly represents an improvement in the quality of life of CEP patients. However, one patient died of an intercurrent cytomegalovirus colitis, emphasizing the high-risk associated with this clinical practice.

The positive results obtained with BMT have provided the framework for hematopoietic stem cell therapy. In this treatment, the patients' own stem cells are stably transduced with vectors containing the *URO-synthase* cDNA. This abrogates the need for compatible human leukocyte antigen donors and minimizes the risk of rejection. The *UROIIIS* gene is incorporated by means of a retroviral vector which acts as a carrier (Mazurier et al., 2001). Recently, a definitive cure of a murine model of CEP was achieved by lentiviral vector-mediated hematopoietic stem cell gene therapy (Richard et al., 2008), and RNA interference has also been successfully employed to downregulate the levels of UROIIIS in cord blood (Robert-Richard et al., 2010). Altogether, these results indicate that current clinical efforts in CEP are focused on the development of hematopoietic stem cell-mediated gene therapy with a considerable degree of success.

### B. Molecular Chaperones and Inhibitors of the Proteasome

It has been shown that UROIIIS is a thermolabile enzyme which suffers from irreversible denaturation over time (see Section II.C). Biophysical analysis of the mutants has clearly shown that thermodynamic destabilization is the main mechanism by which the enzyme loses its activity upon mutation (Fortian et al., 2009). In theory, a recovery of protein stability *in vivo* should be considered as a possible therapy because it would also imply an increase in intracellular enzyme activity. Several works have explored the possibility of manipulating protein stability inside cells by using chemical or molecular chaperones (Loo and Clarke, 2007). Cystic fibrosis, a disease caused by mutations in the CFTR protein, provides a good example in which characterization of the intracellular metabolism of a misfolded protein has allowed successful therapeutic interventions. Seventy percent of cystic fibrosis patients carry the mutation F508del, which moderately affects the activity of the protein but severely affects its stability, causing quick intracellular degradation. Denning et al. (1992) showed that by growing the cells below 30 °C it was possible to increase the amount of the mutant protein, which was also reflected by its better cellular activity. A similar improvement was shown when the cells were grown in the presence of glycerol (Brown et al., 1996) and when another chemical chaperone, 4-phenyl butyrate (4-PBA), was used (Rubenstein et al., 1997), and these formed the basis of a phase I clinical trial in 18 CF-F508del homozygous patients. In this trial, the group of patients treated with 4-PBA showed increased CFTR function in the nasal epithelia, supporting the use of this compound as a potential therapy (Rubenstein and Zeitlin, 1998). Another compound which has shown chaperone features *in vivo* is trimethylamine *N*-oxide, which has been found to be able to ameliorate the effects of mutant aggregating-keratins in cultured cells (Lee et al., 2008). These and other chemical chaperones have shown promising results in other diseases associated with a loss of function due to destabilizing mutations, including retinitis pigmentosa, nephrogenic diabetes insipidus, and hypogonadotropic hypogonadism, all caused by mutations affecting several G-protein-coupled receptors (Conn et al., 2007). Because UROIIIS undergoes a similar mechanism upon mutation, we are currently studying the effect of the abovementioned chemical chaperones on the intracellular stability of UROIIIS.

Once a protein is unfolded inside the cell, it is immediately degraded by the different pathways available. Chemical compounds which act as inhibitors of specific cellular pathways allow the determination of whether or not a specific pathway is involved in the metabolism of a particular protein: MG132 is a well-characterized inhibitor of the ubiquitin-proteasome degradation system (Lee and Goldberg, 1998), whereas chloroquine, leupeptine, and $NH_4Cl$ have been widely used to alter lysosomal activity (Fox and Kang, 1993). Eeyarestatin I (EerI) has been described as a chemical inhibitor that blocks endoplasmic reticulum (ER)-associated protein degradation (ERAD; Wang et al., 2009). Other compounds such us glycerol, trehalose, doxorubicin, cyclopamine phosphate, and sodium 4-phenylbutyrate also stabilize proteins (Fayos et al., 2005), and some of them have also been used to modify protein homeostasis, although their mechanisms of action are not yet clear (Loo and Clarke, 2007).

Our results have shown that C73R UROIIIS is expressed in the cell but it rapidly unfolds and is quickly degraded, resulting in undetectable protein levels in the cell. UROIIIS is processed via the proteosome pathway. Remarkably, this process can be reverted by reversibly inhibiting the proteosome with the aldehyde MG132 (Fortian et al., submitted for publication). Even though the accumulation of the protein in the cell results in partial aggregation, we have shown that, upon MG132 addition, the recovered enzyme retains partial catalytic activity. Thus, even in the absence of a molecular chaperone, partial inhibition of the proteosome may constitute a new therapeutic option for CEP patients. When repeating the experiment with chloroquine, the levels of C73R UROIIIS could not be observed because the enzyme did not degrade via the lysosome. This result explains why early treatments using this inhibitor did not result in a beneficial effect for CEP patients (Moore et al., 1978).

## VI. Concluding Remarks

In this chapter, we have explored the intimate relationship between the fourth enzyme in the heme biosynthetic route (UROIIIS) and the rare autosomal disease, CEP. The main function of UROIIIS is in forcing the substrate to adopt a conformation that impedes spontaneous direct cyclization or the linear tetrapyrrol, also reducing the energy barrier for the cyclization of the D ring with inversion of the configuration. A significant number of residues are implicated in substrate binding, consistent with the large substrate–enzyme interface, in the cleft of the two domains of the protein. Except for Tyr168, none of these residues seem to be directly implicated in the catalytic mechanism. UROIIIS is a kinetically stable enzyme which remains folded, thanks to the large energy barrier between the native and the transition states. With time, the enzyme unfolds and either aggregates (*in vitro*) or is degraded via the proteasome (*in vivo*).

CEP patients suffer from a deficient function of UROIIIS introduced with the mutation (*genetic defect*) combined with the accumulation of derivatives that cannot be catabolized (*metabolic defect*). Current therapy is focused on palliative care of the severe symptoms that the disease causes and the cure of the disease by bone marrow suppression, BMT, and gene therapy, with unequal degrees of success. Missense pathogenic mutations spread out throughout the sequence and most of them retain a significant amount of catalytic activity. Remarkably, most of these mutations decrease kinetic stability, accelerating proteasomal degradation (*stability defect*). The hotspot mutation C73R, found in one-third of CEP patients, decreases the catalytic activity *and* the stability of the enzyme, resulting in a severe CEP phenotype. Thus, any treatment aimed at reversing the stability defect also constitutes a potential line of therapeutic interventions. Two theoretical possibilities have been discussed in the present chapter: the partial inhibition of the proteasomal pathway and the use of molecular chaperones to improve the intracellular stability of the mutant protein.

### Acknowledgments

Support was provided from the Department of Industry, Tourism and Trade of the Government of the Autonomous Community of the Basque Country (Etortek Research Programs 2007/2009), the Innovation Technology Department of the Bizkaia County, the Ministerio de Ciencia y Tecnología (CTQ2006-09101/BQU and CTQ2009-10353/BQU), and the Fundación Ramon Areces and the Ramon y Cajal program (J. F. P.).

## References

Anderson, K. E., Sassa, S., Bishop, D. F., Desnick, R. J. (2001). Disorders of haem biosynthesis: X-linked sideroblatic anemia and the porphyrias. In: The Molecular and Metabolic Bases of Inherited Disease, Scriver, C. S., et al. (Eds.), pp. 2961–3062. McGraw-Hill, New York.

Baker, D., Agard, D. A. (1994). Kinetics versus thermodynamics in protein folding. *Biochemistry* **33**, 7505–7509.

Battersby, A. R. (2000). Tetrapyrroles: the pigments of life. *Nat. Prod. Rep.* **17**, 507–526.

Becktel, W. J., Schellman, J. A. (1987). Protein stability curves. *Biopolymers* **26**, 1859–1877.

Bishop, D. F., Johansson, A., Phelps, R., Shady, A. A., Ramirez, M. C., Yasuda, M., et al. (2006). Uroporphyrinogen III synthase knock-in mice have the human congenital erythropoietic porphyria phenotype, including the characteristic light-induced cutaneous lesions. *Am. J. Hum. Genet.* **78**, 645–658.

Bishop, D. F., Schneider-Yin, X., Clavero, S., Yoo, H. W., Minder, E. I., Desnick, R. J. (2010). Congenital erythropoietic porphyria: a novel uroporphyrinogen III synthase branchpoint mutation reveals underlying wild-type alternatively spliced transcripts. *Blood* **115**, 1062–1069.

Bogorad, L. (1958). The enzymatic synthesis of porphyrins from porphobilinogen. II. Uroporphyrin III. *J. Biol. Chem.* **233**, 510–515.

Boulechfar, S., Da Silva, V., Deybach, J. C., Nordmann, Y., Grandchamp, B., de Verneuil, H. (1992). Heterogeneity of mutations in the uroporphyrinogen III synthase gene in congenital erythropoietic porphyria. *Hum. Genet.* **88**, 320–324.

Brown, C. R., Hong-Brown, L. Q., Biwersi, J., Verkman, A. S., Welch, W. J. (1996). Chemical chaperones correct the mutant phenotype of the delta F508 cystic fibrosis transmembrane conductance regulator protein. *Cell Stress Chaperones* **1**, 117–125.

Castano, D., Millet, O. (2010). Backbone chemical shifts assignments of D-allose binding protein in the free form and in complex with D-allose. *Biomol. NMR Assign.* Advanced online publication (PMID: 20711759).

Conn, P. M., Ulloa-Aguirre, A., Ito, J., Janovick, J. A. (2007). G protein-coupled receptor trafficking in health and disease: lessons learned to prepare for therapeutic mutant rescue in vivo. *Pharmacol. Rev.* **59**, 225–250.

Cunha, L., Kuti, M., Bishop, D. F., Mezei, M., Zeng, L., Zhou, M. M., et al. (2007). Human uroporphyrinogen III synthase: NMR-based mapping of the active site. *Proteins* **71**, 855–873.

Dailey, H. A. (1997). Enzymes of heme biosynthesis. *J. Biol. Inorg. Chem.* **2**, 411–417.

Denning, G. M., Anderson, M. P., Amara, J. F., Marshall, J., Smith, A. E., Welsh, M. J. (1992). Processing of mutant cystic fibrosis transmembrane conductance regulator is temperature-sensitive. *Nature* **358**, 761–764.

Desnick, R. J., Astrin, K. H. (2002). Congenital erythropoietic porphyria: advances in pathogenesis and treatment. *Br. J. Haematol.* **117**, 779–795.

Duncan, R. F. (2005). Inhibition of Hsp90 function delays and impairs recovery from heat shock. *FEBS J.* **272**, 5244–5256.

Duy, C., Fitter, J. (2005). Thermostability of irreversible unfolding alpha-amylases analyzed by unfolding kinetics. *J. Biol. Chem.* **280**, 37360–37365.

Fayos, R., Pons, M., Millet, O. (2005). On the origin of the thermostabilization of proteins induced by sodium phosphate. *J. Am. Chem. Soc.* **127**, 9690–9691.

Fontanellas, A., Bensidhoum, M., Enriquez de Salamanca, R., Moruno Tirado, A., de Verneuil, H., Ged, C. (1996). A systematic analysis of the mutations of the uroporphyrinogen III synthase gene in congenital erythropoietic porphyria. *Eur. J. Hum. Genet.* **4**, 274–282.

Fortian, A., Castano, D., Ortega, G., Lain, A., Pons, M., Millet, O. (2009). Uroporphyrinogen III synthase mutations related to congenital erythropoietic porphyria identify a key helix for protein stability. *Biochemistry* **48**, 454–461.

Fortian, A., Gonzalez, E., Castaño, D., Falcón, J., Millet, O. (2011). Intracellular rescue of the uroporphyrinogen III synthase activity in enzymes carrying the hotspot mutation C73R. *J. Biol. Chem.* online publication.

Fox, R. I., Kang, H. I. (1993). Mechanism of action of antimalarial drugs: inhibition of antigen processing and presentation. *Lupus* **2**(Suppl. 1), S9–S12.

Frank, J., Wang, X., Lam, H. M., Aita, V. M., Jugert, F. K., Goerz, G., et al. (1998). C73R is a hotspot mutation in the uroporphyrinogen III synthase gene in congenital erythropoietic porphyria. *Ann. Hum. Genet.* **62**, 225–230.

Frere, F., Schubert, W. D., Stauffer, F., Frankenberg, N., Neier, R., Jahn, D., et al. (2002). Structure of porphobilinogen synthase from Pseudomonas aeruginosa in complex with 5-fluorolevulinic acid suggests a double Schiff base mechanism. *J. Mol. Biol.* **320**, 237–247.

Ged, C., Megarbane, H., Chouery, E., Lalanne, M., Megarbane, A., de Verneuil, H. (2004). Congenital erythropoietic porphyria: report of a novel mutation with absence of clinical manifestations in a homozygous mutant sibling. *J. Invest. Dermatol.* **123**, 589–591.

Guarini, L., Piomelli, S., Poh-Fitzpatrick, M. B. (1994). Hydroxyurea in congenital erythropoietic porphyria. *N. Engl. J. Med.* **330**, 1091–1092.

Harada, F. A., Shwayder, T. A., Desnick, R. J., Lim, H. W. (2001). Treatment of severe congenital erythropoietic porphyria by bone marrow transplantation. *J. Am. Acad. Dermatol.* **45**, 279–282.

Jaswal, S. S., Sohl, J. L., Davis, J. H., Agard, D. A. (2002). Energetic landscape of alpha-lytic protease optimizes longevity through kinetic stability. *Nature* **415**, 343–346.

Lee, D. H., Goldberg, A. L. (1998). Proteasome inhibitors: valuable new tools for cell biologists. *Trends Cell Biol.* **8**, 397–403.

Lee, D., Santos, D., Al-Rawi, H., McNeill, A. M., Rugg, E. L. (2008). The chemical chaperone trimethylamine N-oxide ameliorates the effects of mutant keratins in cultured cells. *Br. J. Dermatol.* **159**, 252–255.

Leeper, F. J. (1985). The biosynthesis of porphyrins, chlorophylls, and vitamin B12. *Nat. Prod. Rep.* **2**, 19–47.

Loo, T. W., Clarke, D. M. (2007). Chemical and pharmacological chaperones as new therapeutic agents. *Expert Rev. Mol. Med.* **9**, 1–18.

Mathews, M. A., Schubert, H. L., Whitby, F. G., Alexander, K. J., Schadick, K., Bergonia, H. A., et al. (2001). Crystal structure of human uroporphyrinogen III synthase. *EMBO J.* **20**, 5832–5839.

Mazurier, F., Geronimi, F., Lamrissi-Garcia, I., Morel, C., Richard, E., Ged, C., et al. (2001). Correction of deficient CD34+ cells from peripheral blood after mobilization in a patient with congenital erythropoietic porphyria. *Mol. Ther.* **3**, 411–417.

Miles, C. A., Burjanadze, T. V., Bailey, A. J. (1995). The kinetics of the thermal denaturation of collagen in unrestrained rat tail tendon determined by differential scanning calorimetry. *J. Mol. Biol.* **245**, 437–446.

Millet, O., Hudson, R. P., Kay, L. E. (2003). The energetic cost of domain reorientation in maltose-binding protein as studied by NMR and fluorescence spectroscopy. *Proc. Natl. Acad. Sci. USA* **100**, 12700–12705.

Moore, M. R., Thompson, G. G., Goldberg, A., Ippen, H., Seubert, A., Seubert, S. (1978). The biosynthesis of haem in congenital (erythropoietic) porphyria. *Int. J. Biochem.* **9**, 933–938.

Petersen, P. M., Hawker, C. J., Stamford, P. J., Leeper, F. J., Battersby, A. R. (1998). Biosynthesis of porphyrins and related macrocycles. Part 50. Synthesis of the N-formyl-dihydro analogue of the spiro-intermediate and its interaction with uroporphyrinogen III synthase. *J. Chem. Soc. Perkin Trans.* **1**(9), 1531–1539.

Piomelli, S., Poh-Fitzpatrick, M. B., Seaman, C., Skolnick, L. M., Berdon, W. E. (1986). Complete suppression of the symptoms of congenital erythropoietic porphyria by long-term treatment with high-level transfusions. *N. Engl. J. Med.* **314**, 1029–1031.

Plaza del Pino, I. M., Ibarra-Molero, B., Sanchez-Ruiz, J. M. (2000). Lower kinetic limit to protein thermal stability: a proposal regarding protein stability in vivo and its relation with misfolding diseases. *Proteins* **40**, 58–70.

Richard, E., Robert-Richard, E., Ged, C., Moreau-Gaudry, F., de Verneuil, H. (2008). Erythropoietic porphyrias: animal models and update in gene-based therapies. *Curr. Gene Ther.* **8**, 176–186.

Robert-Richard, E., Lalanne, M., Lamrissi-Garcia, I., Guyonnet-Duperat, V., Richard, E., Pitard, V., et al. (2010). Modeling of congenital erythropoietic porphyria by RNA interference: a new tool for preclinical gene therapy evaluation. *J. Gene Med.* **12**, 637–646.

Rodriguez-Larrea, D., Minning, S., Borchert, T. V., Sanchez-Ruiz, J. M. (2006). Role of solvation barriers in protein kinetic stability. *J. Mol. Biol.* **360**, 715–724.

Roessner, C. A., Ponnamperuma, K., Scott, A. I. (2002). Mutagenesis identifies a conserved tyrosine residue important for the activity of uroporphyrinogen III synthase from Anacystis nidulans. *FEBS Lett.* **525**, 25–28.

Rubenstein, R. C., Egan, M. E., Zeitlin, P. L. (1997). In vitro pharmacologic restoration of CFTR-mediated chloride transport with sodium 4-phenylbutyrate in cystic fibrosis epithelial cells containing delta F508-CFTR. *J. Clin. Invest.* **100**, 2457–2465.

Rubenstein, R. C., Zeitlin, P. L. (1998). A pilot clinical trial of oral sodium 4-phenylbutyrate (Buphenyl) in deltaF508-homozygous cystic fibrosis patients: partial restoration of nasal epithelial CFTR function. *Am. J. Respir. Crit. Care Med.* **157**, 484–490.

Schubert, H. L., Phillips, J. D., Heroux, A., Hill, C. P. (2008). Structure and mechanistic implications of a uroporphyrinogen III synthase-product complex. *Biochemistry* **47**, 8648–8655.

Shady, A. A., Colby, B. R., Cunha, L. F., Astrin, K. H., Bishop, D. F., Desnick, R. J. (2002). Congenital erythropoietic porphyria: identification and expression of eight novel mutations in the uroporphyrinogen III synthase gene. *Br. J. Haematol.* **117**, 980–987.

Shaw, P. H., Mancini, A. J., McConnell, J. P., Brown, D., Kletzel, M. (2001). Treatment of congenital erythropoietic porphyria in children by allogeneic stem cell transplantation: a case report and review of the literature. *Bone Marrow Transplant.* **27**, 101–105.

Shoolingin-Jordan, P. M., Leadbeater, R. (1997). Coupled assay for uroporphyrinogen III synthase. *Methods Enzymol.* **281**, 327–336.

Silva, P. J., Ramos, M. J. (2008). Comparative density functional study of models for the reaction mechanism of uroporphyrinogen III synthase. *J. Phys. Chem. B* **112**, 3144–3148.

Stenson, P. D., Ball, E., Howells, K., Phillips, A., Mort, M., Cooper, D. N. (2008). Human Gene Mutation Database: towards a comprehensive central mutation database. *J. Med. Genet.* **45**, 124–126.

Stenson, P. D., Ball, E. V., Mort, M., Phillips, A. D., Shiel, J. A., Thomas, N. S., et al. (2003). Human Gene Mutation Database (HGMD): 2003 update. *Hum. Mutat.* **21**, 577–581.

Tadeo, X., Lopez-Mendez, B., Trigueros, T., Lain, A., Castano, D., Millet, O. (2009). Structural basis for the aminoacid composition of proteins from halophilic archea. *PLoS Biol.* **7**, e1000257.

Tietze, L. F., Geissler, H. (1993). Why is porphobilinogen the biological substrate for the formation of porphyrins - calculations on the conformation of acyclic tetrapyrroles and the acid-catalyzed cyclization of hydroxymethylpyrroles. *Angew. Chemie.* **32**, 1038–1040.

Wang, Q., Mora-Jensen, H., Weniger, M. A., Perez-Galan, P., Wolford, C., Hai, T., et al. (2009). ERAD inhibitors integrate ER stress with an epigenetic mechanism to activate BH3-only protein NOXA in cancer cells. *Proc. Natl. Acad. Sci. USA* **106**, 2200–2205.

Warner, C. A., Yoo, H. W., Roberts, A. G., Desnick, R. J. (1992). Congenital erythropoietic porphyria: identification and expression of exonic mutations in the uroporphyrinogen III synthase gene. *J. Clin. Invest.* **89**, 693–700.

Watson, C. J., Bossenmaier, I., Cardinal, R., Petryka, Z. J. (1974). Repression by hematin of porphyrin biosynthesis in erythrocyte precursors in congenital erythropoietic porphyria. *Proc. Natl. Acad. Sci. USA* **71**, 278–282.

Xu, W., Astrin, K. H., Desnick, R. J. (1995). Congenital erythropoietic porphyria: identification and expression of 10 mutations in the uroporphyrinogen III synthase gene. *J. Clin. Invest.* **95**, 905–912.

# ROLE OF FIBRIN STRUCTURE IN THROMBOSIS AND VASCULAR DISEASE

By AMY L. CILIA LA CORTE, HELEN PHILIPPOU, AND ROBERT A. S. ARIËNS

Division of Cardiovascular and Diabetes Research, Section on Mechanisms of Thrombosis, Leeds Institute for Genetics Health and Therapeutics, Faculty of Medicine and Health, University of Leeds, Leeds, United Kingdom

| | | |
|---|---|---:|
| I. | Introduction | 76 |
| II. | The Coagulation Cascade | 77 |
| III. | Fibrinogen Structure and Function | 81 |
| IV. | Factor XIII Structure and Function | 84 |
| V. | Clot Formation and Function | 86 |
| | A. Cleavage of the Fibrinopeptides | 86 |
| | B. Polymerization of Fibrin Molecules into Fibers | 88 |
| | C. Cross-Linking of Fibrin Molecules by Factor XIIIa | 89 |
| VI. | Fibrinolysis | 91 |
| VII. | Elastic Properties of Fibrin | 94 |
| VIII. | Heterogeneity in Coagulation and Fibrin Structure | 96 |
| IX. | Environmental Factors and Fibrin Structure | 102 |
| X. | Fibrin Density and Thrombosis | 105 |
| XI. | Interactions with Cells and Wound Healing | 107 |
| XII. | Perspectives | 109 |
| | References | 110 |

## Abstract

Fibrin clot formation is a key event in the development of thrombotic disease and is the final step in a multifactor coagulation cascade. Fibrinogen is a large glycoprotein that forms the basis of a fibrin clot. Each fibrinogen molecule is comprised of two sets of Aα, Bβ, and γ polypeptide chains that form a protein containing two distal D regions connected to a central E region by a coiled-coil segment. Fibrin is produced upon cleavage of the fibrinopeptides by thrombin, which can then form double-stranded half staggered oligomers that lengthen into protofibrils. The protofibrils then aggregate and branch, yielding a three-dimensional clot network. Factor XIII, a transglutaminase, cross-links the fibrin stabilizing the clot protecting it from mechanical stress and proteolytic attack. The mechanical properties of the fibrin clot are essential for its function as it must prevent bleeding but still allow the penetration of cells. This viscoelastic property is generated at the level of each individual fiber up to the complete clot. Fibrinolysis is the mechanism of clot removal, and involves a cascade of interacting zymogens and enzymes that act in concert with clot formation to maintain blood flow. Clots vary significantly in structure between individuals due to both genetic and environmental factors and this has an effect on clot stability and susceptibility to lysis. There is increasing evidence that clot structure is a determinant for the development of disease and this review will discuss the determinants for clot structure and the association with thrombosis and vascular disease.

## I. Introduction

Thrombosis and diseases associated with thrombosis are one of the most common causes of morbidity and mortality in the Western world. Thrombotic disease can be divided into two terms; venous thrombosis and arterial thrombosis which reflect whether the thrombosis (blood clot) occurs in a vein under low flow and pressure or the high flow, high pressure artery. The fibrin clot is formed in the final step of the coagulation cascade upon proteolytic cleavage of fibrinogen by thrombin and stabilized by cross-linking reactions catalyzed by activated coagulation factor XIII. Fibrinogen is a large soluble protein that plays a pivotal role in hemostasis as it has many structural elements enabling it to interact with a number of different proteins and cells within the circulation. Fibrinolysis is the process of clot dissolution and this occurs in concert with coagulation to maintain a patent vascular system.

Fibrin clot structure is determined by fiber thickness and distribution of the fibers and branch-points. There are genetic and environmental factors that influence the individual proteins involved in generating the clot network producing heterogeneous clot structures with varying properties. The structure and mechanical properties of fibrin are important determinants for its breakdown and *in vivo* can translate into an increased risk of thrombosis. A number of clinical studies have reported an association between altered fibrin clot structure and thrombotic disease. For example, after a myocardial infarction, fibrin clots have reduced permeability and a lower fiber mass-to-length ratio (Fatah et al., 1992, 1996b) and patients with severe coronary artery disease have more rigid clot structures and an elevated fiber mass-to-length ratio (Greilich et al., 1994). The mechanism via which these structural changes are translated into enhanced vascular risk may in part reside in the reported association between the structure of the fibrin clot and its susceptibility to lysis.

## II. The Coagulation Cascade

Activation of coagulation is stimulated in response to injury to maintain the integrity of the circulation and to stem bleeding (hemostasis). Pathophysiology of this pathway results in inappropriate intravascular coagulation resulting in thrombosis. Historically, the coagulation pathway has been described with an extrinsic and intrinsic pathway. In last couple of decades, the intrinsic pathway was postulated to play a less significant role with respect to initiation of the coagulation system, as stimulation of the extrinsic pathway was considered to play the initial role with respect to activation of coagulation *in vivo*. However, more recently, the intrinsic pathway has been shown to play an important role in sustaining thrombus development *in vivo*.

The coagulation system has been classically described as a cascade, since a small amount of active enzyme is required to activate the subsequent zymogens to generate larger amounts of enzyme. Stimulated by the exposure of tissue factor (TF) following endothelial damage, the extrinsic pathway is triggered to ultimately generate thrombin that is responsible for the conversion of fibrinogen to fibrin (see Fig. 1). The intrinsic and extrinsic pathways converge at the level of Factor (F) IX activation (Osterud and Rapaport, 1977). Later stages of coagulation take place principally on the surface of membrane systems, such as provided by platelets. This localizes the hemostatic response to the site of vascular injury. The main stages of coagulation require the formation of a complex between a proteinase, its substrate, a specific cofactor and calcium ions and phospholipids. For the activation of FX and prothrombin, the cofactors are activated factors (FVIIIa) and V (FVa), respectively. FVIIIa and FVa behave as cofactors by facilitating the formation of enzyme complexes on the membrane surface. Activated factor IX (FIXa) activates FX in a reaction requiring the participation of FVIIIa, calcium ions, and phospholipids. Activated factor X (FXa), FVa, calcium ions and phospholipids form the prothrombinase complex, responsible for the conversion of prothrombin to thrombin.

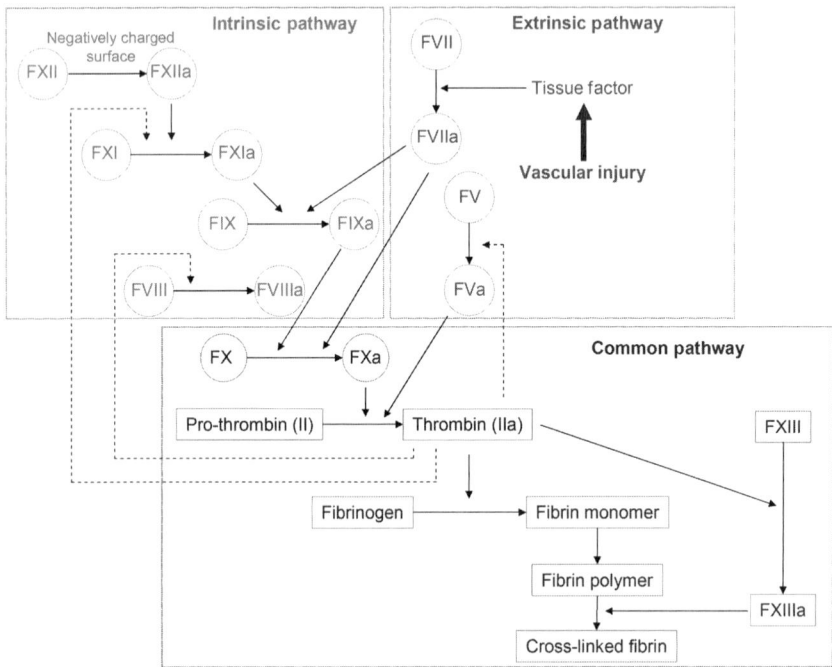

FIG. 1. *Schematic diagram of the blood coagulation pathways.* The intrinsic pathway and extrinsic pathway culminate in the generation of Factor IIa (thrombin) via the common pathway which is then involved in the conversion of fibrinogen to a cross-linked fibrin clot. The dashed arrows demonstrate the positive feedback mechanism exerted by thrombin to consolidate clot formation.

Formation of fibrin delivers the scaffold of a developing clot. Stability of the clot is conferred by the action of a transglutaminase, activated factor XIII (FXIIIa, also generated by the action of thrombin). FXIIIa cross-links the clot itself but also cross-links α2-antiplasmin (α2-AP) to fibrin, a major inhibitor of plasmin which is responsible for clot proteolysis (fibrinolysis). The fibrinolytic pathway is responsible for breaking down the clot (see Section VI). Imbalance between the coagulation and fibrinolytic pathways results in pathological thrombosis or bleeding.

The clotting cascade incorporates a number of positive feedback mechanisms such as the activation of coagulation factors XI, V, and VIII by thrombin that result in further thrombin generation to amplify the generation of fibrin. Because of the involvement of factors XI, V, and VIII,

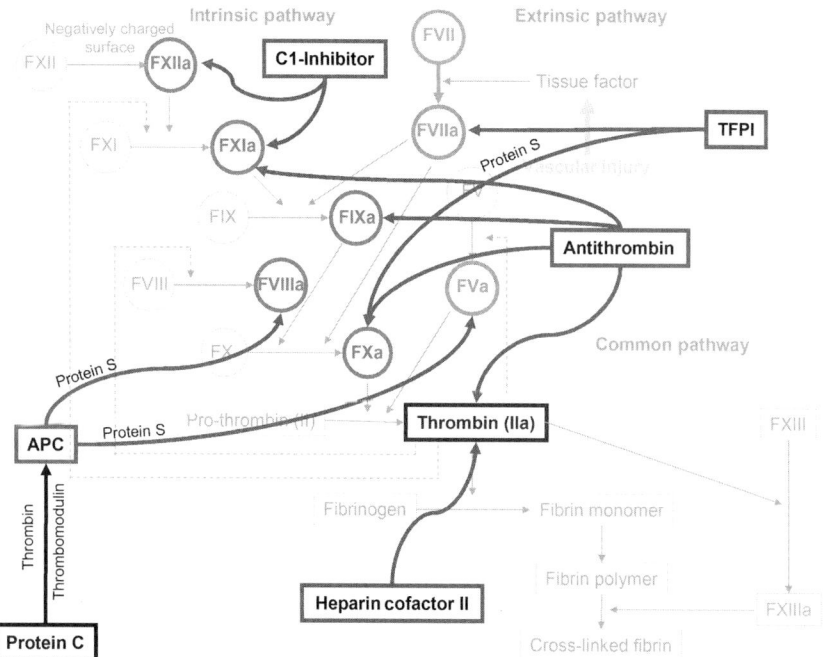

FIG. 2. *Schematic diagram of the natural inhibitors of the blood coagulation pathways.* The physiological regulators of the coagulation cascade prevent uncontrolled thrombin generation by inhibiting multiple factors within each of the three pathways. The bold red arrows indicate which factors are inhibited by C1-inhibitor, antithrombin, tissue factor pathway inhibitor (TFPI), heparin cofactor II, and activated protein C (APC). (For interpretation of the references to color in this figure legend, the reader is referred to the Web version of this chapter.)

many components of the intrinsic pathway are responsible for maintaining thrombin generation following initial triggering of the coagulation system via the exposure of TF triggering the extrinsic pathway.

There are also inhibitory mechanisms that limit procoagulant activity and uncontrolled spread of fibrin formation, throughout the circulation (Fig. 2). The most important inhibitor (antithrombin) is a member of the serine proteinase inhibitor (SERPIN) super family. Antithrombin inhibits thrombin by forming a 1:1 stable thrombin–antithrombin (TAT) complex. Antithrombin also complexes with FXa, FXIIa, FIXa, FXIa, plasmin (Damus et al., 1973; Highsmith and Rosenberg, 1974; Rosenberg et al.,

1975; Stead et al., 1976; Lane and Caso, 1989) and in addition has an inhibitory effect on FVIIa bound to TF (Rao et al., 1993; Rao et al., 1995). The physiological importance of antithrombin is illustrated in individuals with congenital deficiencies that are usually heterozygous and exhibit functional antithrombin levels 4–70% of normal. These individuals have a high probability of developing venous thrombosis (Lane and Caso, 1989). The vascular endothelium produces heparin sulfate proteoglycans which possess heparin-like activity (Marcum and Rosenberg, 1985; Marcum et al., 1986). Heparin accelerates the rate at which antithrombin complexes with serine proteinases (Damus et al., 1973), and hence works as a strong anticoagulant. Additional mechanisms by which heparin and proteoglycans inhibit coagulation involve complex formation with heparin cofactor II and tissue factor pathway inhibitor (TFPI).

A second inhibitory mechanism involves the protein C anticoagulant pathway. Activation of protein C occurs in the presence of calcium ions, following 1:1 complex of thrombin with its cofactor thrombomodulin on endothelial cell surfaces. Protein S acts as a cofactor for activated protein C (APC) in the inactivation of FVIIIa (Vehar and Davie, 1980) and FVa (Comp and Esmon, 1979) by enhancing the binding of APC to phospholipid membranes (Walker, 1981; Harris and Esmon, 1985; Stern et al., 1986). The inactivation of cofactors FVa and FVIIIa by APC inhibits the formation of FXa and thrombin (Dahlback and Stenflo, 1980). The protein C/thrombomodulin pathway is reviewed elsewhere (Esmon, 1983). There is a high incidence of thrombotic disorders in patients with protein C and protein S deficiencies (Griffin et al., 1981; Walker, 1981; Broekmans et al., 1983; Horellou et al., 1984; Suzuki et al., 1984).

Activated protein C resistance (APCR) caused by a *factor V* gene mutation, FV Arg506Gln or FV Leiden, has been identified as the most common inherited risk factor for venous thrombosis. It is found in 20–30% of unselected patients with venous thrombosis and 5% of healthy individuals (Dahlback et al., 1993; Koster et al., 1993; Bertina et al., 1994; Dahlback, 1994). The mutation removes a protein C inactivation site on FV and thus affects the regulation of FVa.

TFPI is mainly produced in endothelial cells (Wun et al., 1988; Girard et al., 1989; Bajaj et al., 1990; Broze Jr. et al., 1990; Werling et al., 1993). TFPI functions *in vivo* to inhibit the FVII/TF catalytic complex in an FXa-dependent manner. This inhibitory process prevents further production of FXa and FIXa by the FVIIa/TF complex. Under these circumstances, additional FXa can only be produced through the intrinsic pathway involving FIXa and FVIII, where FIXa results from the initial action of FVIIa/TF complex and is then produced via FXIa (Broze Jr., 1995). Recently, it has been shown that the inhibitory action of TFPI is enhanced by protein S (Hackeng et al., 2006), a process requiring Kunitz 3-domain of TFPI (Ndonwi et al., 2010). A large pool of TFPI is bound to proteoglycans on the luminal surface of the endothelial cell. Displacement of this pool of TFPI and hence release into the blood occurs following injections of heparin in a dose-dependent fashion. Heparin accelerates the action of TFPI as well as that of antithrombin and heparin cofactor II (Sandset, 1996).

## III. Fibrinogen Structure and Function

A blood clot or thrombus is composed of a network of polymerized fibrin with aggregated platelets. Fibrin is generated by limited proteolysis of the 340 kDa plasma glycoprotein, fibrinogen that is synthesized by hepatocytes. Human fibrinogen is translated from three closely linked genes which are located on chromosome 4q23-32 (Kant et al., 1985). Each gene is transcribed and translated separately and independently translocated to the endoplasmic reticulum where the protein is assembled (Yu et al., 1984). Fibrinogen circulates in the blood at micromolar concentrations and participates in many important biological processes including hemostasis, wound healing, inflammation, angiogenesis, atherosclerosis, and thrombosis.

In 1959, Hall and Slayter identified by electron microscopy that fibrinogen has a trinodular structure comprised of three spherical particles bound together by a thread-like structure (Hall and Slayter, 1959). Subsequent structural studies confirmed the initial low resolution findings and demonstrated the fibrinogen molecule to be composed of two D regions separated from a central E region by two coiled-coil regions (Doolittle et al., 1978). After the publication of many crystal structures of the fibrinogen D region (Spraggon et al., 1997), E region (Madrazo et al., 2001), bovine fibrinogen (Brown et al., 2000), and chicken fibrinogen (Yang et al., 2001), in 2009 the crystal structure of the human fibrinogen molecule was published by Kollman et al. (2009) at approximately 3.3 Å resolution (Fig. 3).

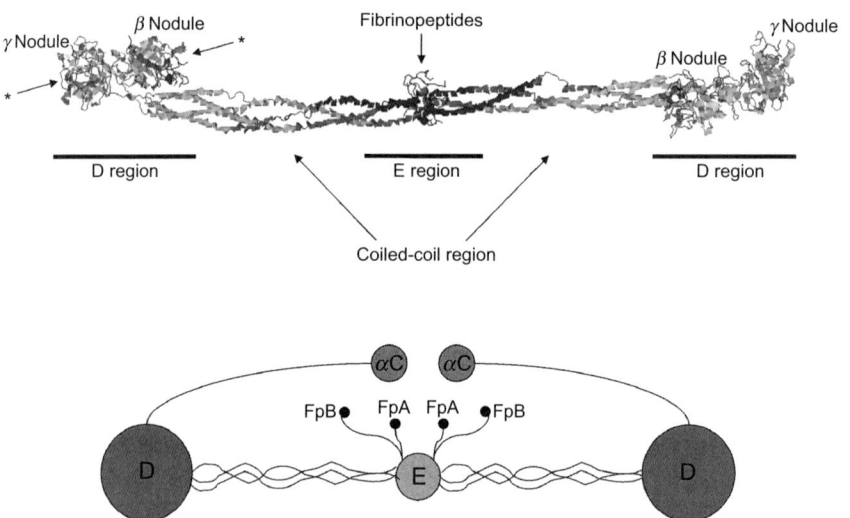

FIG. 3. *The human fibrinogen molecule.* Crystal structure of human fibrinogen determined at 3.3 Å resolution showing distal globular D regions connected by three-stranded coiled-coils to the central E region (top). The model was generated by RasMol using the PDB entry 3GHG (Kollman et al., 2009). The structure is produced with the "group" color scheme (colored according to position in polypeptide chain). The polymerization sites on one side of the molecule are indicated by the asterisks. Schematic representation of the fibrinogen molecule (bottom). Illustrated in the figure are the central E region with the fibrinopeptides (FpA and FpB) connected to the two D regions via the coiled-coils. The αC regions are also depicted.

The 45 nm fibrinogen molecule is comprised of two identical subunits connected by disulfide bonds, each of which contains three polypeptide chains; Aα (610 residues), Bβ (461 residues), and γ (411 residues) (Blomback, 1996). The E region consists of the N-terminal ends of the six polypeptide chains and the D regions are composed of the C-terminal ends of the Bβ- and γ-chains and a portion of the Aα-chain (Brown et al., 2000; Yang et al., 2000b). The E- and D regions are separated by a three-stranded α-helical coiled-coil structure composed of the parallel α-helical regions for the Aα, Bβ, and γ-chains (Doolittle et al., 1978). The D regions are comprised of the β and γ nodules which contain the sites for polymerization (Fig. 3), calcium-binding sites, and also carbohydrate clusters.

Calcium plays a major role in maintaining the structure and stability of fibrinogen. When bound to fibrinogen, calcium has been shown to promote lateral aggregation during fibrin polymerization (Brass et al., 1978), limit the extent of plasmin digestion (Odrljin et al., 1996), protect against denaturation by heat and pH (Ly and Godal, 1973), and limit the extent of disulfide bond reduction by reducing agents (Procyk and Blomback, 1990). There are calcium-binding sites within the γ nodules (between residues 311 and 336 of the γ-chain) and β nodules which would be fully occupied in circulating fibrinogen as the affinity for the binding sites is well below the physiological calcium concentration. Sialic residues have been shown to provide low affinity-binding sites for calcium and may have a regulatory role (Dang et al., 1989). Recently, Averett et al., found that the degree of extension experienced by the γ module in polymerized fibrin prior to the rupture of the bond between two fibrin molecules is dependent on the occupation of one of the γ calcium-binding sites which is located near the polymerization "hole" (Averett et al., 2010).

There are four biantennary carbohydrate clusters linked to fibrinogen by N-glycosyl bonds at residues BβAsn364 and γAsn52. These carbohydrates have been shown to have a significant effect on clot structure as complete removal of the carbohydrates leads to clots with very large fibers (Langer et al., 1988). In contrast, clots from liver cirrhosis patients (have high levels of sialylation on the fibrinogen carbohydrate) are comprised of thin fibers with many branch-points (Martinez et al., 1983a,b). These results led to the suggestion that both the charge and mass of the carbohydrate helps to modulate lateral aggregation (Weisel, 2005).

The C-terminal ends of the Aα-chains (Aα392-610) form two αC regions, each of which are composed of a compact globular domain (αC domain) attached to the rest of the fibrinogen molecule via a flexible N-terminal portion (αC connector; Aα221-391). Nuclear magnetic resonance studies of the αC domains using recombinant bovine fibrinogen and more recently human fibrinogen have demonstrated the presence of two independently folded subdomains; the N-terminal subdomain was shown to be comprised of a mixed parallel/antiparallel β-sheet while the structure of the C-terminal subdomain is less stable and remains unsolved (Burton et al., 2006, 2007; Tsurupa et al., 2009). In the fibrinogen molecule, the αC domains associate with each other and with the central E region (Veklich et al., 1993), most probably by the N-terminal end of the Bβ-chain (Fig. 3). Upon conversion to fibrin and cleavage of the fibrinopeptide B by thrombin (see below), the αC domains dissociate enabling intermolecular αC–αC interactions and leading to lateral aggregation of the protofibrils (Gorkun et al., 1994; Collet et al., 2005a).

## IV. Factor XIII Structure and Function

Factor XIII (FXIII) is a 320 kDa transglutaminase that circulates in blood plasma as a tetramer composed of two A-(possessing catalytic site) and two B-(that act as carrier proteins for the A-subunits) subunits ($A_2B_2$) (Muszbek et al., 2007). The *FXIIIA* subunit gene is located on chromosome 6p24-25 and encodes a mature protein of 731 amino acids (Ichinose and Davie, 1988). The *FXIIIB* gene has been localized to chromosome 1q31-32.1 and is translated into a 641 amino acid protein (Webb et al., 1989; Bottenus et al., 1990). Under normal circumstances, the plasma levels of the FXIIIB subunit are approximately twice that of the A-subunit, with the B-subunit therefore existing as both the $A_2B_2$ tetramer and as a free molecule (Schwartz et al., 1973; Yorifuji et al., 1988).

X-ray crystallography studies have demonstrated that the FXIIIA subunit is a globular protein consisting of well-defined folded domains which include; the activation peptide (residues 1–37), the β-sandwich (residues 38–184), the catalytic core (residues 185–515), barrel 1 (residues 516–628), and barrel 2 (residues 629–730). The A-subunit dimer folds to form a hexagonal globular molecule with the catalytic core in the center (Fig. 4). The activation peptide of each A-subunit occludes the opening of the catalytic site preventing substrate binding to the active site cysteine residue (Cys314) in nonactivated FXIII (Yee et al., 1994). The B-subunit is composed of 10 tandem repeats each approximately 60 amino acids in length called Sushi domains or glycoprotein-1 domains; each Sushi domain contains 2 disulfide bridges that maintain the tertiary structure (Ichinose et al., 1986). The main functions of the B-subunit are the stabilization and transport of the hydrophobic A-subunit and are thought to be of importance in mediating the interaction with other proteins (Siebenlist et al., 1996).

Thrombin cleaves the activation peptide from the A-subunit of FXIII at the Arg37-Gly38 peptide bond generating FXIIIa′. This reaction is enhanced approximately 80-fold by the presence of polymerizing fibrin (Janus et al., 1983). Following cleavage of the activation peptide, the A and B-subunits of FXIII dissociate in the presence of calcium exposing the FXIIIa catalytic triad (Fig. 4). Transglutaminases such as FXIIIa produce a γ-glutamyl-ε-lysine bond between glutamine and lysine residues (Lorand et al., 1968; Folk,

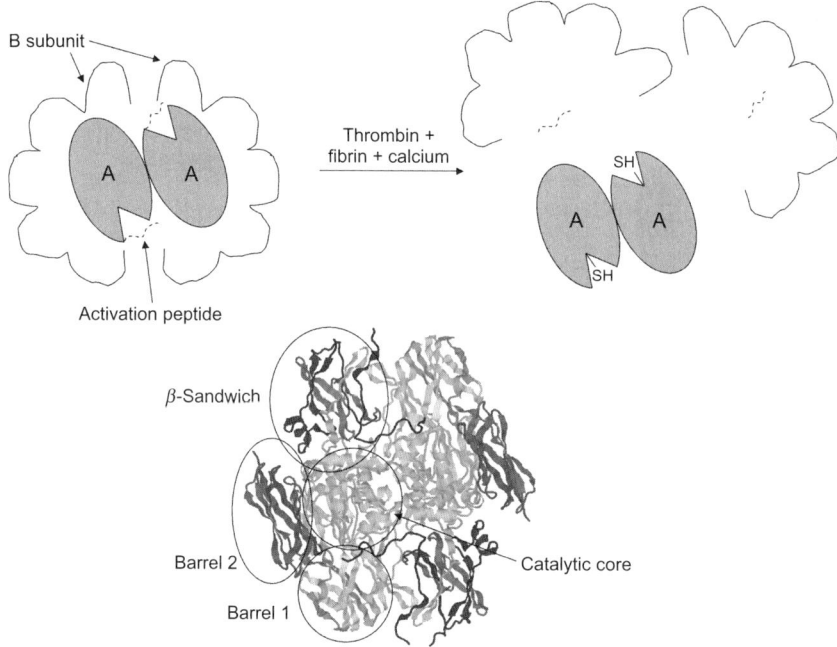

FIG. 4. *Factor XIII structure and activation.* A schematic representation of the Factor XIII tetrameric structure consisting of 2 A-subunits and 2 B-subunits. Upon exposure to fibrin, thrombin, and calcium, the activation peptides are removed from the A-subunit and the B-subunits dissociate from A exposing the active site thiol (−SH) group (top). Crystal structure of the factor XIII A dimer showing the β-sandwich, barrels 1 and 2 and the catalytic core (bottom). The model was generated by RasMol using the PDB entry 1F13 (Weiss et al., 1998).

1983). The enzyme first reacts with glutamine and forms an intermediate from which ammonia is released. In a second step, glutamine is cross-linked to lysine by an isopeptide bond and the active enzyme is released. It is thought that the first step producing the glutamine intermediate drives substrate specificity. In absence of a suitable lysine, the intermediate reacts with water converting glutamine to glutamic acid.

Factor XIIIa cross-links fibrin and other proteins into the clot (Ariëns et al., 2002; Muszbek et al., 2008). Interestingly, as fibrin formation itself considerably enhances activation of FXIII, these proteins together modify the stability of the clot in a very timely mechanistic manner. FXIIIa cross-links the fibrin γ-chains within 5 minutes upon initiation of polymerization whereas α-chain cross-linking typically takes >20 minutes to complete. Within several minutes, FXIIIa cross-links α2-AP to Lys303 of the fibrin α-chain. As α2-AP is the major inhibitor of plasmin (the enzyme responsible for breaking down fibrin clots) by forming a 1:1 complex, the cross-linking of α2-AP into the growing fibrin clot enhances the resistance of the clot to fibrinolysis. FXIIIa cross-links a number of other key proteins to the fibrin clot which influence its susceptibility to lysis (see VI). Fibronectin, Collagen Types I, II, III, and V, von Willebrand factor, vitronectin, and thrombospondin have also been shown to be cross-linked to fibrin and these proteins may be of importance in regulating the mechanical or other biological properties of the clot, such as interactions with cells and other proteins (Mosher and Schad, 1979; Bale et al., 1985; Hada et al., 1986; Sane et al., 1988; Barry and Mosher, 1989).

Studies from Kohler and Catto were the first to indicate a role for FXIII in cardiovascular diseases (Catto et al., 1998, 1999; Kohler and Grant, 1998; Kohler et al., 1998, 1999). Data from our laboratory indicated that cross-linking by FXIIIa influences fibrin clot structure (Ariëns et al., 2000) in a manner that is dependent on fibrinogen concentration (Lim et al., 2003).

## V. Clot Formation and Function

Electron microscopy analysis of polymerized fibrin demonstrated a banding pattern within the fibrin polymers with a repeat distance of 200–240 Å which is equivalent to approximately half the estimated length of an individual fibrin molecule. This observation led to the hypothesis that fibrin monomers polymerized together in a half staggered formation (Ferry, 1952) and that this occurred when sufficiently strong repulsive forces were overcome. Since this first observation, the clotting process has been determined in more detail and is now known to be a multistep process encompassing peptide cleavage events and molecule rearrangements to generate a stable mesh structure (Fig. 5).

### A. Cleavage of the Fibrinopeptides

Initiation of clot formation requires the interaction of thrombin with a funnel-shaped domain within the E region of the fibrinogen molecule (Madrazo et al., 2001). This interaction occurs at three thrombin sites; the

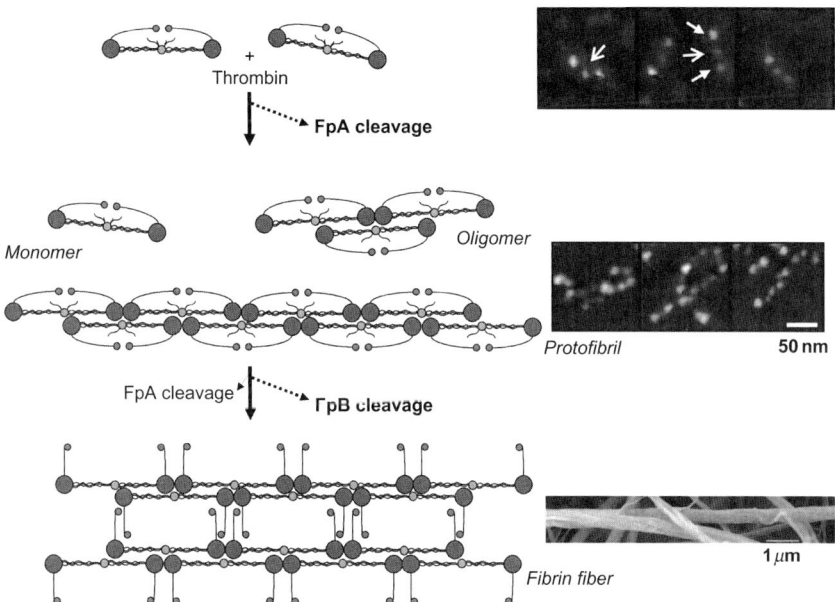

FIG. 5. *Schematic diagram of fibrin polymerization.* The three step fibrin assembly process is depicted; upon thrombin cleavage of fibrinopeptide A (FpA) fibrin monomers, oligomers and half staggered, double-stranded protofibrils form through D–E–D interactions. Thrombin cleavage of fibrinopeptide B (FpB) precedes the release of the αC regions from the central region allowing lateral aggregation of the protofibrils. High resolution noncontact atomic force microscopy (NC-AFM) imaging of the fibrinogen molecule with the trinodular structure identified (central E region: open arrows and two distal D regions: closed arrows) (top). Longitudinal fibrin oligomers visualized by NC-AFM (middle). A fibrin fiber visualized by scanning electron microscopy (bottom).

catalytic domain (residues Asp-His-Ser), an apolar binding pocket, and the fibrinogen recognition site (FRS) which is composed of multiple basic amino acid residues (Binnie and Lord, 1993). The interaction between fibrinogen and the FRS on thrombin is thought to be important for ensuring the correct orientation of thrombin's active site relative to the specific residues within the fibrinogen E region. Residues AαTrp33, AαPhe35, AαAsp38, and AαGlu39; BβAla68 and BβAsp69; γAsp27 and γSer30 were identified as binding to the FRS by co-crystallizing human α-thrombin with the E region of fibrinogen (Pechik et al., 2004). Upon binding, thrombin cleaves a specific arginyl–glycine bond (AαArg16-Gly17) within the amino terminus of the Aα polypeptide chain removing 16 residues from the Aα-chain (fibrinopeptide A, FpA) and at a slower rate thrombin cleaves the BβArg14-Gly15 bond removing 14 residues from the Bβ-chain (fibrinopeptide B, FpB). These cleavage events produce new N-terminal sequences of Gly-Pro-Arg and Gly-His-Arg, respectively, and convert fibrinogen into fibrin monomers which can spontaneously polymerize to form fibrin protofibrils (Fig. 5).

### B. *Polymerization of Fibrin Molecules into Fibers*

Cleavage of the fibrinogen Aα and Bβ polypeptide chains is important in two ways for enabling polymerization of fibrin. Firstly, removal of FpA and FpB alters the charge distribution within the fibrin molecule facilitating electrostatic interactions between adjacent fibrin monomers. At neutral pH, fibrinogen has an overall negative charge which is focused in the E region (Doolittle, 1994). Cleavage of the fibrinopeptides reduces the net negative charge from $-20$ to $-7$ and more importantly the excess charge on the E region is reduced from $-8$ to $+5$. This enables interaction with the fibrin D region of an adjacent monomer as the D region has a slight negative charge $(-3)$ enabling electrostatic interactions to occur.

Secondly, fibrinopeptide cleavage exposes polymerization sites termed $E_A$ and $E_B$ (Aα and Bβ cleaved N-termini, respectively) which are complementary to Da and Db sites that exist in the D regions thus enabling interactions between the D and E regions of adjacent molecules. Da is located between residues γ337 and γ379 of the P (polymerization) domain within the γ nodule and X-ray crystallography studies have demonstrated that these residues form a small hole into which the Gly-Pro-Arg-Val residues of $E_A$ bind (Shimizu et al., 1992; Pratt et al., 1997). The formation of $E_A$:Da associations between fibrin molecules causes the monomers to associate, leading to formation of staggered overlapping double-stranded fibrin fibrils. The exposure of the $E_B$ residues Gly-His-Arg-Pro contributes to lateral association of fibers as the interaction of $E_B$ with the complementary Db site located within the β nodule induces a rearrangement of the β nodule that promotes intermolecular β–β contacts (Yang et al., 2000a). The D:D interaction sites, which also promote self-association of fibrin monomers, are situated between residues 275 and 300 of the γ nodule within the D region. Fibrin clots formed from fibrinogen containing an arginine to cysteine mutation at position 275 in the γ-chain (dysfibrinogenemia: Tokyo II) were found to be completely disordered reflecting inaccurate end-to-end positioning of the fibrin monomers during polymerization resulting from the loss of the normal D:D interactions (Mosesson et al., 1995, 2001).

The linear growth of the protofibrils predominates until they exceed 600 nm in length, at which point lateral aggregation of protofibrils into thicker fibers begins and fibers subsequently branch to form the three-dimensional fibrin network (Mosesson et al., 2001). There are two types of branch junctions; "bilateral junctions," which form when a double-stranded fibril converges laterally with another fibril to form a four-stranded fibril, and "equilateral junctions," which are convergent interaction among three fibrin molecules that give rise to three double-stranded fibrils. It has been observed that equilateral junctions are more frequent when fibrinopeptide cleavage is slow and this leads to fibrin clots that are more branched which in turn mean they are less porous and less susceptible to lysis (Blomback et al., 1994).

The $\alpha$C domain is also involved in fibrin polymerization. In fibrinogen, the $\alpha$C domains are noncovalently associated with the E region but dissociate following FpB cleavage (Veklich et al., 1993; Gorkun et al., 1994). This makes the $\alpha$C domains available for interaction with adjacent $\alpha$C domains thereby promoting lateral fibril associations (Fig. 5). The role of the $\alpha$C domains in polymerization is evidenced by fibrin clots formed from fibrinogen which lacks the C-terminal portions of the $\alpha$C domains. These clots display prolonged thrombin times, reduced turbidity, and generate thinner fibers (Weisel and Medved, 2001).

### C. Cross-Linking of Fibrin Molecules by Factor XIIIa

FXIIIa catalyzes the formation of isopeptide bonds between specific lysine and glutamine residues within the $\gamma$ and $\alpha$ polypeptide chains of fibrin molecules to form $\gamma$–$\gamma$ and $\alpha$–$\alpha$ cross-links, respectively. Polymerization of fibrin is a regulator of the cross-linking process and correct alignment of the fibrin molecules in relation to each other enhances intermolecular catalysis by FXIIIa (Lorand et al., 1998).

The $\gamma$–$\gamma$ cross-links are the first to form with FXIIIa catalyzing the formation of two isopeptide bonds between $\gamma$-Lys406 and either $\gamma$-Gln398 or $\gamma$-Gln399 of adjacent fibrin molecules within the clot (Chen and Doolittle, 1971). The specificity of FXIIIa for the glutamine acyl donor appears to be due to secondary/tertiary structure of the fibrin molecule rather than just the sequence of residues surrounding the donor site, as a synthetic peptide containing the flanking residues was a poor substrate for FXIIIa (Gorman and Folk, 1981). The γ polypeptide chain COOH terminal residues containing the FXIIIa cross-linking sites are flexible, but sufficiently close together to allow cross-linking to proceed rapidly. The precise location of the γ cross-links has been disputed; some studies have suggested that the cross-links are formed transversely between D regions of fibrin molecules on opposing strands (Mosesson et al., 2001), other studies have suggested that the cross-links are formed longitudinally between two fibrin molecules aligned end-to-end (Weisel et al., 1993). It is possible that both occur and that the conditions during clot formation dictate which form predominates; however, recent evidence obtained with atomic force manipulation of fibrin indicates that the γ-chain cross-links are formed predominantly in the longitudinal orientation (Guthold and Carlisle, 2010). This orientation also agrees better with the X-ray crystal structure of the D region from cross-linked fibrin (Spraggon et al., 1997).

Each fibrin molecule contains two γ-chain cross-linking sites (one in each of the two D regions), whereas there are several sites within the fibrin α-chains. Studies using peptide probes containing either glutamine residues (to identify potential lysine acceptor sites) or a primary amine residues (to identify glutamine donor sites) have identified 4 potential glutamine donor sites at residues 221, 237, 328, and 366 and 15 potential lysine acceptor sites at residues 208, 219, 224, 418, 427, 429, 446, 448, 508, 539, 556, 580, 583, 601, and 606 (Fretto et al., 1978; Cottrell et al., 1979; Matsuka et al., 1996; Sobel and Gawinowicz, 1996). However, lysine residues 556 and 580 appear to provide the greater proportion of cross-linking compared to the other sites which only play a minor role. It has been proposed that Lys556 and Lys580 regulate linear growth while the remaining lysine residues are involved in branching (Sobel and Gawinowicz, 1996). The large number of potential cross-linking sites within the Aα-chains are required as each Aα-chain can interact with at least two other Aα-chains within the fibrin clot. These cross-links are slower to form than the γ–γ cross-links, but they play a very important role in stabilizing the molecular assembly of protofibrils within fibers, thereby enhancing the tensile strength of the clot and reducing the susceptibility of the clot to lysis (Gaffney and Whitaker, 1979).

Cross-linking also occurs between α- and γ-chains to produce heterodimers and -polymers, although *in vivo* it is believed that this is due to tissue transglutaminase activity rather than FXIII (Siebenlist and Mosesson, 1996). This was supported by a study investigating the contribution of γ cross-linking to fibrin elastic properties using a recombinant fibrinogen variants with mutations in the γ-chain cross-linking sites (Standeven et al., 2007). Mutation of all three Lys406, Gln398, and Gln399 eliminated all cross-linking of the fibrinogen γ-chain and reduced stiffness of the fibrin fiber, demonstrating the role of γ-chain cross-linking in fibrin elasticity. Elimination of the lys406 site lead to absence of γ-dimer formation, but new γ–α heterodimers and polymers were formed, which were similar to those formed when fibrin is cross-linked by tissue transglutaminase. Interestingly, cross-links that made γ–α heterodimers were not as effective in increasing fibrin stiffness as those producing γ–γ homodimers (Standeven et al., 2007).

## VI. Fibrinolysis

Fibrinolysis or the breakdown of a fibrin clot plays an important role in hemostasis preventing blockage of the lumen of blood vessels and maintaining blood flow. The fibrinolytic system is comprised and tightly regulated by a series of proteins and their specific inhibitors (Fig. 6). Most of the enzymes in the system are serine proteases which contain a catalytic triad of a serine, aspartic acid, and a histidine residue located in the C-terminal protease domain. The inhibitors of the system are known as SERPINS which have a reactive center loop within their C-terminal region. SERPINS effectively act as pseudosubstrates for their target serine proteases and contain a reactive site peptide bond (Arg-X or Lys-X) that mimics the normal substrate. Cleavage of this peptide bond results in the formation of a covalent linkage between the protease and SERPIN which inactivates the enzyme (Collen, 1999; Huntington et al., 2000; Rau et al., 2007).

The central enzyme in the fibrinolytic system is plasmin which is generated from the zymogen plasminogen by two distinct plasminogen activators; tissue plasminogen activator (tPA) and urokinase-type plasminogen activator (uPA) (Fig. 6). Plasminogen is a 92 kDa glycoprotein that circulates in the plasma and following the generation of fibrin, is rapidly converted to plasmin by tPA which is released from endothelial cells.

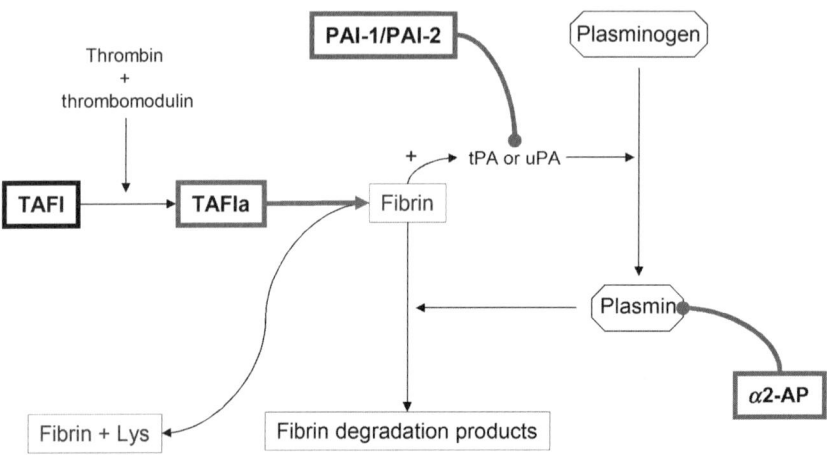

FIG. 6. *Schematic overview of fibrinolysis.* Clot dissolution is mediated by plasmin which is produced by the actions of tissue-type plasminogen activator (tPA) or urokinase plasminogen activator (uPA). The bold red lines indicate where the fibrinolysis pathway is inhibited by plasminogen activator inhibitor-1 and -2 (PAI-1/PAI-2) and α2-antiplasmin (α2-AP). Thrombin-activatable fibrinolysis inhibitor (TAFI) is activated (TAFIa) by the thrombin:thrombomodulin complex and inhibits fibrinolysis by cleaving C-terminal lysine residues on fibrin preventing tPA and plasminogen binding.

Plasminogen contains five homologous kringle domains which contain lysine-binding sites that mediate the binding of plasminogen to fibrin and the interaction of plasmin and its inhibitor α2-AP (Collen, 1999). Fibrin is the major substrate for plasmin and regulates its own degradation by binding both plasminogen and tPA on its surface. In the absence of fibrin, tPA has a weak affinity for plasminogen and is therefore only a weak activator of enzyme; however, in the presence of fibrin, the catalytic efficiency of tPA is increased by 1000-fold (Thorsen, 1992). The reason for this enhancing effect is due the exposure of tPA and plasminogen-binding sites upon the association of individual fibrin monomers into a polymeric structure and thus the colocalization of these proteins on the fibrin surface. These sites mediate the interactions between tPA and plasminogen initiating plasmin generation and clot lysis. Plasmin mediated degradation of fibrin generates new C-terminal lysine residues within the fibrin clot, which can act as new plasminogen-binding sites, leading to a further enhancement in plasmin generation and fibrin degradation. In addition to tPA-induced cleavage of plasminogen, urokinase (uPA) is also able to generate plasmin in a fibrin-independent manner (Stump et al., 1986; Husain, 1993).

Fibrinolysis has to be tightly regulated to prevent bleeding and this is coordinated by the inhibitors; α2-AP, plasminogen activator inhibitor-1 (PAI-1), plasminogen activator inhibitor-2 (PAI-2) and thrombin-activatable fibrinolysis inhibitor (TAFI) (Fig. 6).

Alpha 2-AP is the main physiological inhibitor of plasmin and exists in two forms within the circulation: Met-α2-AP (approximately 30% of the circulating α2-AP) and Asn-α2-AP. The two forms are generated by the cleavage of Met-α2-AP at the Pro12-Asn13 bond by AP-cleaving enzyme (APCE) (Lee et al., 2004). FXIIIa rapidly cross-links α2-AP to Lys303 of the α-chain of fibrin(ogen) via its Gln14 residue within the N-terminus (Kimura and Aoki, 1986). A study by Lee et al., demonstrated that Asn-α2-AP was cross-linked to fibrin approximately 12-fold faster than Met-α2-AP due to the availability of the Gln14 residue (Gln2 in Asn-α2-AP) but both forms display the same inhibitory activities (Lee et al., 2007). Cross-linking of α2-AP localizes the plasmin inhibitor to the fibrin clot protecting the clot from lysis. Alpha 2-AP inhibits plasmin by forming an irreversible enzyme–inhibitor complex which is mediated by the cleavage of the Arg364-Met365 bond within the C-terminal region of α2-AP by plasmin (Collen, 1980). The binding of plasmin to fibrin protects it from inactivation by α2-AP but not completely. Clots made from α2-AP deficient plasma were found to lyse extremely quickly and subsequent addition of α2-AP to the clot increased lysis time, but did not restore the normal lysis rate. Only the addition of α2-AP before clot formation corrected the lysis times, indicating the importance of cross-linkage of α2-AP to the clot for regulating its fibrinolytic susceptibility (Sakata and Aoki, 1982).

PAI-1 and PAI-2 are inhibitors of tPA and uPA; however, PAI-1 is considered the major inhibitor as significant plasma levels of PAI-2 are only found in humans during pregnancy. Like other serpins, PAI-1 forms a stable acyl–enzmye complex with tPA or uPA inactivating them and inhibiting plasmin generation. PAI-1 is synthesized in endothelial cells and platelets and therefore, it has a high localized concentration in platelet-rich thrombi protecting the blood clot from premature lysis (Fay et al., 1994).

TAFI is expressed by the liver and is present in platelets (Nesheim, 2003). Thrombin cleaves TAFI at Arg92-Ala93 bond generating an activation peptide (Phe1-Arg92) and the catalytic portion (Ala93-Val401) which is a zinc-dependent carboxypeptidase with specificity for carboxy terminal Arg and Lys residues. However, thrombin on its own is a relatively weak activator of TAFI. It is only in the presence of thrombomodulin that the activation of TAFI by thrombin is increased more that 1000-fold (Bajzar et al., 1996; Nesheim et al., 1997). The proteolytic activity of activated TAFI (TAFIa) is unstable under physiological conditions (half-life about 10 min). TAFI is cross-linked to fibrin by FXIIIa via its activation peptide (Valnickova and Enghild, 1998). As Arg and Lys residues are the binding site for plasminogen and tPA on fibrin, TAFIa can inhibit lysis by preventing the binding of these factors thus reducing plasmin generation (Redlitz et al., 1995). High concentrations of TAFIa have also been shown to directly inhibit plasmin (Rijken and Sakharov, 2001).

The interplay between the enzymes and inhibitors in the circulation and at the fibrin surface produces a complex interaction regulating coagulation and fibrinolysis. For effective digestion, plasmin has to degrade fibrin in the coiled-coil region between the D and E regions, a process that is interfered with by the presence of α-chain cross-links (Francis et al., 1980). This indicates that clot structure can affect the efficiency of fibrinolysis. Studies investigating this have demonstrated that clots with thick fibers and large pores lyse at a faster rate than tight fiber networks consisting of thin fibers. This is potentially due to the three-dimensional structure of the clot and the ability of the fibrinolytic enzymes to be able to permeate the clot (Collet et al., 2000, 2003). However, it is likely that other mechanisms also play a role, including changes in the affinity between the fibrinolytic proteins and fibrin.

## VII. Elastic Properties of Fibrin

The fibrin network that builds up a clot shows remarkable elastic properties. Fibrin demonstrates both viscous and elastic characteristics. Its viscoelastic properties are unique among other (biological and nonbiological) polymers. While fibrin elasticity can be similar to that of rubber, its structure is vastly different (Weisel, 2008). The elastic properties of fibrin can be attributed to fibrin structures at different scales, ranging from molecular and microscopic to macroscopic, with contributions from the individual fibrin molecules, fibrin fibers, and the three-dimensional network that constitute the clot (Brown et al., 2009).

At a molecular level, fibrin molecules can extend their molecular length under mechanical stress firstly due to unfolding the two alpha-helical coiled-coil regions in each molecule, accounting for a large proportion of the stretch of fibrin. Secondary areas of the molecule that have been reported to unfold under stress are the E regions (Lim et al., 2008), and the D regions (Liu et al., 2010). Forces that drive refolding of the fibrin domains during relaxation contribute to fibrin elasticity.

At a fiber level, fibrin demonstrates a remarkable elasticity and ability to stretch. Fibers are stiffer for stretch than for flexion (Collet et al., 2005b). Individual fibers can be stretched an extraordinary 3.2-fold before the fiber breaks. After cross-linking with FXIIIa, this extensibility of the fibrin fiber is further increased to a 4.3-fold stretch without rupture. Fibrin fiber extensibility is the largest observed among biological polymers (Liu et al., 2006). Electron microscopy studies have shown that the protofibrils in the fibrin fiber twist round one another. Outer protofibrils have to stretch when the fiber grows in diameter, limiting fiber growth to a diameter in which the thermodynamic energy involved in the stretch of the twisting protofibrils balance that of the molecular interaction forces (Weisel et al., 1987). Partial unwinding of the twisted protofibrils is likely to contribute to elasticity of the individual fibrin fiber, and its ability to stretch (Weisel, 2004; Upmanyu et al., 2008). The complicated nature of fibrin structure and its multiple components that impart mechanical properties to the fibrin fiber including protein unfolding and protofibril twisting lead to a nonlinear elastic response when fibrin is exposed to mechanical stress. One typical nonlinear elastic response is strain-hardening that is observed when a fibrin fiber is stretched. During strain-hardening, stiffness of the fibrin fiber increases with increasing strain (Kang et al., 2009). Protein unfolding events during stretching of the fiber play a possible role in strain-hardening, but the precise mechanism(s) accounting for this phenomenon are currently unknown and yet to be elucidated. Cross-linking by FXIIIa, while increasing overall fiber stiffness, reduces the degree of strain-hardening of fibrin (Liu et al., 2010), suggesting that the introduction of isopeptide bonds reduces the role of either the $\gamma$- or the $\alpha$-chains in strain-hardening.

The formation of a three-dimensional fibrin network introduces additional characteristics that contribute to overall clot elasticity. A major characteristic of the fibrin architecture is the number of branch-points that occur within the network. The relative strength of these branch-points contributes to the elastic properties of the fibrin clot. In a study of manipulation of the branch-points with the tip of an atomic force microscope, it was observed that 70% of branched fiber that are stretched to rupture break at the site of a branch-point and 30% of along the fibrin fiber. This response was inversed by cross-linking of the fibrin by FXIIIa, after which only 40% of the branch-points in cross-linked clots ruptured under strain as opposed to 60% of ruptures occurring along the fibrin fiber (Carlisle et al., 2010), suggesting that cross-linking by FXIIIa preferentially strengthens fiber branch-points. A second feature of the three-dimensional fibrin network is that the fibers of the network align when mechanically stretched (Brown et al., 2009; Kang et al., 2009). Such fiber alignment during stretch is accompanied by a dramatic reduction in clot volume and by water displacement (Brown et al., 2009).

Several factors have been reported to modulate the elasticity of the fibrin clot. The incorporation of red blood cells in the fibrin network has been reported to decrease the stiffness and increase the viscous component of fibrin clot (Gersh et al., 2009a). Homocysteine, a by-product of the methionine metabolism pathway, has been reported to increase the stiffness of the fibrin network, thereby possibly contributing to the reported cardiovascular risk associated with high homocysteine levels (Rojas et al., 2009). Proteolysis of N-terminus of the fibrinogen β-chain by a snake venom enzyme from the American diamondback rattlesnake produces a fibrin clot that is much reduced in its elastic integrity, with an approximately eightfold reduction of the elastic modulus (Abou-Saleh et al., 2009), demonstrating the role of the N-terminal β-chain in fibrin elasticity. Also the C-terminal region of the β-chain is involved in fibrin elasticity as demonstrated by the threefold higher stiffness of a commonly occurring variant of fibrinogen, βLys448 (Ajjan et al., 2008a).

## VIII. Heterogeneity in Coagulation and Fibrin Structure

Clot structure is heterogeneous, being influenced by both genetic and environmental factors. These factors can affect levels of proteins, affect the structure of the proteins which can then modify their function, or alter their interactions with other proteins. Examples of both genetic and environmental factors that influence fibrin structure are discussed below.

Dysfibrinogenemias are the result of structural aberrations in the fibrinogen molecule which can occur due to single base mutations that lead to substitutions of a single amino acid, or base additions/deletions. Alterations in amino acid sequence can have a profound effect on protein structure and in the case of fibrinogen can cause thrombosis or bleeding. Mutations that result in dysfibrinogenemias are rare with carriers generally being heterozygous for the mutation and therefore asymptomatic. Currently there are 570 dysfibrinogenemias documented (www.geht.org, release 33) and discussion of these is beyond the scope of this chapter, see reviews for further information (Asselta et al., 2006; Acharya and Dimichele, 2008).

Genetic polymorphisms (common genetic mutations, generally defined with a frequency of 1% or greater) within the fibrinogen genes generally occur in noncoding regions which potentially affect the quantity of circulating fibrinogen. This in itself can affect clot structure as clots formed with high levels of fibrinogen have more fibers and the fibers are longer in length. The maximum rate of assembly is also increased; however, at very high fibrinogen concentrations the opposite effect may occur (Weisel and Nagaswami, 1992). The two most studied polymorphisms are a *BclI* variation in the 3′ region of the *Bβ* gene, and the -455G/A polymorphism in the 5′ promoter region of the *Bβ* gene both of which have been shown to be associated with increased fibrinogen levels (Humphries et al., 1987; Behague et al., 1996; Zito et al., 1997). Polymorphisms in the coding regions also occur and they can affect the concentration and the quality of the circulating fibrinogen or both. Two nonsynonymous polymorphisms will be discussed below with further examples shown in Table I.

An A to G transition in the gene coding for the fibrinogen α-chain results in the substitution of a threonine residue at position 312 with alanine with an allele frequency of 0.24 (Baumann and Henschen, 1993). This polymorphism has been shown to be associated with increased poststroke mortality in subjects with atrial fibrillation and an increased risk of pulmonary embolism in subjects with venous thrombosis (Carter et al., 1999, 2000). The polymorphism lies in close proximity to the α–α cross-linking site at residue 328 (Muszbek et al., 1996) and, therefore, could influence the strength of the cross-linking of the clot (Standeven et al., 2003). Additionally, it lies within the region if the α-chain which plays a role in promoting the dissociation of the FXIIIA and B subunits and enhances FXIII activity (Credo et al., 1981).

A second polymorphism results in a change from a lysine residue at position 448 in the Bβ-chain to an arginine residue. This polymorphism has an allele frequency of 0.15 (Baumann and Henschen, 1993) and while this is a modest change in amino acid side chain, this polymorphism has

TABLE I
Genetic Variants of Fibrinogen and Factor XIII

| Gene | Variant | Location | Function | Relation to disease | References |
|---|---|---|---|---|---|
| *FGA* | AαThr312Ala | Close to FXIII cross-linking site | Changes fibrin structure/function and FXIII cross-linking | Atrial fibrillation/pulmonary embolism | See Section VIII |
|  | *TaqI* | 3′ UTR of α-chain | Unknown | None | Remijn et al. (2001) |
| *FGB* | −455G/A | Promoter of β-chain | Increased fibrinogen concentration | More common in CAD | Carter et al. (1996) |
|  | BβArg448Lys | Close to β-chain polymerization site/calcium-binding site | Changes fibrin structure/function | Macrovascular disease | See Section VIII |
|  | *BclI* | 3′ UTR of β-chain | Increased fibrinogen concentration | More common in CAD | See Section VIII |
| *FXIIIA* | Val34Leu | Activation Peptide | Increases activation rate; affects fibrin structure | CAD, stroke, DVT | See Section VIII |
|  | Tyr204Phe | Catalytic Core | Unknown | Stroke | Pruissen et al. (2008) |
|  | Pro564Leu | Barrel 1 | Unknown | Stroke | Reiner et al. (2002) |
|  | Val650Ile | Barrel 2 | None | None |  |
|  | Glu651Gln | Barrel 2 | None | None |  |
| *FXIIIB* | His95Arg | Second Sushi domain | Increases subunit dissociation | Venous thrombosis, aneurysmal subarachnoid hemorrhage | See Section VIII |

CAD, coronary artery disease; DVT, deep vein thrombosis; UTR, untranslated region.

been associated with coronary artery disease (Behague et al., 1996; Carter et al., 1997b) and a predisposition for stroke in females (Carter et al., 1997a). The location of the BβArg448Lys polymorphism is relatively close to the proposed β-chain polymerization site and a calcium-binding site. It is therefore possible that this polymorphism could play a role in fibrin polymerization. Analysis of clots formed from plasma and purified BβLys448 fibrinogen displayed different clot properties to BβArg448 fibrinogen (Lim et al., 2003). Subsequent analysis of clot structure using recombinant BβLys448 fibrinogen demonstrated that the clots had thin fibers and small pores, show increased stiffness, and are more resistant to fibrinolysis (Ajjan et al., 2008a).

Many mutations have been reported in the *FXIIIA* and *FXIIIB* genes with the majority causing deficiency of the subunit due to either truncations of the proteins or altering enzyme activity leading to bleeding disorders. Key lessons have been learned with regards to FXIII structure and function and we refer to the following excellent reviews for detailed discussion on this topic (Hsieh and Nugent, 2008; Muszbek et al., 2008).

There are several polymorphisms within the *FXIII* genes (Table I), two of which have been studied in detail and shown to alter the susceptibility of the carrier to thrombosis. The Val34Leu polymorphism in the FXIIIA subunit was identified by Mikkola et al. and is the result of a G to T transition in exon 2 (Mikkola et al., 1994). The polymorphism is relatively common with an allele frequency of 0.27 and is located 3 amino acid residues away from the thrombin activation site of the FXIIIA subunit. Clinical studies have suggested that possession of the Leu allele at this locus might be protective against the development of myocardial infarction, stroke and deep vein thrombosis (Catto et al., 1998, 1999; Kohler et al., 1998; Franco et al., 1999; Wartiovaara et al., 1999). A study from our laboratory demonstrated that possession of the Leu allele enhanced the efficiency of FXIII activation peptide cleavage leading to alterations in clot structure (Ariëns et al., 2000). Investigation of the interaction between thrombin and FXIII has shown that the bulkier Leu side-chain alters the substrate peptide conformation increasing the catalytic efficiency of thrombin (Trumbo and Maurer, 2000, 2003). Thrombin-induced cleavage of the FXIII34Val activation peptides occurs with the cleavage of FpB but cleavage of the FXIII34Leu activation peptides coincides with FpA cleavage from fibrinogen producing clots with thinner fibers and smaller intrinsic pores (Ariëns et al., 2000). The changes in clot structure were proposed to be due to accelerated rates of cross-link formation due to the enhanced rate of FXIII activation. Earlier formation of cross-links could act to inhibit lateral aggregation of protofibrils, resulting in thinner fiber formation compared to slower cross-linking by FXIII34Val which would allow more lateral aggregation and therefore thicker fibers to form.

Komanasin et al. identified an A to G transition at position 8259 within the *FXIIIB* gene which results in a histidine to arginine substitution in the second Sushi domain. Functional studies demonstrated that this polymorphism is associated with an increased dissociation rate of the factor XIII $A_2B_2$ tetramer following activation by thrombin and that the arginine 95 variant is associated with an increased risk for venous thrombosis (Komanasin et al., 2005) and more recently, aneurysmal subarachnoid hemorrhage (Ruigrok et al., 2010).

Fibrinogen $\gamma'$ occurs due to alternative splicing and polyadenylation of the *FGG* mRNA transcript. The $\gamma'$-chain which is present in approximately 8–15% of plasma fibrinogen molecules is longer than the main $\gamma$-chain ($\gamma A$) containing 427 residues compared to 411 of $\gamma$-chain. The extension of $\gamma A$ occurs due to utilization of an alternative polyadenylation site located in intron 9. This causes intron 9 to remain unspliced leading to translation of a unique 20 amino acid extension ($\gamma'$408-427: VRPEHPAE-TEYDSLYPEDDL) and loss of 4 amino acids ($\gamma$408-411: AGDV) translated from exon 10 in $\gamma A$ (Wolfenstein-Todel and Mosesson, 1981). The 20 amino acid extension is anionic in nature partly due to sulfonation of the two tyrosine residues in the sequence. This property along with its larger size enables separation of the $\gamma A/\gamma'$-chains by anion-exchange chromatography (Mosesson and Finlayson, 1963a,b). The folding of the $\gamma'$-chain during clot formation is unknown; however, it has potential to interfere with formation of cross-linked fibrin. The $\gamma'$ extension is located in the D region and could potentially interfere with the D–D interface during initial polymerization or stretch to the E region of a neighboring molecule of the D–E–D complex (Uitte de Willige et al., 2009). Many studies have been carried out to determine the effect of the fibrinogen $\gamma'$-chain on clot structure and in the main it has been observed that the presence of the $\gamma'$-chain produces clots with thinner fibers and smaller pores (Cooper et al., 2003; Siebenlist et al., 2005; Gersh et al., 2009b). These structural characteristics indicate that clots with high levels of $\gamma'$ may increase the risk for thrombosis as clots with this structure have been shown to be stiffer and more resistant to lysis (Gersh et al., 2009b).

In addition to the structural changes, γA/γ′ has been shown to modulate thrombin activity. Thrombin binds to the γ′-chain with high affinity via exosite II, this affinity is increased further when thrombin is simultaneously bound to the E region via exosite I (Pospisil et al., 2003). Binding of thrombin to γ′ inhibits thrombin activity and may act as an inhibitor of the intrinsic pathway for thrombin generation in the solution phase. It has also been proposed that fibrinogen γ′ could act as a storage molecule for active thrombin in the fibrin clot.

Fibrinogen γ′ has also been proposed to bind Factor XIII via its B subunits (Mosesson and Finlayson, 1963a; Siebenlist et al., 1996) with a 20-fold higher affinity compared to γA/γA (Siebenlist et al., 1996); however, this has been recently disputed by Gersh et al. who demonstrated no difference in affinity between γA/γA, γA/γ′ or γ′/γ′, and FXIII (Gersh and Lord, 2006).

Finally, fibrinogen γ′ does not promote platelet aggregation unlike fibrinogen γA due to the lack of the integrin-binding site for αIIbβ3 which is translated from exon 10 (Peerschke et al., 1986; Farrell and Thiagarajan, 1994). In addition, the γ′-chain has been shown to hinder thrombin binding to GpIbα and thrombin cleavage of PAR1 on platelets indicating that γ′ significantly modifies platelet activation *in vivo* (Lancellotti et al., 2008; Lovely et al., 2008).

Epidemiological studies have shown that elevated levels of γ′ are associated with arterial thrombosis (coronary artery disease, myocardial infarction, and stroke) (Lovely et al., 2002; Mannila et al., 2007; Cheung et al., 2008), while reduced levels have been correlated with venous thrombosis (deep vein thrombosis, thrombotic microangiopathy) (Uitte de Willige et al., 2005; Mosesson et al., 2007). It is generally accepted that fibrin clot structure can directly contribute to disease and the functional effects of γA/γ′ fibrinogen described above could go some way to explaining the altered thrombotic risk associated with γ′ fibrinogen.

In addition to changes in the γ-chain, the α-chain has also been shown to be alternatively spliced. Fibrinogen 420 (Fu and Grieninger, 1994) accounts for approximately 1% of circulating fibrinogen molecules and has an extended C-terminus (237 amino acids), which folds to form a globular domain termed $α_EC$. This extension is due to translation of Exon VI, which only occurs when the gene is appropriately spliced, leading to a mass increase of 50% compared to the main α-chain form (Fu et al., 1992). The $α_EC$ region is similar in structure to the β and γ nodules; it contains a calcium-binding site but no polymerization pocket. Clots formed using purified fibrinogen 420 assembled in the same way as fibrinogen 340 (more common form); however, the $\alpha_E C$ regions in fibrinogen 420 occupy the fiber surface and slow lateral aggregation resulting in clots with thinner more branched fibers (Mosesson et al., 2004). The functional consequences of fibrinogen 420 are still largely unknown; however, there is evidence to suggest a role in cellular adhesion (Fu et al., 1998), and a recent study suggested that $\alpha_E C$ has chaperone-like activity (Tang et al., 2009).

Alternative splicing occurs in many genes generating different protein isoforms from the same gene that can have a profound effect on protein structure as evidenced above. Allele specific alternative splicing (asAS), where a nucleotide polymorphism alters splicing of all transcripts of the gene so that each allele produces just one form of the transcript, is not as common. The FXIIIB subunit is an example of a heterogeneous protein generated by asAS (Iwata et al., 2009).

FXIIIB is noncatalytic but is involved as a carrier molecule stabilizing the catalytic FXIIIA subunit. In addition to being a carrier molecule, its levels are also important for regulating the plasma levels of FXIIIA subunit. FXIIIB was first demonstrated to be heterogeneous by Curtis et al. using isoelectric focusing, thus enabling the separation of different variants according to pI (Curtis et al., 1974). This heterogeneity has been classified into three major population associated phenotypes, FXIIIB*1, FXIIIB*2, and FXIIIB*3 for European, African, and Asian populations, respectively (Kamboh and Ferrell, 1986). Recently, the molecular basis for these phenotypes was identified by sequencing all 12 exons of the FXIIIB gene in individuals of whom the phenotypes were known (Iwata et al., 2009; Ryan et al., 2009).

## IX. Environmental Factors and Fibrin Structure

As a consequence of the initial studies that have indicated the existence of a relationship between fibrin structure and thrombosis (see below), an increasing number of studies have aimed at elucidating the effects of environmental and modifiable factors, such as dietary and treatment related agents on fibrin clot structure. The modifiable factors that have been studied can be classed as posttranslational modifications caused by disease mechanisms or drug treatment, changes related to inflammatory response, or other environmental factors. Several studies have reported an effect of metabolite homocysteine on fibrin clot structure. Homocysteine is an intermediate product from the conversion of methionine to cysteine and glutathione during folate- and vitamin B6- and B12-dependent metabolic reactions. Accumulation of homocysteine occurs in vitamin B6 deficiency and in carriers of a mutation in methylenetetrahydrofolate reductase (MTHFR). High levels of homocysteine have been associated with an increased risk of cardiovascular disease. Homocysteine has been reported to lead to a fibrin structure with reduced pores and thinner fibers (Lauricella et al., 2002), leading to a reduced susceptibility of the fibrin clot for fibrinolysis induced by tPA (Sauls et al., 2006; Undas et al., 2006). These effects were also observed *in vivo* in a rabbit model of hyperhomocysteinuria (Sauls et al., 2003). The fibrin clot effects of homocysteine together with the effect of homocysteine on clot stiffness likely contribute to the increased vascular risk associated with the abnormal accumulation of this metabolite.

High glucose levels in diabetes associate with protein modifications due to nonenzymatic glycation through the formation of Schiff bases and rearrangements to irreversible Amadori products. Glycation of a plasma protein depends on the levels of hyperglycemia of the patient, the duration of hyperglycemia, susceptibility of the protein to glycation and the half-life of the protein. Lysine and arginine residues are most sensitive to glycation, particularly, if located in the vicinity of an imidazole group of a nearby histidine residue (Shilton and Walton, 1991). It has been shown that clots made from fibrinogen that was purified from the plasma of diabetic patients with poor glycemic control showed a different structure than control clots with smaller pores and increased fiber branching, and that these changes correlated with HbA1c, a sensitive marker of glycemic control (Dunn et al., 2005). Furthermore, changes to fibrin in patients with diabetes lead to a reduced rate of fibrinolysis due to impaired plasmin generation and reduced binding of both plasminogen and tPA to the fibrin clot (Dunn et al., 2006). Fibrin structure and fibrinolysis rates returned to normal after improvement of glycemic control (Pieters et al., 2008). Hyperglycemia was also associated with abnormal fibrin structure and reduced lysis rates in patients with coronary artery disease without prior diagnosis of diabetes (Undas et al., 2008b). Glycation in patients with diabetes has been reported not to change the function of TAFI, suggesting that a normally functioning TAFI in patients with impaired glucose control may tip the balance toward hypofibrinolysis (Verkleij et al., 2009).

Lysine residues may also be subject to another posttranslational modification induced by acetyl-salicylic acid (ASA or aspirin). The acetylation of lysines by ASA has been reported for many proteins such as prostaglandin synthase (Roth et al., 1975; Loll et al., 1995), antithrombin (Villanueva and Allen, 1986), platelet and red blood cell membrane proteins (Watala and Gwozdzinski, 1993), and hemoglobin (Xu et al., 2000). Acetylation of albumin and hemoglobin has been reported to inhibit glycation of these proteins (Rendell et al., 1986). Fibrinogen can also be acetylated by ASA (Caspary, 1956), leading to modifications of the $\alpha$-, $\beta$- and $\gamma$-chains, with a 1–1 stoichiometry of one acetyl group per chain (Bjornsson et al., 1989). Recent data using an *in vitro* fibrinogen expression system in Chinese hamster ovary cells showed that acetylation of fibrin(ogen) leads to antithrombotic alterations in fibrin structure and function, whereby the clot is made less dense and easier to be broken down by plasmin (Ajjan et al., 2008b). In this cellular system, acetylation of fibrinogen appeared to mainly occur on the $\alpha$-chain. Similar effects of ASA on fibrin structure and function are observed *in vitro* (He et al., 2009) as well as *in vivo* (Antovic et al., 2005) and in patients (Fatah et al., 1996a). In addition, aspirin reduces FXIII activation, especially, in carriers of the common Leu34 allele (Undas et al., 2003). Hence, acetylation by aspirin appears to counteract the effects on fibrin structure induced by hyperglycemia, and therefore aspirin treatment may help to reduce the risk of thrombosis in diabetes. The effects of aspirin on fibrin structure and its susceptibility to lysis are additional to the well-described inhibition of platelet aggregation by aspirin, which is due inactivation of the platelet COX-1 enzyme through the acetylation of Ser529, and subsequent reduced synthesis of thromboxane A2, a potent platelet agonist and vasoconstrictor (Undas et al., 2007).

Many other environmental and modifiable factors have been shown to modulate fibrin structure and function. Oxidative stress as occurring in patients with cardiovascular disease and diabetes has been shown to associate with prothrombotic changes in fibrin clot structure and fibrinolysis rates (Undas et al., 2008a). Thrombin and prothrombin concentrations as well as fibrinogen concentrations also modulate fibrin clot structure. Higher thrombin levels lead to a denser fibrin clot structure with thinner fibers, which is more difficult to break down by the fibrinolytic system (Wolberg et al., 2003). Elevated fibrinogen modulates fibrin structure by increasing clot density and reducing susceptibility to lysis

(Weisel and Nagaswami, 1992; Scott et al., 2004). Since fibrinogen is an acute phase protein and its levels are strongly and rapidly upregulated during the inflammatory response, its effects on clot structure provide another link between cardiovascular disease and inflammation. Whereas above mentioned modifiable factors are to some degree influenced by both environmental and genetic regulation, there are also examples of purely environmental factors that change fibrin structure. For example, recent evidence suggests that fibrin clot structure may be affected by urban particulate matter and air pollution. Exposure to air pollution has been linked to cardiovascular disease, with an increase in both long-term risk (Pope III et al., 2004; Miller et al., 2007) but also immediate risk of the disease, with effects noticeable as soon as 1–2 h after exposure (Peters et al., 2004; Pope III et al., 2006). *In vitro* studies showed that particulate matter increased fibrin network heterogeneity. These effects were attributable mainly to the ultrafine fraction of the particulate matter, with diameter less than 200 nm, and were partly reversed by a reactive oxygen species scavenger, suggesting that oxidation may play a role in the mechanism by which the changes occur (Metassan et al., 2010a). However, no effects of moderate short-term transient exposure to particulate matter were observed in young, healthy volunteers, indicating that at least in these individuals the fibrin clot was not altered by city-level of air pollution (Metassan et al., 2010b). Further studies will be required to investigate the effects of air pollution on fibrin clot structure in patients with vascular or pulmonary disease and in chronic exposure.

## X. Fibrin Density and Thrombosis

Thrombosis is characterized by the formation of an occlusive clot within a blood vessel, leading to ischemia or tissue death due to starvation and oxygen deprivation downstream of the occlusion. Thrombosis can occur in either the arterial or the venous circulation. Arterial thrombosis in the arterial circulation is a complex disease, involving inflammation of the vessel wall and the formation of an atherosclerotic plaque in the vessel (Fuster et al., 2005). The formation of the plaque is a relatively slow process, which can take many years. Once the plaque is fully developed, it becomes unstable and ruptures, triggering thrombosis and precipitating the disease in the acute phase. Arterial thrombosis manifests itself most dramatically in two main locations, in the arteries that supply blood to the heart (also called coronary arteries) and in the arteries that supply blood to the brain, causing myocardial infarction and ischemic stroke, respectively. Thrombosis in the venous circulation manifests itself as thrombosis of the deep veins (veins in the arms or legs) or as pulmonary embolism, where parts of the clot break off and block smaller veins downstream in the lung. Relative to arterial thrombosis, venous thrombosis is phenotypically more homogeneous, and is generally caused by an imbalance between coagulation and anticoagulation.

A first indication that alterations in fibrin clot structure associate with thrombosis came from studies in patients with thrombosis caused by dysfibrinogenemia. Certain functional abnormalities of the fibrinogen molecule are not associated with bleeding as may be expected from the deficiency of a protein that is central to hemostasis, but with a thrombotic tendency instead. Studies have indicated that alterations in fibrin clot structure and reduced susceptibility of the fibrin to fibrinolysis are likely a key mechanism by which the risk of thrombosis is increased in these cases (Soria et al., 1991). However, dysfibrinogenemias are rare, and it was not until fibrin structure was investigated systematically in patients with cardiovascular disease that a clear and consistent pattern emerged linking fibrin clot structure and its susceptibility to fibrinolysis to the risk of thrombosis.

Studies in patients with myocardial infarction showed a dense fibrin network with reduced porosity (Fatah et al., 1992, 1996b). These findings were confirmed in a study of patients with premature coronary artery disease and the clots were found to be stiffer and slower to lyse than control clots (Collet et al., 2006). In patients with acute coronary disease, the changes in fibrin clot structure are even more pronounced than in patients with stable coronary disease, with a further decrease in porosity and increase in lysis times (Undas et al., 2008a). The first-degree relatives of patients with coronary artery disease also have changes in fibrin clot structure suggesting that there is a familial basis for fibrin structure abnormalities predisposing to the disease (Mills et al., 2002).

Other thrombotic diseases are also associated with changes in fibrin clots structure, generally comparable to those described above for coronary artery disease. Stroke is characterized by either intracranial bleeding (hemorrhagic stroke) or thrombosis (ischemic stroke). The etiology of ischemic stroke is similar to that of coronary artery disease, involving atherosclerosis of the arteries and the formation of a plaque. Plaque rupture provides the trigger for thrombosis, leading to death of brain tissue with potentially devastating effects. Patients with ischemic stroke were found to produce clots with increased fiber branching, reduced porosity, and increased resistance to fibrinolysis (Undas et al., 2009a, 2010). These data suggest that more aggressive fibrinolytic therapies may be needed to treat ischemic stroke. However, tPA, the most widely used fibrinolytic agent, crosses the blood–brain barrier and has strong neurotoxic effects, limiting the effectiveness of this treatment. Recent data with tPA linked to larger particles such as red blood cells that prevent its leakage into the brain and limits its activity to newly developing blood clots have been promising (Danielyan et al., 2008).

Patients with peripheral vascular disease and their first-degree relatives show prothrombotic changes in fibrin structure (Bhasin et al., 2008), indicating a role for fibrin in the thrombotic response in this disease of the larger vessels in the lower extremities. Other data suggest a role for fibrin structure in abdominal aortic aneurysm (Scott D. J. A., unpublished), which is characterized by a dilatation of the aorta just below the renal arteries, with the formation of a (nonobstructive) intraluminal thrombus and an increased risk for other vascular disease. A recent study has shown that fibrin clot structure is altered in patients with deep vein thrombosis and their first-degree relatives (Undas et al., 2009b). The fibrin structure changes are prothrombotic in nature and similar to those observed in other thrombotic diseases. Interestingly, the fibrin clot characteristics of patients with pulmonary embolism were different to those with deep vein thrombosis alone, suggesting that certain fibrin characteristics may be involved in embolization of the thrombus. The general consistency of the fibrin clot architectural characteristics between patients with different types of arterial as well as venous thrombosis may be the strongest indicator yet of a causal role for fibrin structure in the pathogenesis of thrombosis.

## XI. Interactions with Cells and Wound Healing

Fibrinogen and fibrin interact with vascular cells via two main motives, one in the γ-chain and one in the α-chain of the molecule. The C-terminal Ala-Gly-Asp sequence (γA408-410) of the fibrinogen γA-chain binds primarily to platelets (Kloczewiak et al., 1984). The integrin on the platelet membrane responsible for this interaction is αIIbβIII (Kamata et al., 1996; Arnaout, 2002). Pharmacological inhibition of the interaction between the αIIbβIII and the fibrin(ogen) γA408-410 Ala-Gly-Asp motive remains one of the most potent treatments for arterial thrombosis currently in use (Varon and Spectre, 2009). The Ala-Gly-Asp motive is lacking in the fibrinogen γ' splice variant so that this variant does not bind platelets. Fibrinogen contained in the α-granules of platelets consists only of γA fibrinogen (Francis et al., 1984; Mosesson et al., 1984). The mechanism behind the lack of γ' fibrinogen in the platelet α-granules is currently unknown, but it could involve selective uptake of γA fibrinogen from plasma through the formation of a tertiary complex of 2 αIIbβIII integrins and 1 γA/γA fibrinogen homodimer molecule, followed by internalization of the tertiary fibrinogen–integrin complex into α-granules. It has also been reported, however, that platelets are capable of producing protein via a DNA-independent mechanism (platelets are anucleate cells), from mRNA and translation machinery packaged into the cells during shedding of platelets from the megakaryocyte (Denis et al., 2005; Schwertz et al., 2006). It cannot be excluded therefore that the γA fibrinogen present in the platelet α-granule is the exclusive product of this expression machinery.

A second motive on the fibrinogen α-chain (α572-574 Arg-Gly-Asp) is primarily responsible for the interaction with endothelial cells (Cheresh et al., 1989; Tranqui et al., 1989; Suehiro et al., 2000). Cross-linking of the α-chain C-terminal region by FXIIIa enhances the interaction between fibrinogen and the endothelial cell (Belkin et al., 2005). The integrin on the endothelial cell membrane responsible for fibrinogen binding is αVβ3 (Charo et al., 1987; Cheresh, 1987). The interactions of fibrinogen with endothelial cells play an important role during thrombosis where the thrombus interacts with the intact endothelium surrounding the site of injury, and also during subsequent processes of wound-healing and angiogenesis, involving cell adhesion to the fibrin matrix and cell migration. Fibrin(ogen) can also interact with the red blood cell, via a β3-dependent integrin, with an interaction force that is similar to that between fibrinogen and platelets (Carvalho et al., 2010). The interaction between fibrinogen and red blood cells may play an important role in determination of blood viscosity and fibrin viscoelastic properties (Gersh et al., 2009a).

After a blood clot is formed at a site of vascular lesion and the stemming of bleeding is achieved, the fibrin clot continues to play an important role in the process of wound healing (Laurens et al., 2006). The clearest evidence for a role of fibrin in wound healing comes from observations made in fibrinogen deficient mice, in which distinct differences were observed in wound healing mechanisms, with altered epithelial cell migration patterns, epithelial hyperplasia, reduced granulation tissue to close the wound gap, and reduced tensile strength of the wound (Drew et al., 2001). The formation of new blood vessels in the extracellular matrix, or angiogenesis, plays an important part of the wound healing process. Angiogenesis is characterized by the sprouting of tubules of migrating endothelial cell into the matrix. The endothelial cells at the tip of the tubule secrete matrix metalloproteases which locally dissolve the matrix for the tubules to grow (van Hinsbergh and Koolwijk, 2008). Endothelial tubule formation in a fibrin matrix is dependent on both $\alpha V\beta 3$ and $\alpha 5\beta 1$ integrins (Laurens et al., 2009). The structure of the fibrin clot may also modulate angiogenesis. A fibrin network with densely packed fibers may be more difficult to penetrate for the endothelial cell, than a network that is less densely packed and with larger pores. Fibrin in plasma exists in a high molecular weight and a low molecular weight form, the latter of which is partially degraded at the C-terminal end of the $\alpha$-chain. It has been shown that angiogenesis is delayed in clots made from low molecular weight fibrinogen, which may be caused by a combination of a denser clot structure that is observed for this variant and the lack of the integrin binding motive Arg572-Gly573-Asp574 in the $\alpha$-chain (Kaijzel et al., 2006).

## XII. Perspectives

The central role of fibrinogen in hemostasis and thrombosis has long been recognized, and although fibrinogen was the first coagulation factor to be described, we are still discovering new aspects of it structure and function and its role in disease. An invaluably large body of knowledge of the protein and gene structure has been, and still is, obtained from fibrinogen heterogeneity in the form of dys-, hypo-, and afibrinogenaemias. In addition, more common fibrinogen variants which are products of polymorphisms and alternative splicing provide important lessons for protein structure and function, and may also be relevant for complex diseases such as cardiovascular disease. An increasing number of environmental factors are also known to regulate the structure and function of the fibrin clot. These environmental factors hold the promise of treatment of disease, as some of them are modifiable by diet, exposure, or pharmacological intervention.

Fibrinogen holds a pivotal position in regulating both coagulation and fibrinolysis. Recent data shows that, in addition to fibrinogen heterogeneity the three-dimensional structure of the fibrin clot produced from fibrinogen by thrombin is crucially important. The modification of fibrin by FXIIIa-dependent cross-linking clearly stabilizes the clot and increases its resistance to mechanical and proteolytic insults. The literature indicates that cross-linking by FXIIIa may be important in vascular disease and thrombosis, but the mechanisms by which this occurs is unknown. We are only beginning to understand the mechanisms that regulate fibrin elasticity and how FXIII modulates this. Further studies are required to identify how the viscoelastic properties of fibrin may relate to thrombosis in the high-pressure, arterial circulation or in the low-pressure venous circulation, and its role in clot embolization.

The biochemical regulation of fibrinolysis has been extensively characterized, but recent studies show that complex interplay exists between the structure of the fibrin clot and the susceptibility of the clot to lysis. This interplay is more than just the speed at which the lytic enzymes permeate the clot. Binding of tPA and plasminogen may be modulated by fibrin structure and the mechanisms behind this are still unknown. Perhaps, exposure of lysine residues may be different in fibrin clots with altered structure, and the role of shear stress or stretch of the fibrin clot on its dissolution by the plasmin system could also be an important factor *in vivo*.

Fibrinogen is one of the most abundant proteins in the human body, and while many aspects of its structure and function are known, the advent of new technologies has provided insight into some of the unknowns of this intriguing molecule. Future studies will show whether it will be possible to modulate fibrin clot structure and use this to our advantage in the treatment of disease.

## References

Abou-Saleh, R. H., Connell, S. D., Harrand, R., Ajjan, R. A., Mosesson, M. W., Smith, D. A., et al. (2009). Nanoscale probing reveals that reduced stiffness of clots from fibrinogen lacking 42 N-terminal Bbeta-chain residues is due to the formation of abnormal oligomers. *Biophys. J.* **96**, 2415–2427.

Acharya, S. S., Dimichele, D. M. (2008). Rare inherited disorders of fibrinogen. *Haemophilia* **14**, 1151–1158.

Ajjan, R., Lim, B. C., Standeven, K. F., Harrand, R., Dolling, S., Phoenix, F., et al. (2008). Common variation in the C-terminal region of the fibrinogen beta-chain: effects on fibrin structure, fibrinolysis and clot rigidity. *Blood* **111**, 643–650.

Ajjan, R., Storey, R. F., Grant, P. J. (2008). Aspirin resistance and diabetes mellitus. *Diabetologia* **51**, 385–390.

Antovic, A., Perneby, C., Ekman, G. J., Wallen, H. N., Hjemdahl, P., Blomback, M., et al. (2005). Marked increase of fibrin gel permeability with very low dose ASA treatment. *Thromb. Res.* **116**, 509–517.

Ariëns, R. A., Lai, T. S., Weisel, J. W., Greenberg, C. S., Grant, P. J. (2002). Role of factor XIII in fibrin clot formation and effects of genetic polymorphisms. *Blood* **100**, 743–754.

Ariëns, R. A., Philippou, H., Nagaswami, C., Weisel, J. W., Lane, D. A., Grant, P. J. (2000). The factor XIII V34L polymorphism accelerates thrombin activation of factor XIII and affects cross-linked fibrin structure. *Blood* **96**, 988–995.

Arnaout, M. A. (2002). Integrin structure: new twists and turns in dynamic cell adhesion. *Immunol. Rev.* **186**, 125–140.

Asselta, R., Duga, S., Tenchini, M. L. (2006). The molecular basis of quantitative fibrinogen disorders. *J. Thromb. Haemost.* **4**, 2115–2129.

Averett, L. E., Akhremitchev, B. B., Schoenfisch, M. H., Gorkun, O. V. (2010). Calcium dependence of fibrin nanomechanics: the gamma1 calcium mediates the unfolding of fibrinogen induced by force applied to the "A–a" bond. *Langmuir* **26**, 14716–14722.

Bajaj, M. S., Kuppuswamy, M. N., Saito, H., Spitzer, S. G., Bajaj, S. P. (1990). Cultured normal human hepatocytes do not synthesize lipoprotein-associated coagulation inhibitor: evidence that endothelium is the principal site of its synthesis. *Proc. Natl. Acad. Sci. USA* **87**, 8869–8873.

Bajzar, L., Morser, J., Nesheim, M. (1996). TAFI, or plasma procarboxypeptidase B, couples the coagulation and fibrinolytic cascades through the thrombin–thrombomodulin complex. *J. Biol. Chem.* **271**, 16603–16608.

Bale, M. D., Westrick, L. G., Mosher, D. F. (1985). Incorporation of thrombospondin into fibrin clots. *J. Biol. Chem.* **260**, 7502–7508.

Barry, E. L., Mosher, D. F. (1989). Factor XIIIa-mediated cross-linking of fibronectin in fibroblast cell layers. Cross-linking of cellular and plasma fibronectin and of amino-terminal fibronectin fragments. *J. Biol. Chem.* **264**, 4179–4185.

Baumann, R. E., Henschen, A. H. (1993). Human fibrinogen polymorphic site analysis by restriction endonuclease digestion and allele-specific polymerase chain reaction amplification: identification of polymorphisms at positions A alpha 312 and B beta 448. *Blood* **82**, 2117–2124.

Behague, I., Poirier, O., Nicaud, V., Evans, A., Arveiler, D., Luc, G., et al. (1996). Beta fibrinogen gene polymorphisms are associated with plasma fibrinogen and coronary artery disease in patients with myocardial infarction. The ECTIM Study. Etude Cas-Temoins sur l'Infarctus du Myocarde. *Circulation* **93**, 440–449.

Belkin, A. M., Tsurupa, G., Zemskov, E., Veklich, Y., Weisel, J. W., Medved, L. (2005). Transglutaminase-mediated oligomerization of the fibrin(ogen) alphaC domains promotes integrin-dependent cell adhesion and signaling. *Blood* **105**, 3561–3568.

Bertina, R. M., Koeleman, B. P., Koster, T., Rosendaal, F. R., Dirven, R. J., de Ronde, H., et al. (1994). Mutation in blood coagulation factor V associated with resistance to activated protein C. *Nature* **369**, 64–67.

Bhasin, N., Ariëns, R. A., West, R. M., Parry, D. J., Grant, P. J., Scott, D. J. (2008). Altered fibrin clot structure and function in the healthy first-degree relatives of subjects with intermittent claudication. *J. Vasc. Surg.* **48**, 1497–1503.

Binnie, C. G., Lord, S. T. (1993). The fibrinogen sequences that interact with thrombin. *Blood* **81**, 3186–3192.

Bjornsson, T. D., Schneider, D. E., Berger, H., Jr. (1989). Aspirin acetylates fibrinogen and enhances fibrinolysis. Fibrinolytic effect is independent of changes in plasminogen activator levels. *J. Pharmacol. Exp. Ther.* **250**, 154–161.

Blombäck, B. (1996). Fibrinogen and fibrin–proteins with complex roles in hemostasis and thrombosis. *Thromb. Res.* **83**, 1–75.

Blombäck, B., Carlsson, K., Fatah, K., Hessel, B., Procyk, R. (1994). Fibrin in human plasma: gel architectures governed by rate and nature of fibrinogen activation. *Thromb. Res.* **75**, 521–538.

Bottenus, R. E., Ichinose, A., Davie, E. W. (1990). Nucleotide sequence of the gene for the b subunit of human factor XIII. *Biochemistry* **29**, 11195–11209.

Brass, E. P., Forman, W. B., Edwards, R. V., Lindan, O. (1978). Fibrin formation: effect of calcium ions. *Blood* **52**, 654–658.

Broekmans, A. W., Veltkamp, J. J., Bertina, R. M. (1983). Congenital protein C deficiency and venous thromboembolism. A study of three Dutch families. *N. Engl. J. Med.* **309**, 340–344.

Brown, A. E., Litvinov, R. I., Discher, D. E., Purohit, P. K., Weisel, J. W. (2009). Multiscale mechanics of fibrin polymer: gel stretching with protein unfolding and loss of water. *Science* **325**, 741–744.

Brown, J. H., Volkmann, N., Jun, G., Henschen-Edman, A. H., Cohen, C. (2000). The crystal structure of modified bovine fibrinogen. *Proc. Natl. Acad. Sci. USA* **97**, 85–90.

Broze, G. J., Jr. (1995). Tissue factor pathway inhibitor and the revised theory of coagulation. *Annu. Rev. Med.* **46**, 103–112.

Broze, G. J., Jr., Girard, T. J., Novotny, W. F. (1990). Regulation of coagulation by a multivalent Kunitz-type inhibitor. *Biochemistry* **29**, 7539–7546.

Burton, R. A., Tsurupa, G., Hantgan, R. R., Tjandra, N., Medved, L. (2007). NMR solution structure, stability, and interaction of the recombinant bovine fibrinogen alphaC-domain fragment. *Biochemistry* **46**, 8550–8560.

Burton, R. A., Tsurupa, G., Medved, L., Tjandra, N. (2006). Identification of an ordered compact structure within the recombinant bovine fibrinogen alphaC-domain fragment by NMR. *Biochemistry* **45**, 2257–2266.

Carlisle, C. R., Sparks, E. A., Der, L. C., Guthold, M. (2010). Strength and failure of fibrin fiber branchpoints. *J. Thromb. Haemost.* **8**, 1135–1138.

Carter, A. M., Catto, A. J., Bamford, J. M., Grant, P. J. (1997). Gender-specific associations of the fibrinogen B beta 448 polymorphism, fibrinogen levels, and acute cerebrovascular disease. *Arterioscler. Thromb. Vasc. Biol.* **17**, 589–594.

Carter, A. M., Catto, A. J., Grant, P. J. (1999). Association of the alpha-fibrinogen Thr312Ala polymorphism with poststroke mortality in subjects with atrial fibrillation. *Circulation* **99**, 2423–2426.

Carter, A. M., Catto, A. J., Kohler, H. P., Ariëns, R. A., Stickland, M. H., Grant, P. J. (2000). Alpha-fibrinogen Thr312Ala polymorphism and venous thromboembolism. *Blood* **96**, 1177–1179.

Carter, A. M., Mansfield, M. W., Stickland, M. H., Grant, P. J. (1996). Beta-fibrinogen gene-455 G/A polymorphism and fibrinogen levels. Risk factors for coronary artery disease in subjects with NIDDM. *Diabetes Care.* **19**, 1265–1268.

Carter, A. M., Ossei-Gerning, N., Wilson, I. J., Grant, P. J. (1997). Association of the platelet Pl(A) polymorphism of glycoprotein IIb/IIIa and the fibrinogen Bbeta 448 polymorphism with myocardial infarction and extent of coronary artery disease. *Circulation* **96**, 1424–1431.

Carvalho. F. A., Connell, S., Miltenberger-Miltenyi, G., Pereira, S. V., Tavares, A., Ariëns, R. A., et al. (2010). Atomic force microscopy-based molecular recognition of a fibrinogen receptor on human erythrocytes. *ACS Nano* **4**, 4609–4620.

Caspary, E. A. (1956). Studies on the acetylation of human fibrinogen. *Biochem. J.* **62**, 507–512.

Catto, A. J., Kohler, H. P., Bannan, S., Stickland, M., Carter, A., Grant, P. J. (1998). Factor XIII Val 34 Leu: a novel association with primary intracerebral hemorrhage. *Stroke* **29**, 813–816.

Catto, A. J., Kohler, H. P., Coore, J., Mansfield, M. W., Stickland, M. H., Grant, P. J. (1999). Association of a common polymorphism in the factor XIII gene with venous thrombosis. *Blood* **93**, 906–908.

Charo, I. F., Bekeart, L. S., Phillips, D. R. (1987). Platelet glycoprotein IIb-IIIa-like proteins mediate endothelial cell attachment to adhesive proteins and the extracellular matrix. *J. Biol. Chem.* **262**, 9935–9938.

Chen, R., Doolittle, R. F. (1971). Gamma–gamma cross-linking sites in human and bovine fibrin. *Biochemistry* **10**, 4487–4491.

Cheresh, D. A. (1987). Human endothelial cells synthesize and express an Arg-Gly-Asp-directed adhesion receptor involved in attachment to fibrinogen and von Willebrand factor. *Proc. Natl. Acad. Sci. USA* **84**, 6471–6475.

Cheresh, D. A., Berliner, S. A., Vicente, V., Ruggeri, Z. M. (1989). Recognition of distinct adhesive sites on fibrinogen by related integrins on platelets and endothelial cells. *Cell* **58**, 945–953.

Cheung, E. Y., Uitte, d.W., Vos, H. L., Leebeek, F. W., Dippel, D. W., Bertina, R. M., et al. (2008). Fibrinogen gamma' in ischemic stroke: a case–control study. *Stroke* **39**, 1033–1035.

Collen, D. (1980). Natural inhibitors of fibrinolysis. *J. Clin. Pathol. Suppl. (R. Coll. Pathol.)* 24–30.

Collen, D. (1999). The plasminogen (fibrinolytic) system. *Thromb. Haemost.* **82**, 259–270.

Collet, J. P., Allali, Y., Lesty, C., Tanguy, M. L., Silvain, J., Ankri, A., et al. (2006). Altered fibrin architecture is associated with hypofibrinolysis and premature coronary atherothrombosis. *Arterioscler. Thromb. Vasc. Biol.* **26**, 2567–2573.

Collet, J. P., Lesty, C., Montalescot, G., Weisel, J. W. (2003). Dynamic changes of fibrin architecture during fibrin formation and intrinsic fibrinolysis of fibrin-rich clots. *J. Biol. Chem.* **278**, 21331–21335.

Collet, J. P., Moen, J. L., Veklich, Y. I., Gorkun, O. V., Lord, S. T., Montalescot, G., et al. (2005). The alphaC domains of fibrinogen affect the structure of the fibrin clot, its physical properties, and its susceptibility to fibrinolysis. *Blood* **106**, 3824–3830.

Collet, J. P., Park, D., Lesty, C., Soria, J., Soria, C., Montalescot, G., et al. (2000). Influence of fibrin network conformation and fibrin fiber diameter on fibrinolysis speed: dynamic and structural approaches by confocal microscopy. *Arterioscler. Thromb. Vasc. Biol.* **20**, 1354–1361.

Collet, J. P., Shuman, H., Ledger, R. E., Lee, S., Weisel, J. W. (2005). The elasticity of an individual fibrin fiber in a clot. *Proc. Natl. Acad. Sci. USA* **102**, 9133–9137.

Comp, P. C., Esmon, C. T. (1979). Activated protein C inhibits platelet prothrombin-converting activity. *Blood* **54**, 1272–1281.

Cooper, A. V., Standeven, K. F., Ariens, R. A. (2003). Fibrinogen gamma-chain splice variant gamma' alters fibrin formation and structure. *Blood* **102**, 535–540.

Cottrell, B. A., Strong, D. D., Watt, K. W., Doolittle, R. F. (1979). Amino acid sequence studies on the alpha chain of human fibrinogen. Exact location of cross-linking acceptor sites. *Biochemistry* **18**, 5405–5410.

Credo, R. B., Curtis, C. G., Lorand, L. (1981). Alpha-chain domain of fibrinogen controls generation of fibrinoligase (coagulation factor XIIIa). Calcium ion regulatory aspects. *Biochemistry* **20**, 3770–3778.

Curtis, C. G., Brown, K. L., Credo, R. B., Domanik, R. A., Gray, A., Stenberg, P., et al. (1974). Calcium-dependent unmasking of active center cysteine during activation of fibrin stabilizing factor. *Biochemistry* **13**, 3774–3780.

Dahlback, B. (1994). Inherited resistance to activated protein C, a major cause of venous thrombosis, is due to a mutation in the factor V gene. *Haemostasis* **24**, 139–151.

Dahlback, B., Carlsson, M., Svensson, P. J. (1993). Familial thrombophilia due to a previously unrecognized mechanism characterized by poor anticoagulant response to activated protein C: prediction of a cofactor to activated protein C. *Proc. Natl. Acad. Sci. USA* **90**, 1004–1008.

Dahlback, B., Stenflo, J. (1980). Inhibitory effect of activated protein C on activation of prothrombin by platelet-bound factor Xa. *Eur. J. Biochem.* **107**, 331–335.

Damus, P. S., Hicks, M., Rosenberg, R. D. (1973). Anticoagulant action of heparin. *Nature* **246**, 355–357.

Dang, C. V., Shin, C. K., Bell, W. R., Nagaswami, C., Weisel, J. W. (1989). Fibrinogen sialic acid residues are low affinity calcium-binding sites that influence fibrin assembly. *J. Biol. Chem.* **264**, 15104–15108.

Danielyan, K., Ganguly, K., Ding, B. S., Atochin, D., Zaitsev, S., Murciano, J. C., et al. (2008). Cerebrovascular thromboprophylaxis in mice by erythrocyte-coupled tissue-type plasminogen activator. *Circulation* **118**, 1442–1449.

Denis, M. M., Tolley, N. D., Bunting, M., Schwertz, H., Jiang, H., Lindemann, S., et al. (2005). Escaping the nuclear confines: signal-dependent pre-mRNA splicing in anucleate platelets. *Cell* **122**, 379–391.

Doolittle, R. F. (1994). Fibrinogen and fibrin. In: Haemostasis and Thrombosis, Bloom, A., Forbes, C., Thomas, D., Tuddenham, E (Eds), pp. 491–513. Churchill Livingston, Edinburgh.

Doolittle, R. F., Goldbaum, D. M., Doolittle, L. R. (1978). Designation of sequences involved in the "coiled-coil" interdominal connections in fibrinogen: constructions of an atomic scale model. *J. Mol. Biol.* **120**, 311–325.

Drew, A. F., Liu, H., Davidson, J. M., Daugherty, C. C., Degen, J. L. (2001). Wound-healing defects in mice lacking fibrinogen. *Blood* **97**, 3691–3698.

Dunn, E. J., Ariëns, R. A., Grant, P. J. (2005). The influence of type 2 diabetes on fibrin structure and function. *Diabetologia* **48**, 1198–1206.

Dunn, E. J., Philippou, H., Ariëns, R. A., Grant, P. J. (2006). Molecular mechanisms involved in the resistance of fibrin to clot lysis by plasmin in subjects with type 2 diabetes mellitus. *Diabetologia* **49**, 1071–1080.

Esmon, C. T. (1983). Protein-C: biochemistry, physiology, and clinical implications. *Blood* **62**, 1155–1158.

Farrell, D. H., Thiagarajan, P. (1994). Binding of recombinant fibrinogen mutants to platelets. *J. Biol. Chem.* **269**, 226–231.

Fatah, K., Beving, H., Albage, A., Ivert, T., Blomback, M. (1996). Acetylsalicylic acid may protect the patient by increasing fibrin gel porosity. Is withdrawing of treatment harmful to the patient? *Eur. Heart J.* **17**, 1362–1366.

Fatah, K., Hamsten, A., Blomback, B., Blomback, M. (1992). Fibrin gel network characteristics and coronary heart disease: relations to plasma fibrinogen concentration, acute phase protein, serum lipoproteins and coronary atherosclerosis. *Thromb. Haemost.* **68**, 130–135.

Fatah, K., Silveira, A., Tornvall, P., Karpe, F., Blomback, M., Hamsten, A. (1996). Proneness to formation of tight and rigid fibrin gel structures in men with myocardial infarction at a young age. *Thromb. Haemost.* **76**, 535–540.

Fay, W. P., Eitzman, D. T., Shapiro, A. D., Madison, E. L., Ginsburg, D. (1994). Platelets inhibit fibrinolysis in vitro by both plasminogen activator inhibitor-1-dependent and -independent mechanisms. *Blood* **83**, 351–356.

Ferry, J. D. (1952). The mechanism of polymerization of fibrinogen. *Proc. Natl. Acad. Sci. USA* **38**, 566–569.

Folk, J. E. (1983). Mechanism and basis for specificity of transglutaminase-catalyzed epsilon-(gamma-glutamyl) lysine bond formation. *Adv. Enzymol. Relat. Areas Mol. Biol.* **54**, 1–56.

Francis, C. W., Marder, V. J., Martin, S. E. (1980). Plasmic degradation of crosslinked fibrin. I. Structural analysis of the particulate clot and identification of new macromolecular-soluble complexes. *Blood* **56**, 456–464.

Francis, C. W., Nachman, R. L., Marder, V. J. (1984). Plasma and platelet fibrinogen differ in gamma chain content. *Thromb. Haemost.* **51**, 84–88.

Franco, R. F., Reitsma, P. H., Lourenco, D., Maffei, F. H., Morelli, V., Tavella, M. H., et al. (1999). Factor XIII Val34Leu is a genetic factor involved in the etiology of venous thrombosis. *Thromb. Haemost.* **81**, 676–679.

Fretto, L. J., Ferguson, E. W., Steinman, H. M., McKee, P. A. (1978). Localization of the alpha-chain cross-link acceptor sites of human fibrin. *J. Biol. Chem.* **253**, 2184–2195.

Fu, Y., Grieninger, G. (1994). Fib420: a normal human variant of fibrinogen with two extended alpha chains. *Proc. Natl. Acad. Sci. USA* **91**, 2625–2628.

Fu, Y., Weissbach, L., Plant, P. W., Oddoux, C., Cao, Y., Liang, T. J., et al. (1992). Carboxy-terminal-extended variant of the human fibrinogen alpha subunit: a novel exon conferring marked homology to beta and gamma subunits. *Biochemistry* **31**, 11968–11972.

Fu, Y., Zhang, J. Z., Redman, C. M., Grieninger, G. (1998). Formation of the human fibrinogen subclass fib420: disulfide bonds and glycosylation in its unique (alphaE chain) domains. *Blood* **92**, 3302–3308.

Fuster, V., Moreno, P. R., Fayad, Z. A., Corti, R., Badimon, J. J. (2005). Atherothrombosis and high-risk plaque: part I: evolving concepts. *J. Am. Coll. Cardiol.* **46**, 937–954.

Gaffney, P. J., Whitaker, A. N. (1979). Fibrin crosslinks and lysis rates. *Thromb. Res.* **14**, 85–94.

Gersh, K. C., Lord, S. T. (2006). An investigation of Factor XIII binding to recombinant gamma'/gamma' and gamma/gamma' fibrinogen. *Blood* **108**, 1705.

Gersh, K. C., Nagaswami, C., Weisel, J. W. (2009a). Fibrin network structure and clot mechanical properties are altered by incorporation of erythrocytes. *Thromb. Haemost.* **102**, 1169–1175.

Gersh, K. C., Nagaswami, C., Weisel, J. W., Lord, S. T. (2009b). The presence of gamma' chain impairs fibrin polymerization. *Thromb. Res.* **124**, 356–363.

Girard, T. J., Warren, L. A., Novotny, W. F., Likert, K. M., Brown, S. G., Miletich, J. P., et al. (1989). Functional significance of the Kunitz-type inhibitory domains of lipoprotein-associated coagulation inhibitor. *Nature* **338**, 518–520.

Gorkun, O. V., Veklich, Y. I., Medved, L. V., Henschen, A. H., Weisel, J. W. (1994). Role of the alpha C domains of fibrin in clot formation. *Biochemistry* **33**, 6986–6997.

Gorman, J. J., Folk, J. E. (1981). Structural features of glutamine substrates for transglutaminases. Specificities of human plasma factor XIIIa and the guinea pig liver enzyme toward synthetic peptides. *J. Biol. Chem.* **256**, 2712–2715.

Greilich, P. E., Carr, M. E., Zekert, S. L., Dent, R. M. (1994). Quantitative assessment of platelet function and clot structure in patients with severe coronary artery disease. *Am. J. Med. Sci.* **307**, 15–20.

Griffin, J. H., Evatt, B., Zimmerman, T. S., Kleiss, A. J., Wideman, C. (1981). Deficiency of protein C in congenital thrombotic disease. *J. Clin. Invest.* **68**, 1370–1373.

Guthold, M., Carlisle, C. (2010). Single fibrin fiber experiments suggest longitudinal rather than transverse crosslinking. *J. Thromb, Haemost.*

Hackeng, T. M., Sere, K. M., Tans, G., Rosing, J. (2006). Protein S stimulates inhibition of the tissue factor pathway by tissue factor pathway inhibitor. *Proc. Natl. Acad. Sci. USA* **103**, 3106–3111.

Hada, M., Kaminski, M., Bockenstedt, P., McDonagh, J. (1986). Covalent crosslinking of von Willebrand factor to fibrin. *Blood* **68**, 95–101.

Hall, C. E., Slayter, H. S. (1959). The fibrinogen molecule: its size, shape, and mode of polymerization. *J. Biophys. Biochem. Cytol.* **5**, 11–16.

Harris, K. W., Esmon, C. T. (1985). Protein S is required for bovine platelets to support activated protein C binding and activity. *J. Biol. Chem.* **260**, 2007–2010.

He, S., Bark, N., Wang, H., Svensson, J., Blomback, M. (2009). Effects of acetylsalicylic acid on increase of fibrin network porosity and the consequent upregulation of fibrinolysis. *J. Cardiovasc. Pharmacol.* **53**, 24–29.

Highsmith, R. F., Rosenberg, R. D. (1974). The inhibition of human plasmin by human antithrombin-heparin cofactor. *J. Biol. Chem.* **249**, 4335–4338.
Horellou, M. H., Conard, J., Bertina, R. M., Samama, M. (1984). Congenital protein C deficiency and thrombotic disease in nine French families. *Br. Med. J. Clin. Res. Ed.* **289**, 1285–1287.
Hsieh, L., Nugent, D. (2008). Factor XIII deficiency. *Haemophilia* **14**, 1190–1200.
Humphries, S. E., Cook, M., Dubowitz, M., Stirling, Y., Meade, T. W. (1987). Role of genetic variation at the fibrinogen locus in determination of plasma fibrinogen concentrations. *Lancet* **1**, 1452–1455.
Huntington, J. A., Read, R. J., Carrell, R. W. (2000). Structure of a serpin–protease complex shows inhibition by deformation. *Nature* **407**, 923–926.
Husain, S. S. (1993). Fibrin affinity of urokinase-type plasminogen activator. Evidence that $Zn^{2+}$ mediates strong and specific interaction of single-chain urokinase with fibrin. *J. Biol. Chem.* **268**, 8574–8579.
Ichinose, A., Davie, E. W. (1988). Characterization of the gene for the a subunit of human factor XIII (plasma transglutaminase), a blood coagulation factor. *Proc. Natl. Acad. Sci. USA* **85**, 5829–5833.
Ichinose, A., McMullen, B. A., Fujikawa, K., Davie, E. W. (1986). Amino acid sequence of the b subunit of human factor XIII, a protein composed of ten repetitive segments. *Biochemistry* **25**, 4633–4638.
Iwata, H., Kitano, T., Umetsu, K., Yuasa, I., Yamazaki, K., Kemkes-Matthes, B., et al. (2009). Distinct C-terminus of the B subunit of factor XIII in a population-associated major phenotype: the first case of complete allele-specific alternative splicing products in the coagulation and fibrinolytic systems. *J. Thromb. Haemost.* **7**, 1084–1091.
Janus, T. J., Lewis, S. D., Lorand, L., Shafer, J. A. (1983). Promotion of thrombin-catalyzed activation of factor XIII by fibrinogen. *Biochemistry* **22**, 6269–6272.
Kaijzel, E. L., Koolwijk, P., van Erck, M. G., van Hinsbergh, V. W., de Maat, M. P. (2006). Molecular weight fibrinogen variants determine angiogenesis rate in a fibrin matrix in vitro and in vivo. *J. Thromb. Haemost.* **4**, 1975–1981.
Kamata, T., Irie, A., Tokuhira, M., Takada, Y. (1996). Critical residues of integrin alphaIIb subunit for binding of alphaIIbbeta3 (glycoprotein IIb-IIIa) to fibrinogen and ligand-mimetic antibodies (PAC-1, OP-G2, and LJ-CP3). *J. Biol. Chem.* **271**, 18610–18615.
Kamboh, M. I., Ferrell, R. E. (1986). Genetic studies of low abundance human plasma proteins. II. Population genetics of coagulation factor XIIIB. *Am. J. Hum. Genet.* **39**, 817–825.
Kang, H., Wen, Q., Janmey, P. A., Tang, J. X., Conti, E., MacKintosh, F. C. (2009). Nonlinear elasticity of stiff filament networks: strain stiffening, negative normal stress, and filament alignment in fibrin gels. *J. Phys. Chem. B* **113**, 3799–3805.
Kant, J. A., Fornace, A. J., Jr., Saxe, D., Simon, M. I., McBride, O. W., Crabtree, G. R. (1985). Evolution and organization of the fibrinogen locus on chromosome 4: gene duplication accompanied by transposition and inversion. *Proc. Natl. Acad. Sci. USA* **82**, 2344–2348.

Kimura, S., Aoki, N. (1986). Cross-linking site in fibrinogen for alpha 2-plasmin inhibitor. *J. Biol. Chem.* **261**, 15591–15595.
Kloczewiak, M., Timmons, S., Lukas, T. J., Hawiger, J. (1984). Platelet receptor recognition site on human fibrinogen. Synthesis and structure-function relationship of peptides corresponding to the carboxy-terminal segment of the gamma chain. *Biochemistry* **23**, 1767–1774.
Kohler, H. P., Futers, T. S., Grant, P. J. (1999). Prevalence of three common polymorphisms in the A-subunit gene of factor XIII in patients with coronary artery disease. *Thromb. Haemost.* **81**, 511–515.
Kohler, H. P., Grant, P. J. (1998). Clustering of haemostatic risk factors with FXIII-Val34Leu in patients with myocardial infarction. *Thromb. Haemost.* **80**, 862.
Kohler, H. P., Stickland, M. H., Ossei-Gerning, N., Carter, A., Mikkola, H., Grant, P. J. (1998). Association of a common polymorphism in the factor XIII gene with myocardial infarction. *Thromb. Haemost.* **79**, 8–13.
Kollman, J. M., Pandi, L., Sawaya, M. R., Riley, M., Doolittle, R. F. (2009). Crystal structure of human fibrinogen. *Biochemistry* **48**, 3877–3886.
Komanasin, N., Catto, A. J., Futers, T. S., van Hylckama Vlieg, A., Rosendaal, F. R., Ariens, R. A. (2005). A novel polymorphism in the factor XIII B-subunit (His95Arg): relationship to subunit dissociation and venous thrombosis. *J. Thromb. Haemost.* **3**, 2487–2496.
Koster, T., Rosendaal, F. R., de Ronde, H., Briët, E., Vandenbroucke, J. P., Bertina, R. M. (1993). Venous thrombosis due to poor anticoagulant response to activated protein C: Leiden Thrombophilia Study. *Lancet* **342**, 1503–1506.
Lancellotti, S., Rutella, S., De Filippis, V., Pozzi, N., Rocca, B., De Cristofaro, R. (2008). Fibrinogen-elongated gamma chain inhibits thrombin-induced platelet response, hindering the interaction with different receptors. *J. Biol. Chem.* **283**, 30193–30204.
Lane, D. A., Caso, R. (1989). Antithrombin: structure, genomic organization, function and inherited deficiency. *Baillières Clin. Haematol.* **2**, 961–998.
Langer, B. G., Weisel, J. W., Dinauer, P. A., Nagaswami, C., Bell, W. R. (1988). Deglycosylation of fibrinogen accelerates polymerization and increases lateral aggregation of fibrin fibers. *J. Biol. Chem.* **263**, 15056–15063.
Laurens, N., Engelse, M. A., Jungerius, C., Lowik, C. W., van Hinsbergh, V. W., Koolwijk, P. (2009). Single and combined effects of alphavbeta3- and alpha5-beta1-integrins on capillary tube formation in a human fibrinous matrix. *Angiogenesis* **12**, 275–285.
Laurens, N., Koolwijk, P., de Maat, M. P. (2006). Fibrin structure and wound healing. *J. Thromb. Haemost.* **4**, 932–939.
Lauricella, A. M., Quintana, I. L., Kordich, L. C. (2002). Effects of homocysteine thiol group on fibrin networks: another possible mechanism of harm. *Thromb. Res.* **107**, 75–79.
Lee, K. N., Jackson, K. W., Christiansen, V. J., Chung, K. H., McKee, P. A. (2004). A novel plasma proteinase potentiates alpha2-antiplasmin inhibition of fibrin digestion. *Blood* **103**, 3783–3788.
Lee, K. N., Jackson, K. W., Christiansen, V. J., Lee, C. S., Chun, J. G., McKee, P. A. (2007). Why alpha-antiplasmin must be converted to a derivative form for optimal function. *J. Thromb. Haemost.* **5**, 2095–2104.

Lim, B. C., Ariëns, R. A., Carter, A. M., Weisel, J. W., Grant, P. J. (2003). Genetic regulation of fibrin structure and function: complex gene–environment interactions may modulate vascular risk. *Lancet* **361**, 1424–1431.

Lim, B. B., Lee, E. H., Sotomayor, M., Schulten, K. (2008). Molecular basis of fibrin clot elasticity. *Structure* **16**, 449–459.

Liu, W., Carlisle, C. R., Sparks, E. A., Guthold, M. (2010). The mechanical properties of single fibrin fibers. *J. Thromb. Haemost.* **8**, 1030–1036.

Liu, W., Jawerth, L. M., Sparks, E. A., Falvo, M. R., Hantgan, R. R., Superfine, R., et al. (2006). Fibrin fibers have extraordinary extensibility and elasticity. *Science* **313**, 634.

Loll, P. J., Picot, D., Garavito, R. M. (1995). The structural basis of aspirin activity inferred from the crystal structure of inactivated prostaglandin H2 synthase. *Nat. Struct. Biol.* **2**, 637–643.

Lorand, L., Downey, J., Gotoh, T., Jacobsen, A., Tokura, S. (1968). The transpeptidase system which crosslinks fibrin by gamma-glutamyle-episilon-lysine bonds. *Biochem. Biophys. Res. Commun.* **31**, 222–230.

Lorand, L., Parameswaran, K. N., Murthy, S. N. (1998). A double-headed Gly-Pro-Arg-Pro ligand mimics the functions of the E domain of fibrin for promoting the end-to-end crosslinking of gamma chains by factor XIIIa. *Proc. Natl. Acad. Sci. USA* **95**, 537–541.

Lovely, R. S., Falls, L. A., Al Mondhiry, H. A., Chambers, C. E., Sexton, G. J., Ni, H., et al. (2002). Association of gammaA/gamma' fibrinogen levels and coronary artery disease. *Thromb. Haemost.* **88**, 26–31.

Lovely, R. S., Rein, C. M., White, T. C., Jouihan, S. A., Boshkov, L. K., Bakke, A. C., et al. (2008). GammaA/gamma' fibrinogen inhibits thrombin-induced platelet aggregation. *Thromb. Haemost.* **100**, 837–846.

Ly, B., Godal, H. C. (1973). Denaturation of fibrinogen: the protective effect of calcium. *Haemostasis* **1**, 204–209.

Madrazo, J., Brown, J. H., Litvinovich, S., Dominguez, R., Yakovlev, S., Medved, L., et al. (2001). Crystal structure of the central region of bovine fibrinogen (E5 fragment) at 1.4-A resolution. *Proc. Natl. Acad. Sci. USA* **98**, 11967–11972.

Mannila, M. N., Lovely, R. S., Kazmierczak, S. C., Eriksson, P., Samnegard, A., Farrell, D. H., et al. (2007). Elevated plasma fibrinogen gamma' concentration is associated with myocardial infarction: effects of variation in fibrinogen genes and environmental factors. *J. Thromb. Haemost.* **5**, 766–773.

Marcum, J. A., McKenney, J. B., Galli, S. J., Jackman, R. W., Rosenberg, R. D. (1986). Anticoagulantly active heparin-like molecules from mast cell-deficient mice. *Am. J. Physiol.* **250**, H879–H888.

Marcum, J. A., Rosenberg, R. D. (1985). Heparinlike molecules with anticoagulant activity are synthesized by cultured endothelial cells. *Biochem. Biophys. Res. Commun.* **126**, 365–372.

Martinez, J., Keane, P. M., Gilman, P. B., Palascak, J. E. (1983). The abnormal carbohydrate composition of the dysfibrinogenemia associated with liver disease. *Ann. NY Acad. Sci.* **408**, 388–396.

Martinez, J., MacDonald, K. A., Palascak, J. E. (1983). The role of sialic acid in the dysfibrinogenemia associated with liver disease: distribution of sialic acid on the constituent chains. *Blood* **61**, 1196–1202.

Matsuka, Y. V., Medved, L. V., Migliorini, M. M., Ingham, K. C. (1996). Factor XIIIa-catalyzed cross-linking of recombinant alpha C fragments of human fibrinogen. *Biochemistry* **35**, 5810–5816.

Metassan, S., Charlton, A. J., Routledge, M. N., Scott, D. J., Ariëns, R. A. (2010). Alteration of fibrin clot properties by ultrafine particulate matter. *Thromb. Haemost.* **103**, 103–113.

Metassan, S., Routledge, M. N., Lucking, A. J., de Willige, S. U., Philippou, H., Mills, N. L., et al. (2010). Fibrin clot structure remains unaffected in young, healthy individuals after transient exposure to diesel exhaust. *Part Fibre. Toxicol.* **7**, 17.

Mikkola, H., Syrjala, M., Rasi, V., Vahtera, E., Hamalainen, E., Peltonen, L., et al. (1994). Deficiency in the A-subunit of coagulation factor XIII: two novel point mutations demonstrate different effects on transcript levels. *Blood* **84**, 517–525.

Miller, K. A., Siscovick, D. S., Sheppard, L., Shepherd, K., Sullivan, J. H., Anderson, G. L., et al. (2007). Long-term exposure to air pollution and incidence of cardiovascular events in women. *N. Engl. J. Med.* **356**, 447–458.

Mills, J. D., Ariëns, R. A., Mansfield, M. W., Grant, P. J. (2002). Altered fibrin clot structure in the healthy relatives of patients with premature coronary artery disease. *Circulation* **106**, 1938–1942.

Mosesson, M. W., DiOrio, J. P., Hernandez, I., Hainfeld, J. F., Wall, J. S., Grieninger, G. (2004). The ultrastructure of fibrinogen-420 and the fibrin-420 clot. *Biophys. Chem.* **112**, 209–214.

Mosesson, M. W., Finlayson, J. S. (1963a). Biochemical and chromatographic studies of certain activities associated with human fibrinogen preparations. *J. Clin. Invest.* **42**, 747–755.

Mosesson, M. W., Finlayson, J. S. (1963b). Subfractions of human fibrinogen; preparation and analysis. *J. Lab. Clin. Med.* **62**, 663–674.

Mosesson, M. W., Hernandez, I., Raife, T. J., Medved, L., Yakovlev, S., Simpson-Haidaris, P. J., et al. (2007). Plasma fibrinogen gamma' chain content in the thrombotic microangiopathy syndrome. *J. Thromb. Haemost.* **5**, 62–69.

Mosesson, M. W., Homandberg, G. A., Amrani, D. L. (1984). Human platelet fibrinogen gamma chain structure. *Blood* **63**, 990–995.

Mosesson, M. W., Siebenlist, K. R., DiOrio, J. P., Matsuda, M., Hainfeld, J. F., Wall, J. S. (1995). The role of fibrinogen D domain intermolecular association sites in the polymerization of fibrin and fibrinogen Tokyo II (gamma 275 Arg->Cys). *J. Clin. Invest.* **96**, 1053–1058.

Mosesson, M. W., Siebenlist, K. R., Meh, D. A. (2001). The structure and biological features of fibrinogen and fibrin. *Ann. NY Acad. Sci.* **936**, 11–30.

Mosher, D. F., Schad, P. E. (1979). Cross-linking of fibronectin to collagen by blood coagulation Factor XIIIa. *J. Clin. Invest.* **64**, 781–787.

Muszbek, L., Adany, R., Mikkola, H. (1996). Novel aspects of blood coagulation factor XIII. I. Structure, distribution, activation, and function. *Crit. Rev. Clin. Lab. Sci.* **33**, 357–421.

Muszbek, L., Ariëns, R. A., Ichinose, A. (2007). Factor XIII: recommended terms and abbreviations. *J. Thromb. Haemost.* **5**, 181–183.

Muszbek, L., Bagoly, Z., Bereczky, Z., Katona, E. (2008). The involvement of blood coagulation factor XIII in fibrinolysis and thrombosis. *Cardiovasc. Hematol. Agents Med. Chem.* **6**, 190–205.
Ndonwi, M., Tuley, E. A., Broze, G. J., Jr. (2010). The Kunitz-3 domain of TFPI-alpha is required for protein S-dependent enhancement of factor Xa inhibition. *Blood* **116**, 1344–1351.
Nesheim, M. (2003). Thrombin and fibrinolysis. *Chest* **124**, 33S–39S.
Nesheim, M., Wang, W., Boffa, M., Nagashima, M., Morser, J., Bajzar, L. (1997). Thrombin, thrombomodulin and TAFI in the molecular link between coagulation and fibrinolysis. *Thromb. Haemost.* **78**, 386–391.
Odrljin, T. M., Rybarczyk, B. J., Francis, C. W., Lawrence, S. O., Hamaguchi, M., Simpson-Haidaris, P. J. (1996). Calcium modulates plasmin cleavage of the fibrinogen D fragment gamma chain N-terminus: mapping of monoclonal antibody J88B to a plasmin sensitive domain of the gamma chain. *Biochim. Biophys. Acta* **1298**, 69–77.
Osterud, B., Rapaport, S. I. (1977). Activation of factor IX by the reaction product of tissue factor and factor VII: additional pathway for initiating blood coagulation. *Proc. Natl. Acad. Sci. USA* **74**, 5260–5264.
Pechik, I., Madrazo, J., Mosesson, M. W., Hernandez, I., Gilliland, G. L., Medved, L. (2004). Crystal structure of the complex between thrombin and the central "E" region of fibrin. *Proc. Natl. Acad. Sci. USA* **101**, 2718–2723.
Peerschke, E. I., Francis, C. W., Marder, V. J. (1986). Fibrinogen binding to human blood platelets: effect of gamma chain carboxyterminal structure and length. *Blood* **67**, 385–390.
Peters, A., von Klot, S., Heier, M., Trentinaglia, I., Hormann, A., Wichmann, H. E., et al. (2004). Exposure to traffic and the onset of myocardial infarction. *N. Engl. J. Med.* **351**, 1721–1730.
Pieters, M., Covic, N., van der Westhuizen, F. H., Nagaswami, C., Baras, Y., Toit, L. D., et al. (2008). Glycaemic control improves fibrin network characteristics in type 2 diabetes—a purified fibrinogen model. *Thromb. Haemost.* **99**, 691–700.
Pope, C. A., III, Burnett, R. T., Thurston, G. D., Thun, M. J., Calle, E. E., Krewski, D., et al. (2004). Cardiovascular mortality and long-term exposure to particulate air pollution: epidemiological evidence of general pathophysiological pathways of disease. *Circulation* **109**, 71–77.
Pope, C. A., III, Muhlestein, J. B., May, H. T., Renlund, D. G., Anderson, J. L., Horne, B. D. (2006). Ischemic heart disease events triggered by short-term exposure to fine particulate air pollution. *Circulation* **114**, 2443–2448.
Pospisil, C. H., Stafford, A. R., Fredenburgh, J. C., Weitz, J. I. (2003). Evidence that both exosites on thrombin participate in its high affinity interaction with fibrin. *J. Biol. Chem.* **278**, 21584–21591.
Pratt, K. P., Cote, H. C., Chung, D. W., Stenkamp, R. E., Davie, E. W. (1997). The primary fibrin polymerization pocket: three-dimensional structure of a 30-kDa C-terminal gamma chain fragment complexed with the peptide Gly-Pro-Arg-Pro. *Proc. Natl. Acad. Sci. USA* **94**, 7176–7181.

Procyk, R., Blomback, B. (1990). Disulfide bond reduction in fibrinogen: calcium protection and effect on clottability. *Biochemistry* **29**, 1501–1507.
Pruissen, D. M., Slooter, A. J., Rosendaal, F. R., van der Graaf, Y., Algra, A. (2008). Coagulation factor XIII gene variation, oral contraceptives, and risk of ischemic stroke. *Blood* **111**(3), 1282–1286.
Rao, L. V., Nordfang, O., Hoang, A. D., Pendurthi, U. R. (1995). Mechanism of antithrombin III inhibition of factor VIIa/tissue factor activity on cell surfaces. Comparison with tissue factor pathway inhibitor/factor Xa-induced inhibition of factor VIIa/tissue factor activity. *Blood* **85**, 121–129.
Rao, L. V., Rapaport, S. I., Hoang, A. D. (1993). Binding of factor VIIa to tissue factor permits rapid antithrombin III/heparin inhibition of factor VIIa. *Blood* **81**, 2600–2607.
Rau, J. C., Beaulieu, L. M., Huntington, J. A., Church, F. C. (2007). Serpins in thrombosis, hemostasis and fibrinolysis. *J. Thromb. Haemost.* **5**(Suppl 1), 102–115.
Redlitz, A., Tan, A. K., Eaton, D. L., Plow, E. F. (1995). Plasma carboxypeptidases as regulators of the plasminogen system. *J. Clin. Invest.* **96**, 2534–2538.
Reiner, A. P., Frank, M. B., Schwartz, S. M., Linenberger, M. L., Longstreth, W. T., Teramura, G., Rosendaal, F. R., Psaty, B. M., Siscovick, D. S. (2002). Coagulation factor XIII polymorphisms and the risk of myocardial infarction and ischaemic stroke in young women. *Br. J. Haematol.* **116**(2), 376–382.
Remijn, J. A., van Wijk, R., de Groot, P. G., van Solinge, W. W. (2001). Nature of the fibrinogen Aα gene TaqI polymorphism. *Thromb. Haemost.* **86**(3), 935–936.
Rendell, M., Nierenberg, J., Brannan, C., Valentine, J. L., Stephen, P. M., Dodds, S., et al. (1986). Inhibition of glycation of albumin and hemoglobin by acetylation in vitro and in vivo. *J. Lab. Clin. Med.* **108**, 286–293.
Rijken, D. C., Sakharov, D. V. (2001). Basic principles in thrombolysis: regulatory role of plasminogen. *Thromb. Res.* **103**(Suppl 1), S41–S49.
Rojas, A. M., Kordich, L., Lauricella, A. M. (2009). Homocysteine modifies fibrin clot deformability: another possible explanation of harm. *Biorheology* **46**, 379–387.
Rosenberg, J. S., McKenna, P. W., Rosenberg, R. D. (1975). Inhibition of human factor IXa by human antithrombin. *J. Biol. Chem.* **250**, 8883–8888.
Roth, G. J., Stanford, N., Majerus, P. W. (1975). Acetylation of prostaglandin synthase by aspirin. *Proc. Natl. Acad. Sci. USA* **72**, 3073–3076.
Ruigrok, Y. M., Slooter, A. J., Rinkel, G. J., Wijmenga, C., Rosendaal, F. R. (2010). Genes influencing coagulation and the risk of aneurysmal subarachnoid hemorrhage, and subsequent complications of secondary cerebral ischemia and rebleeding. *Acta Neurochir. (Wien.)* **152**, 257–262.
Ryan, A. W., Hughes, D. A., Tang, K., Kelleher, D. P., Ryan, T., McManus, R., et al. (2009). Natural selection and the molecular basis of electrophoretic variation at the coagulation F13B locus. *Eur. J. Hum. Genet.* **17**, 219–227.
Sakata, Y., Aoki, N. (1982). Significance of cross-linking of alpha 2-plasmin inhibitor to fibrin in inhibition of fibrinolysis and in hemostasis. *J. Clin. Invest.* **69**, 536–542.
Sandset, P. M. (1996). Tissue factor pathway inhibitor (TFPI)–an update. *Haemostasis* **26** (Suppl 4), 154–165.

Sane, D. C., Moser, T. L., Pippen, A. M., Parker, C. J., Achyuthan, K. E., Greenberg, C. S. (1988). Vitronectin is a substrate for transglutaminases. *Biochem. Biophys. Res. Commun.* **157**, 115–120.

Sauls, D. L., Lockhart, E., Warren, M. E., Lenkowski, A., Wilhelm, S. E., Hoffman, M. (2006). Modification of fibrinogen by homocysteine thiolactone increases resistance to fibrinolysis: a potential mechanism of the thrombotic tendency in hyperhomocysteinemia. *Biochemistry* **45**, 2480–2487.

Sauls, D. L., Wolberg, A. S., Hoffman, M. (2003). Elevated plasma homocysteine leads to alterations in fibrin clot structure and stability: implications for the mechanism of thrombosis in hyperhomocysteinemia. *J. Thromb. Haemost.* **1**, 300–306.

Schwartz, M. L., Pizzo, S. V., Hill, R. L., McKee, P. A. (1973). Human Factor XIII from plasma and platelets. Molecular weights, subunit structures, proteolytic activation, and cross-linking of fibrinogen and fibrin. *J. Biol. Chem.* **248**, 1395–1407.

Schwertz, H., Tolley, N. D., Foulks, J. M., Denis, M. M., Risenmay, B. W., Buerke, M., et al. (2006). Signal-dependent splicing of tissue factor pre-mRNA modulates the thrombogenicity of human platelets. *J. Exp. Med.* **203**, 2433–2440.

Scott, E. M., Ariëns, R. A., Grant, P. J. (2004). Genetic and environmental determinants of fibrin structure and function: relevance to clinical disease. *Arterioscler. Thromb. Vasc. Biol.* **24**, 1558–1566.

Shilton, B. H., Walton, D. J. (1991). Sites of glycation of human and horse liver alcohol dehydrogenase in vivo. *J. Biol. Chem.* **266**, 5587–5592.

Shimizu, A., Nagel, G. M., Doolittle, R. F. (1992). Photoaffinity labeling of the primary fibrin polymerization site: isolation and characterization of a labeled cyanogen bromide fragment corresponding to gamma-chain residues 337–379. *Proc. Natl. Acad. Sci. USA* **89**, 2888–2892.

Siebenlist, K. R., Meh, D. A., Mosesson, M. W. (1996). Plasma factor XIII binds specifically to fibrinogen molecules containing gamma chains. *Biochemistry* **35**, 10448–10453.

Siebenlist, K. R., Mosesson, M. W. (1996). Evidence of intramolecular cross-linked A alpha.gamma chain heterodimers in plasma fibrinogen. *Biochemistry* **35**, 5817–5821.

Siebenlist, K. R., Mosesson, M. W., Hernandez, I., Bush, L. A., Di Cera, E., Shainoff, J. R., et al. (2005). Studies on the basis for the properties of fibrin produced from fibrinogen-containing gamma' chains. *Blood* **106**, 2730–2736.

Sobel, J. H., Gawinowicz, M. A. (1996). Identification of the alpha chain lysine donor sites involved in factor XIIIa fibrin cross-linking. *J. Biol. Chem.* **271**, 19288–19297.

Soria, J., Soria, C., Collet, J. P., Mirishahi, M., Lu, H., Caen, J. P. (1991). Dysfibrinogenemia and thrombosis. *Nouv. Rev. Fr. Hématol.* **33**, 457–459.

Spraggon, G., Everse, S. J., Doolittle, R. F. (1997). Crystal structures of fragment D from human fibrinogen and its crosslinked counterpart from fibrin. *Nature* **389**, 455–462.

Standeven, K. F., Carter, A. M., Grant, P. J., Weisel, J. W., Chernysh, I., Masova, L., et al. (2007). Functional analysis of fibrin {gamma}-chain cross-linking by activated factor XIII: determination of a cross-linking pattern that maximizes clot stiffness. *Blood* **110**, 902–907.

Standeven, K. F., Grant, P. J., Carter, A. M., Scheiner, T., Weisel, J. W., Ariëns, R. A. (2003). Functional analysis of the fibrinogen Aalpha Thr312Ala polymorphism: effects on fibrin structure and function. *Circulation* **107**, 2326–2330.

Stead, N., Kaplan, A. P., Rosenberg, R. D. (1976). Inhibition of activated factor XII by antithrombin-heparin cofactor. *J. Biol. Chem.* **251**, 6481–6488.

Stern, D. M., Nawroth, P. P., Harris, K., Esmon, C. T. (1986). Cultured bovine aortic endothelial cells promote activated protein C-protein S-mediated inactivation of factor Va. *J. Biol. Chem.* **261**, 713–718.

Stump, D. C., Lijnen, H. R., Collen, D. (1986). Purification and characterization of single-chain urokinase-type plasminogen activator from human cell cultures. *J. Biol. Chem.* **261**, 1274–1278.

Suehiro, K., Mizuguchi, J., Nishiyama, K., Iwanaga, S., Farrell, D. H., Ohtaki, S. (2000). Fibrinogen binds to integrin alpha(5)beta(1) via the carboxyl-terminal RGD site of the Aalpha-chain. *J. Biochem.* **128**, 705–710.

Suzuki, K., Nishioka, J., Matsuda, M., Murayama, H., Hashimoto, S. (1984). Protein S is essential for the activated protein C-catalyzed inactivation of platelet-associated factor Va. *J. Biochem.* **96**, 455–460.

Tang, H., Fu, Y., Zhan, S., Luo, Y. (2009). Alpha(E)C, the C-terminal extension of fibrinogen, has chaperone-like activity. *Biochemistry* **48**, 3967–3976.

Thorsen, S. (1992). The mechanism of plasminogen activation and the variability of the fibrin effector during tissue-type plasminogen activator-mediated fibrinolysis. *Ann. NY Acad. Sci.* **667**, 52–63.

Tranqui, L., Andrieux, A., Hudry-Clergeon, G., Ryckewaert, J. J., Soyez, S., Chapel, A., et al. (1989). Differential structural requirements for fibrinogen binding to platelets and to endothelial cells. *J. Cell Biol.* **108**, 2519–2527.

Trumbo, T. A., Maurer, M. C. (2000). Examining thrombin hydrolysis of the factor XIII activation peptide segment leads to a proposal for explaining the cardioprotective effects observed with the factor XIII V34L mutation. *J. Biol. Chem.* **275**, 20627–20631.

Trumbo, T. A., Maurer, M. C. (2003). V34I and V34A substitutions within the factor XIII activation peptide segment (28–41) affect interactions with the thrombin active site. *Thromb. Haemost.* **89**, 647–653.

Tsurupa, G., Hantgan, R. R., Burton, R. A., Pechik, I., Tjandra, N., Medved, L. (2009). Structure, stability, and interaction of the fibrin(ogen) alphaC-domains. *Biochemistry* **48**, 12191–12201.

Uitte de Willige, S., De Visser, M. C., Houwing-Duistermaat, J. J., Rosendaal, F. R., Vos, H. L., Bertina, R. M. (2005). Genetic variation in the fibrinogen gamma gene increases the risk for deep venous thrombosis by reducing plasma fibrinogen gamma' levels. *Blood* **106**, 4176–4183.

Uitte de Willige, S., Standeven, K. F., Philippou, H., Ariëns, R. A. (2009). The pleiotropic role of the fibrinogen gamma' chain in hemostasis. *Blood* **114**, 3994–4001.

Undas, A., Brozek, J., Jankowski, M., Siudak, Z., Szczeklik, A., Jakubowski, H. (2006). Plasma homocysteine affects fibrin clot permeability and resistance to lysis in human subjects. *Arterioscler. Thromb. Vasc. Biol.* **26**, 1397–1404.

Undas, A., Brummel-Ziedins, K. E., Mann, K. G. (2007). Antithrombotic properties of aspirin and resistance to aspirin: beyond strictly antiplatelet actions. *Blood* **109**, 2285–2292.
Undas, A., Podolec, P., Zawilska, K., Pieculewicz, M., Jedlinski, I., Stepien, E., et al. (2009). Altered fibrin clot structure/function in patients with cryptogenic ischemic stroke. *Stroke* **40**, 1499–1501.
Undas, A., Slowik, A., Wolkow, P., Szczudlik, A., Tracz, W. (2010). Fibrin clot properties in acute ischemic stroke: relation to neurological deficit. *Thromb. Res.* **125**, 357–361.
Undas, A., Sydor, W. J., Brummel, K., Musial, J., Mann, K. G., Szczeklik, A. (2003). Aspirin alters the cardioprotective effects of the factor XIII Val34Leu polymorphism. *Circulation* **107**, 17–20.
Undas, A., Szuldrzynski, K., Stepien, E., Zalewski, J., Godlewski, J., Tracz, W., et al. (2008). Reduced clot permeability and susceptibility to lysis in patients with acute coronary syndrome: effects of inflammation and oxidative stress. *Atherosclerosis* **196**, 551–557.
Undas, A., Wiek, I., Stepien, E., Zmudka, K., Tracz, W. (2008). Hyperglycemia is associated with enhanced thrombin formation, platelet activation, and fibrin clot resistance to lysis in patients with acute coronary syndrome. *Diab. Care* **31**, 1590–1595.
Undas, A., Zawilska, K., Ciesla-Dul, M., Lehmann-Kopydlowska, A., Skubiszak, A., Ciepluch, K., et al. (2009). Altered fibrin clot structure/function in patients with idiopathic venous thromboembolism and in their relatives. *Blood* **114**, 4272–4278.
Upmanyu, M., Wang, H. L., Liang, H. Y., Mahajan, R. (2008). Strain-dependent twist-stretch elasticity in chiral filaments. *J. R. Soc. Interface* **5**, 303–310.
Valnickova, Z., Enghild, J. J. (1998). Human procarboxypeptidase U, or thrombin-activatable fibrinolysis inhibitor, is a substrate for transglutaminases. Evidence for transglutaminase-catalyzed cross-linking to fibrin. *J. Biol. Chem.* **273**, 27220–27224.
van Hinsbergh, V. W., Koolwijk, P. (2008). Endothelial sprouting and angiogenesis: matrix metalloproteinases in the lead. *Cardiovasc. Res.* **78**, 203–212.
Varon, D., Spectre, G. (2009). Antiplatelet agents. *Hematology. Am. Soc. Hematol. Educ. Program.* 267–272.
Vehar, G. A., Davie, E. W. (1980). Preparation and properties of bovine factor VIII (antihemophilic factor). *Biochemistry* **19**, 401–410.
Veklich, Y. I., Gorkun, O. V., Medved, L. V., Nieuwenhuizen, W., Weisel, J. W. (1993). Carboxyl-terminal portions of the alpha chains of fibrinogen and fibrin. Localization by electron microscopy and the effects of isolated alpha C fragments on polymerization. *J. Biol. Chem.* **268**, 13577–13585.
Verkleij, C. J., Nieuwdorp, M., Gerdes, V. E., Morgelin, M., Meijers, J. C., Marx, P. F. (2009). The effects of hyperglycaemia on thrombin-activatable fibrinolysis inhibitor. *Thromb. Haemost.* **102**, 460–468.
Villanueva, G. B., Allen, N. (1986). Acetylation of antithrombin III by aspirin. *Semin. Thromb. Hemost.* **12**, 213–215.
Walker, F. J. (1981). Regulation of activated protein C by protein S. The role of phospholipid in factor Va inactivation. *J. Biol. Chem.* **256**, 11128–11131.

Wartiovaara, U., Perola, M., Mikkola, H., Totterman, K., Savolainen, V., Penttila, A., et al. (1999). Association of FXIII Val34Leu with decreased risk of myocardial infarction in Finnish males. *Atherosclerosis* **142**, 295–300.

Watala, C., Gwozdzinski, K. (1993). Effect of aspirin on conformation and dynamics of membrane proteins in platelets and erythrocytes. *Biochem. Pharmacol.* **45**, 1343–1349.

Webb, G. C., Coggan, M., Ichinose, A., Board, P. G. (1989). Localization of the coagulation factor XIII B subunit gene (F13B) to chromosome bands 1q31-32.1 and restriction fragment length polymorphism at the locus. *Hum. Genet.* **81**, 157–160.

Weisel, J. W. (2004). The mechanical properties of fibrin for basic scientists and clinicians. *Biophys. Chem.* **112**, 267–276.

Weisel, J. W. (2005). Fibrinogen and fibrin. *Adv. Protein Chem.* **70**, 247–299.

Weisel, J. W. (2008). Biophysics. Enigmas of blood clot elasticity. *Science* **320**, 456–457.

Weisel, J. W., Francis, C. W., Nagaswami, C., Marder, V. J. (1993). Determination of the topology of factor XIIIa-induced fibrin gamma-chain cross-links by electron microscopy of ligated fragments. *J. Biol. Chem.* **268**, 26618–26624.

Weisel, J. W., Medved, L. (2001). The structure and function of the alpha C domains of fibrinogen. *Ann. NY Acad. Sci.* **936**, 312–327.

Weisel, J. W., Nagaswami, C. (1992). Computer modeling of fibrin polymerization kinetics correlated with electron microscope and turbidity observations: clot structure and assembly are kinetically controlled. *Biophys. J.* **63**, 111–128.

Weisel, J. W., Nagaswami, C., Makowski, L. (1987). Twisting of fibrin fibers limits their radial growth. *Proc. Natl. Acad. Sci. USA* **84**, 8991–8995.

Weiss, M. S., Metzner, H. J., Hilgenfeld, R. (1998). Two non-proline cis peptide bonds may be important for factor XIII function. *FEBS Lett.* **423**, 291–296.

Werling, R. W., Zacharski, L. R., Kisiel, W., Bajaj, S. P., Memoli, V. A., Rousseau, S. M. (1993). Distribution of tissue factor pathway inhibitor in normal and malignant human tissues. *Thromb. Haemost.* **69**, 366–369.

Wolberg, A. S., Monroe, D. M., Roberts, H. R., Hoffman, M. (2003). Elevated prothrombin results in clots with an altered fiber structure: a possible mechanism of the increased thrombotic risk. *Blood* **101**, 3008–3013.

Wolfenstein-Todel, C., Mosesson, M. W. (1981). Carboxy-terminal amino acid sequence of a human fibrinogen gamma-chain variant (gamma'). *Biochemistry* **20**, 6146–6149.

Wun, T. C., Kretzmer, K. K., Girard, T. J., Miletich, J. P., Broze, G. J., Jr. (1988). Cloning and characterization of a cDNA coding for the lipoprotein-associated coagulation inhibitor shows that it consists of three tandem Kunitz-type inhibitory domains. *J. Biol. Chem.* **263**, 6001–6004.

Xu, A. S., Ohba, Y., Vida, L., Labotka, R. J., London, R. E. (2000). Aspirin acetylation of betaLys-82 of human hemoglobin. NMR study of acetylated hemoglobin Tsurumai. *Biochem. Pharmacol.* **60**, 917–922.

Yang, Z., Kollman, J. M., Pandi, L., Doolittle, R. F. (2001). Crystal structure of native chicken fibrinogen at 2.7 A resolution. *Biochemistry* **40**, 12515–12523.

Yang, Z., Mochalkin, I., Doolittle, R. F. (2000). A model of fibrin formation based on crystal structures of fibrinogen and fibrin fragments complexed with synthetic peptides. *Proc. Natl. Acad. Sci. USA* **97**, 14156–14161.

Yang, Z., Mochalkin, I., Veerapandian, L., Riley, M., Doolittle, R. F. (2000). Crystal structure of native chicken fibrinogen at 5.5-A resolution. *Proc. Natl. Acad. Sci. USA* **97**, 3907–3912.

Yee, V. C., Pedersen, L. C., Le Trong, I., Bishop, P. D., Stenkamp, R. E., Teller, D. C. (1994). Three-dimensional structure of a transglutaminase: human blood coagulation factor XIII. *Proc. Natl. Acad. Sci. USA* **91**, 7296–7300.

Yorifuji, H., Anderson, K., Lynch, G. W., Van de Water, L., McDonagh, J. (1988). B protein of factor XIII: differentiation between free B and complexed B. *Blood* **72**, 1645–1650.

Yu, S., Sher, B., Kudryk, B., Redman, C. M. (1984). Fibrinogen precursors. Order of assembly of fibrinogen chains. *J. Biol. Chem.* **259**, 10574–10581.

Zito, F., Di Castelnuovo, A., Amore, C., D'Orazio, A., Donati, M. B., Iacoviello, L. (1997). Bcl I polymorphism in the fibrinogen beta-chain gene is associated with the risk of familial myocardial infarction by increasing plasma fibrinogen levels. A case–control study in a sample of GISSI-2 patients. *Arterioscler. Thromb. Vasc. Biol.* **17**, 3489–3494.

# STRUCTURAL, DYNAMIC, AND FUNCTIONAL ASPECTS OF HELIX ASSOCIATION IN MEMBRANES: A COMPUTATIONAL VIEW

By ANTON A. POLYANSKY,[*,†] PAVEL E. VOLYNSKY,[*] AND ROMAN G. EFREMOV[*]

[*]M. M. Shemyakin and Yu. A. Ovchinnikov Institute of Bioorganic Chemistry, Russian Academy of Sciences, Moscow, Russia
[†]Max F. Perutz Laboratories, GmbH Campus Vienna Biocenter 5, Vienna, Austria

| | | |
|---|---|---|
| I. | Introduction | 130 |
| II. | Prediction of Structure of HH Dimers | 133 |
| | A. Dimerization Motifs in Sequences of TM Helices | 134 |
| | B. Helix Packing Based on Conformational Search | 136 |
| | C. Helix Packing Based on Complementarity of Their Surface Properties | 137 |
| III. | Free Energy of TM Helix–Helix Association | 141 |
| | A. Association in the Implicit Membrane | 141 |
| | B. Application of Atomistic Models | 142 |
| | C. Application of CG Models | 144 |
| IV. | Combination of Modeling and Experimental Techniques | 146 |
| V. | Limitations and Shortcomings of the Computational Methods | 148 |
| VI. | Comparison with NMR Models: How Reliable Is The Reference? | 150 |
| VII. | Is an Accurate Structure Prediction Always Required? | 151 |
| VIII. | What Makes TM Helices Suitable Pharmacological Targets? | 152 |
| IX. | Modeling of TM Helical Dimers as a First Step Toward 3D Structure of Polytopic MPs | 152 |
| X. | From Structure and Thermodynamics to Function and Design | 153 |
| XI. | Conclusion | 154 |
| | References | 155 |

## Abbreviations

| | |
|---|---|
| MP | membrane protein |
| TM | transmembrane |
| GPCR | G protein-coupled receptor |
| RTK | receptor tyrosine kinase |
| NMR | nuclear magnetic resonance |
| HH dimer | dimer of helices |
| MD | molecular dynamics |
| CG | coarse-grained |

| | |
|---|---|
| GpA | transmembrane segment of the human Glycophorin A |
| DPC | dodecylphosphocholine |
| DPPC | 1,2-dipalmitoyl-*sn*-glycero-3-phosphocholine |
| DMPC | 1,2-dimyristoyl-*sn*-glycero-3-phosphocholine |
| MC | Monte Carlo |
| RMSD | root-mean-square deviation |
| ErbB2 | epidermal growth factor receptor 2 |
| MHP | molecular hydrophobicity potential |
| MCP | bacteriophage M13 major coat protein |
| PMF | potential of mean force |
| WHAM | weighted histogram analysis method |
| DLPC | 1,2-dilauroyl-*sn*-glycero-3-phosphocholine |
| DOPC | 1,2-dioleoyl-*sn*-glycero-3-phosphocholine |
| EphA1 | ephrin type-A receptor 1 |
| DHPC | 1,2-dihexanoyl-*sn*-glycero-3-phosphocholine |
| BRh | bacteriorhodopsin |
| LESDs | low-energy states of dimers |

## Abstract

This review surveys recent achievements of molecular computer modeling in understanding spatial structure, dynamics, and mechanisms of functioning of transmembrane $\alpha$-helical dimers in membranes. The factors driving self-association of hydrophobic helices in the membrane milieu are considered with examples of their applications to biologically relevant problems. The emphasis is made on the recent results, which help to understand important aspects of structure–function relations for these systems and their biological activity. Limitations and shortcomings of the methods, along with their perspectives in design of new membrane active agents, are discussed.

## I. Introduction

Oligomerization of membrane proteins (MPs) represents a basis of various functions in the living cell (e.g., reception of extracellular signals). Transmembrane (TM) fragments of a large variety of MPs are usually formed by bundles of $\alpha$-helices (polytopic proteins) or just by single helices (bitopic proteins). Hydrophobic $\alpha$-helices therefore represent a dominant structural motif found in membrane-spanning domains of proteins. Apart from the structural importance, self-association of TM helices often determines functional activity of MPs—many of such proteins accomplish their function only in oligomeric state (Wallin and von Heijne, 1998; Matthews et al., 2006). Among them are integral membrane receptors (e.g., proteins from the GPCR family, receptors of hormones, sensory systems; Lohse, 2010), ion pumps and channels (Green and Millar, 1995), integrins (Cox et al., 2010), amyloid precursor protein (Richter et al., 2010), class II major histocompatibility complex (King and Dixon, 2010), and many others (Ubarretxena-Belandia and Engelman, 2001). For example, recently it has been shown that dimerization of a wide class of receptor tyrosine kinases (RTK) is required for signal transduction, and defective RTKs stimulate a cell to constant growth, which could be the cause of tumor appearing (Kolibaba and Druker, 1997; Cymer and Schneider, 2010). In a series of recent papers, the dimerization events were interrelated with TM domains of RTKs (Bennasroune et al., 2004, 2005; Li and Hristova, 2006, 2010). It is important that single point mutations in TM regions of such bitopic proteins can modulate efficiency of dimerization and thus lead to dramatic changes of their biological function, including development of oncologic disorders (Roskoski, 2004; Moore et al., 2008). Understanding the interactions involved in helix–helix recognition inside the membrane is therefore an important challenge. First, this problem has a fundamental impact because the factors driving helix packing in heterogeneous membrane milieu are still poorly understood (Bowie, 2005). Second, because in cell membranes many of the aforementioned key receptors are activated only upon dimerization (oligomerization) of their TM segments, solving the problem would be invaluable in the optimization of these molecules' behavior for pharmaceutical applications. For example, one of the intriguing perspectives in drug design consists in development of new molecules—artificial hydrophobic peptides, peptide-mimetics, and low-molecular-weight compounds, which would mediate dimerization (oligomerization) of RTKs, GPCRs, ABC-transporters, etc. via concurrent binding to their TM domains. Such molecules can therefore be called "TM peptides-interceptors." For instance, knowledge of the detailed molecular mechanisms of helix association is crucial for development of a new generation of RTKs inhibitors. First such studies have recently appeared (Tarasova et al., 1999, 2005; Bennasroune et al., 2004, 2005; Gerber et al., 2004; Yin et al., 2007; Cymer and Schneider, 2010) and preliminary results look very promising.

Interaction of individual α-helices with membranes and membrane mimics has been extensively studied in experiments (see Liang, 2002; MacKenzie, 2006; Rath et al., 2007; Schneider et al., 2007, for reviews). Further, about 100 high-resolution X-ray structures of nonhomologous α-helical proteins (at 40% identity cut-off) have been obtained (November 2010; http://blanco.biomol.uci.edu/Membrane_Proteins_xtal.html). Most of them are integral MPs, whose membrane bound domains represent α-helical bundles. Their analysis revealed a number of common principles of helix packing in membranes (Eilers et al., 2002; DeGrado et al., 2003; Walters and DeGrado, 2006; Chugunov et al., 2007a,b). On the other hand, their atomic structural/functional details resist easy experimental characterization—such systems are large and highly dynamic. Therefore, they are heavily susceptible to analysis via X-ray or NMR spectroscopy. The former technique experiences problems with crystallization of MPs, while the second one meets difficulties dealing with big and low-mobile protein–lipid complexes. In addition, the crystallographic methods, which recently allowed obtaining high-resolution structure of multispan GPCRs (Cherezov et al., 2007), cannot be directly translated to bitopic flexible receptors like RTKs, integrins, and so forth. Therefore, characterization of the extracellular, cytoplasmic, and membrane parts of such MPs are still made separately.

That is why, detail characterization of structural-dynamic properties of noncovalently bonded protein oligomers in the membrane represents a big challenge. Up to date, just a few of dimeric structures of TM domains of bitopic proteins have been experimentally obtained. Therefore, elaboration of independent techniques is especially timely. Molecular modeling gives promising alternative, which considerably extends and complements traditional structural biology tools. Several strategies to resolve this problem by various theoretical and physicochemical methods and their combination are currently available, thus providing structural-dynamic information about atomic-scale details of TM helix–helix and helix–membrane interactions. Although the primary objective is structure elucidation, experimental high-resolution structure obtained in a particular membrane-mimicking environment usually corresponds to only one of possible homo- or heterodimeric states of TM domains, which are apparently realized *in vivo* in the course of bitopic protein activity. Molecular modeling, in its turn, allows predicting all possible alternative dimerization interfaces of the bitopic protein TM domains, existence of which *in vivo* should be verified in experiment. Application of *in silico* methods to TM fragments of bitopic proteins gives not only insight into spatial organization of a dimer of helices (HH dimer) but also provides opportunities to explore its dynamics, to estimate free energy gain upon dimerization, and to quantify contribution of the membrane environment. Apparently, thorough understanding of all the aspects of TM helix–helix interactions can only be achieved with multidisciplinary approach based on a comprehensive set of modeling, biochemical, and biophysical tools.

Although experimental (Liang, 2002; Bocharov et al., 2010a; Cymer and Schneider, 2010; Li and Hristova, 2010) and theoretical (see Forrest and Sansom, 2000; Efremov et al., 2004; Psachoulia et al., 2010, for reviews) aspects of TM helix–helix interactions have been addressed in several recent reviews, some computational aspects of the aforementioned issues demand a special consideration. This review surveys recent achievements of molecular computer modeling in understanding spatial structure, dynamics, and mechanisms of functioning of TM helical dimers in membranes. The factors driving self-association of hydrophobic helices in the membrane milieu are considered with examples of their applications to biologically relevant problems. The emphasis is made on the recent results, which help to understand important aspects of structure–function relations for these systems and their biological activity. Limitations and shortcomings of the methods, along with their perspectives in design of new membrane active agents, are discussed.

## II. Prediction of Structure of HH Dimers

A small number of dimeric structures of TM helices of bitopic proteins obtained from NMR studies underlies an importance of theoretical prediction of spatial configuration of such oligomers. The most "realistic" computational study would be direct observation of helices association via simulations of their molecular dynamics (MD) in the explicit membrane environment. However, high complexity of such multicomponent system (peptides, lipids, solvent, ions) and big scale (usually, about $10^5$ atoms) provide obvious limitations to employ full-atomic MD simulations on the biologically relevant timescales (ms). Simplified representation of the membrane environment like water-octanol slab (Stockner et al., 2004) or detergents micelles (Braun et al., 2004) gives yet opportunities to apply full-atomic model for observing association of the helices upon MD on the nanosecond timescale. Thus, Stockner et al. (2004) observed formation of a left-handed dimer stabilized by interhelical H-bonds for model MS1 peptide upon its 45-ns MD simulations in the water-octanol-water slab, starting from monomeric states of the helices. Application of different coarse-grained (CG) representations of molecules significantly increases efficiency of the modeling. For instance, Bond and Sansom (2006) observed insertion and dimerization of TM fragments of human Glycophorin A (GpA) in dodecylphosphocholine (DPC) micelles and dipalmitoyl-phosphocholine (DPPC) bilayer using CG MD simulation. It was shown that CG representation allows prediction of the native-like structure of the dimer, which was found to be in a dynamic equilibrium with monomeric GpA inside the lipid bilayer. Subsequent combination in MD of CG and full-atomic representations gives a multiscale view of the dimerization process: dimer self-assembling in a membrane is modeled at low CG resolution, and resulting configurations of the system are further subjected to full-atomic relaxation. Similar technique was recently applied to observe assembling of homo- (GpA, syndecan-2 receptor) and heterodimers (alphaIIb/beta3 integrin) in right- or left-handed conformations (Psachoulia et al., 2010). An alternative and more common approach is supposed to use predicted structure of a dimer as initial configuration of the system that simplifies exploration of a huge number of possible conformational degrees of freedom with various mutual orientations of helices. The latter issue can be addressed with employment of different prediction methods (see below).

### A. *Dimerization Motifs in Sequences of TM Helices*

The simplest prediction technique is based on statistical analysis of frequency of amino acid residues and their combinations, which form interhelical contacts. This analysis reveals a number of *dimerization motifs* one can find in the primary structure of a protein (Lemmon et al., 1992; MacKenzie et al., 1997; Gurezka et al., 1999). A prediction of dimerization interface in this case is based on the assumption that if two helices possess such sequence patterns, they would form a dimer with close-packed contact in those particular regions. For instance, "glycophorin-like" motif (at the first time observed in the dimer of human GpA) represents GVxxxGV pattern, were x can be any amino acid residue (Fleming et al., 1997; MacKenzie et al., 1997). This motif and its modifications like S'L'xxxS'L' (where S' and L' correspond to residues with short and long side chains, respectively) and (G, A,S)xxxGxxxG—the so-called glycine zipper motif (Kim et al., 2005)—are often found on the interface of TM helices oligomers. In most cases, these packing interactions give pairs of α-helices with a right-handed tilt. Other TM helices interactions involve a leucine zipper type of side chain packing providing formation of dimers with a left-handed helical twist. Within leucine zipper, the interacting residues form repeated heptad (*abcdefg*) motifs, where residues at *a*- and *d*-positions constitute the hydrophobic core of the interfaces (Gurezka et al., 1999). Along with hydrophobic aliphatic residues, all three aromatic amino acids promote association of TM helices (Sal-Man et al., 2007). The largest effects on stabilization of TM helix dimers were found for the WXXW motif (where X is any residue)—this sequence pattern is over-represented within a broad set of TM protein domains, especially in bacteria. Molecular modeling (Sal-Man et al., 2007) reveals that the strength of association of the WXXW pair is moderate compared with another common motif formed by polar residues Gln and Ser—QXXS (Sal-Man et al., 2005). The later one was shown to enable homo- and heteroassociation of TM helices *in vivo*. Most of polar residues (Asn, Asp, Glu, His, Ser, Thr) were found to form various dimerization motifs (see MacKenzie, 2006, for a review).

Sequence-based prediction of TM dimers with implicit usage of geometrical features of helices can be considered as a particular and simple case of coiled-coil structure identification, where numerous approaches and programs have been developed (McDonnell et al., 2006; Apgar et al., 2008). These methods engage knowledge-based scoring functions and incorporate experimental data about relevant heptad pairs and pairwise interaction statistics and/or energies. However, potentially significant atomic details necessary for modeling orientation specificity might be obscured by using of strong assumptions about the independence of pairwise interactions (Apgar et al., 2008). Other limitation of the implicit structural methods is prediction of dimeric structure for helices simultaneously containing several dimerization motifs (e.g., TM fragments of RTK) (Gerber et al., 2004; Bocharov et al., 2008a,b, 2010b; Moore et al., 2008; Mineev et al., 2010). In this ambiguous case, only information about dimerization motifs found in sequences of TM helices is not sufficient for the prediction of possible interhelical contact region. More details about 3D structure of TM helix–helix dimers can be obtained via energy-based methods exploring conformational space of the system.

## B. Helix Packing Based on Conformational Search

Various techniques of conformational search of dimeric configuration for TM helices were developed. Kim et al. used only the van der Waals energy term, and, starting from a set of random orientations of a pair of helices, optimized the parameters by a Monte Carlo (MC) search (Kim et al., 2003). Based on a rather simple description of the lipid environment, implicit membrane models were proved very useful for fast and efficient prediction of TM dimeric structures (Ducarme et al., 2000; Im et al., 2003; Vereshaga et al., 2005, 2007; Efremov et al., 2006; Zhang and Lazaridis, 2009). Such methods are based on scanning of conformational space via a series of consecutive MC runs in implicit membrane, which is often described using atomic solvation parameters derived for gas-cyclohexane and gas-water transfer (e.g., Nolde et al., 1997; Efremov et al., 2001). The set of the lowest-energy structures is clustered and the representative structures are considered as possible dimeric models. We used this approach in studies of GpA dimerization (Vereshaga et al., 2005; Efremov et al., 2006). Simulations gave a series of dimeric structures, one of which was in a good agreement with the experimental one. This model provides a large set of possibilities to study helix–helix dimerization, including the role of membrane thickness and its polarity, presence of TM potential, etc. Predicted models can be optimized via following MD simulations in full-atom membrane. Such a combination of approaches was applied in studies of dimerization of TM-fragment of proapoptotic protein Bnip3 (Vereshaga et al., 2007). All the resulting MC-models displayed good agreement with the mutagenesis data and one of them closely resembled the spatial structure obtained by NMR spectroscopy in lipid bicelles (Bocharov et al., 2007). MD relaxation of the models in explicit dimyristoyl-phosphatidylcholine (DMPC) membrane was used to optimize protein–lipid contacts and avoid models with significantly discrepant geometry as worst stable.

Simulations in multicanonical ensembles (Mitsutake et al., 2001; Sugita, 2009) can significantly increase a prediction efficiency of the force-field based approaches with simplified membrane description. The replica-exchange MC method was also successfully used to reproduce the NMR structure of GpA using long-term simulation starting from a single conformation of the pair of helices (Kokubo and Okamoto, 2004). Gervais et al. (2009) developed two-step MC procedure and applied it to assess dimerization of GpA. The protocol included the following steps: (i) estimation of the energy density of states of the system by the Wang–Landau (Wang and Landau, 2001) algorithm; (ii) production run, during which various energetically and structural observables are sampled to get thermodynamic parameters of the system (e.g., energy contribution of different residues of the motif in dimerization). Replica-exchange MD simulations in the implicit membrane were also successfully used for prediction of GpA dimer (Bu et al., 2007). Miyashita et al. (2009) employed similar techniques to predict dimeric structures of the wild type and the mutant forms of the amyloid precursor protein fragment ($A\beta_{23-55}$). In particular, it was shown that the mutation induces drastic changes in the dimer structures and thus can be responsible for amyloid formation.

Although the methods based on conformational search strategies were proved to be powerful in elaboration of correct 3D models of TM helical dimers, often, important information can be obtained using relatively more simple and fast methods. Many of them are based on complementarity of surface properties of the interacting helices and/or on statistical data on helix packing derived from known spatial structures of proteins.

### C. Helix Packing Based on Complementarity of Their Surface Properties

It is known that periodicity of TM helices structure underlies particular organization of their surfaces. For instance, glycine residues of the glycophorin motif form on the surface a groove surrounded by ridges created by neighboring valines. These spatial properties determine the well-known "knob-to-holes" or "ridge-to-groove" packing rules for TM helices (Langosch and Heringa, 1998), which was first postulated more than 50 years ago by Crick (1953). Knob-to-holes packing in coiled-coil structures was well characterized for many proteins with this particular fold using SOCKET software (Walshaw and Woolfson, 2001), where simple distances cut-offs algorithms are employed. This packing rule works for both globular and membrane helical oligomeres, with one exception that helices in the membrane seem to pack generally more tightly (Eilers et al., 2000). Geometric complementarity of TM helices along with the knowledge of important inter-residue interactions stabilizing dimeric structure (H-bonds (Lee and Im, 2008), cation–pi interactions (Johnson et al., 2007), etc.) provide a basis to rationalize prediction of TM dimers based on application of scoring functions (e.g., Fleishman and Ben-Tal, 2002). Methods developed up-to-date differ in the type of scoring function they use to select the "best" dimeric structure among a large variety of conformations. The principal element of all these procedures is docking of flexible helices. Quality of the predicted structures is estimated using a number of force-field based (van der Waals, torsional, H-bonding, electrostatic, and solvation potentials) and/or knowledge-based (e.g., statistics of pairwise residue–residue contacts on the interface) terms. In most cases, accuracy of the approaches is tested via comparisons of the predicted models of GpA with that obtained by NMR (MacKenzie et al., 1997), while other TM dimers extracted from polytopic MPs can also be considered. GpA structure is usually predicted within 2 Å $C_\alpha$ RMSD from the NMR model. Park et al. (2004) described a scoring function designed to maximize the number of contacts between interface residues and to penalize the burial of large side chains, which they used to perform an exhaustive search in the configurational space with 6 degrees of freedom determining mutual spatial dispositions of two helices. Yin et al. (2007) applied simple algorithm of dimer reconstruction using backbone helical scaffolds extracted from the set of TM protein structures to design "interceptor" peptides modulating biological function of integrins. In a simple case of helices containing single dimerization motif and forming homodimer (e.g., GpA, Bnip3), its structure was shown to be efficiently reconstructed by global conformational search followed by MD relaxation (Samna Soumana et al., 2007) and ROSETTA algorithm (Barth et al., 2007). However, in more complicated cases recognition of a correct conformation among a huge number of decoys still represents an important issue for the latter approach. Sodt and Head-Gordon (2010) designed novel statistical contact potential based on solved structures of TM helical bundles, which would help to improve standard ROSETTA energy function. Global search procedure based on knowledge-based scoring functions was also used by Fleishman et al. (2002) to propose putative mechanism of conformational-switch activation for RTK ErbB2. In addition to the force-field approaches, intermediate-resolution electron microscopy map were shown to make available reconstruction of 3D structure of TM helical oligomers (Kovacs et al., 2007). Recently, Nugent and Jones (2010) described a novel method of prediction of lipid exposure and residue contacts on the helix–helix interface using support vector machine algorithms. This approach is based on statistics of lipid exposure of residues and residue–residue contacts obtained based on a series of CG MD simulations of MPs and on the analysis of their experimental structures. Alternatively, lipid faced and helix–helix interface regions of TM helix can be detected using molecular hydrophobicity potential (MHP) approach (Efremov et al., 1992a,b, 2007). Based on MHP analysis of solvent accessible surface of a helix, recently we developed a novel approach (so-called PREDDIMER) for *de novo* reconstructions of 3D models for dimers of TM helices from their primary structures. More specifically, the prediction of dimerization interfaces is based on evaluating complementarity of landscapes (surface roughness) and MHP properties for given helices. PREDDIMER algorithm includes several consecutive stages (Fig. 1): (i) building "ideal" helices from TM-sequence; (ii) calculation of their surface and mapping hydrophobic/hydrophilic properties

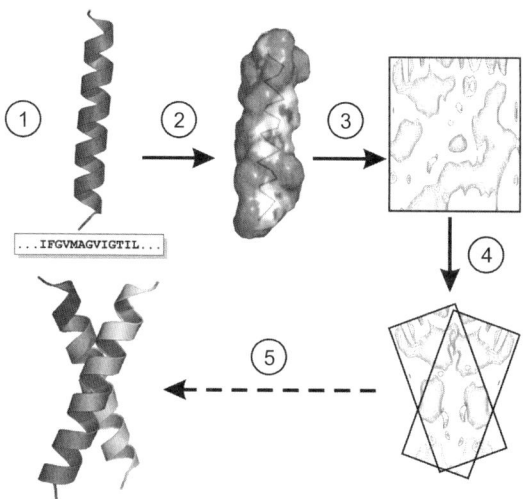

FIG. 1. PREDDIMER—a computational algorithm for prediction of 3D models of transmembrane (TM) helical dimers. The method is based on alignment of the peptides' surfaces in order to achieve the best complementarity of hydrophobic (molecular hydrophobicity potential, MHP) and landscape properties (expressed in terms of 2D maps) of the monomers. Key steps are numbered: 1, building ideal helices from their TM-sequences; 2, calculation and mapping of helices' surfaces (MHP + landscape); 3, projection of surface properties onto the cylinder and building of 2D maps; 4, mutual superimposition of the maps by varying rotation and tilt angles of the helices, and their pairwise comparison using scoring functions; 5, reconstruction of 3D structure of the dimer. Solvent accessible surface of a helix as well as 2D maps are colored according to MHP values. Orange, white, and blue regions correspond to hydrophobic, neutral, and hydrophilic surface patches, respectively. (See color plate 3).

using conception of the MHP; (iii) identification of the dimerization interface by evaluating complementary of landscapes and MHP properties on the helix surface; (iv) reconstruction of 3D model of the dimer and estimation of its quality according to scoring functions and free energy of dimerization obtained via original simulation approach. The approach was tested on the database of dimeric structures of bitopic proteins experimentally obtained via NMR spectroscopy in the membrane-like environment (MacKenzie et al., 1997; Bocharov et al., 2007, 2008a,b, 2010b; Mineev et al., 2010). The computational results demonstrated high efficiency and accuracy of the method. In all cases, backbone RMSDs between the predicted and the experimental structures do not exceed 2 Å (Table I). The effect of single point mutations for several proteins (e.g., GpA) was also investigated. Thus, "silent" mutations slightly affected the dimer structure, whereas the critical ones resulted in formation of structures with weak packing properties.

Standard prediction approaches are usually lacking realistic evaluation criteria for the structure accuracy. For instance, application of scoring functions based on the potential energy usually does not allow distinction between two close dimeric structures (e.g., left- or right-handed dimers differing in sign of the tilt angle between the helix axes). This problem

TABLE I
3D Structure Predictions for a Series of Transmembrane Helix–Helix Dimers Using MHP and Landscape Complementarity Approach

| Structures | Bnip3[a] | Epha1[b] | Epha2[c] | Erbb1/Erbb2[d] | Gpa[e] |
|---|---|---|---|---|---|
| Predicted (tilt angle, °)[f] | −30.3 | −40.9 | 21.1 | −38 | −43.8 |
| Predicted (RMSD, Å) | 1.9/1.7[g] | 2.1/1.3 | 1.6/1.5 | 2.0/2.1 | 1.6/1.2 |
| NMR (tilt angle, °) | −35.9 | −36.8 | 26.8 | −35.8 | -37.3 |

Comparison of the predicted models with those obtained by NMR.
[a]Bocharov et al. (2007).
[b]Bocharov et al. (2008b).
[c]Bocharov et al. (2010b).
[d]Mineev et al. (2010).
[e]MacKenzie et al. (1997).
[f]Only data for predicted models of a dimer displaying lowest RMSD from NMR are given.
[g]Backbone RMSD of predicted structure (before/after MD relaxation in the water-cyclohexane-water slab) from the NMR model.

can be avoided by introducing into the prediction algorithm some additional criteria, especially ones based on the free energy values for helix–helix association.

### III. Free Energy of TM Helix–Helix Association

Estimation of the free energy of dimerization is indispensable for correct prediction of the spatial structure of helical oligomers and for subsequent computer-aided molecular design of new biologically active compounds acting on MPs in cells (e.g., MacKenzie, 2006). First of all, it allows selection of the most favorable conformation from the set of structural models. This significantly increases efficiency of dimeric structure prediction. On other hand, comparison of such values for several different dimers can help to range them according to their dimerization strength. Finally, comparison of dimers with similar sequences (for instance, some wild-type TM protein fragments and their single point mutants) can give important information about the role of different residues in dimerization. Particularly, consideration of residues in different protonation states is capable of getting important insight in pH effects of dimerization. All these data provide important fundamental basis concerning dimerization of TM helices, which can be further employed in *de novo* construction of peptides and/or chemical compounds specifically dimerized with a particular target.

#### A. Association in the Implicit Membrane

Zhang and Lazaridis (2006) assessed the free energy of dimerization ($\Delta\Delta G_{ass}$) as a potential energy of interactions between monomers (enthalpic contribution) with applied entropic corrections. Such a model was used to calculate the value of $\Delta\Delta G_{ass}$ for GpA. Translational and rotational entropy of association was calculated from the probability distribution of the corresponding degrees of freedom obtained from MD simulations. The side chain conformational entropy of association was estimated from the probability distribution calculated by rotation of all side chain as rigid bodies. The calculated standard association free energy of GpA in $N$-DPC micelles was in good agreement with the experimental value. The proposed approach was also applied to study the role of different residues in dimerization of GpA-Tox hybrid, TM helices of bacteriophage M13 major coat protein

(MCP) and MCP-GpA, MCP-Tox hybrids (Zhang and Lazaridis, 2009). It was shown that besides residues lying on the dimerization interface, other residues may also significantly affect the dimerization energy. Such noninterfacial residues in the putative TM domain can affect the association affinity. Further, the extent of their burial in the membrane could be different in monomers and dimers (particularly for charged residues). Flanking polar residues can affect association affinity via direct interactions. Although calculations of $\Delta\Delta G_{ass.}$ in the implicit membrane have obvious limitations related to ignoring of the effects of "membrane response" accompanying dimerization process (solvent entropy). These effects seem to be very important and in several cases, where they even represent the main driving force of dimerization (see below).

## B. Application of Atomistic Models

In explicit membrane environment, $\Delta\Delta G_{ass.}$ can be calculated through the potential of mean force (PMF). PMF provides a complete description of thermodynamic properties along the selected degrees of freedom, that is, reaction coordinate. PMF calculations still remain a challenging task because it is hard to explore rare events that are unlikely to occur on the time scales of regular MD simulations. At the same time, PMF can be efficiently calculated using umbrella-sampling techniques in which a restraint potential function is employed to sample the conformations along the reaction coordinate. In this case, reconstruction of PFM can be performed by application of the weighted histogram analysis method (WHAM) (Kumar et al., 1992; Souaille and Roux, 2001) or the mean force integration method (Henin et al., 2005; Lee and Im, 2008).

One of the pioneering works in application of this technique to study TM helix association was dedicated to GpA and several its mutants (Henin et al., 2005). Conformational space was explored using MD simulations in the membrane-like environment (explicit water-dodecane-water slab). The choice of dodecane as a membrane mimetic was dictated by slow relaxation times of collective motions in phospholipid bilayers. The total PMF decomposition into helix–helix and helix–solvent contributions has greatly improved understanding of the recognition and association mechanism of the GpA TM domain, although the calculated association free energy was overestimated compared with experimental data.

Recently, we applied a simplified membrane model to design highthroughput scheme of the free energy estimation in order to increase accuracy of selection of the best conformation among the predicted dimer models. The water-cyclohexane-water slab was adopted as the membrane mimic. Cyclohexane is smaller than dodecan that significantly saves CPU costs of the simulations. We also used only TM protein fragments without long extramembrane regions. PMF statistics were collected over several independent constrained MD runs with gradually increasing interhelical distances. Apart from efficient distinction of correct predicted models, the proposed approach also allows estimation of the dimerization efficiency. The results obtained for two proteins with known experimental dimeric structure and different association constants are shown in Fig. 2. For illustration, we selected Bnip3 (Fig. 2A, B) and ErbB2 (Fig. 2C, D), which form strong and medium-strength dimer, respectively. The structure of Bnip3

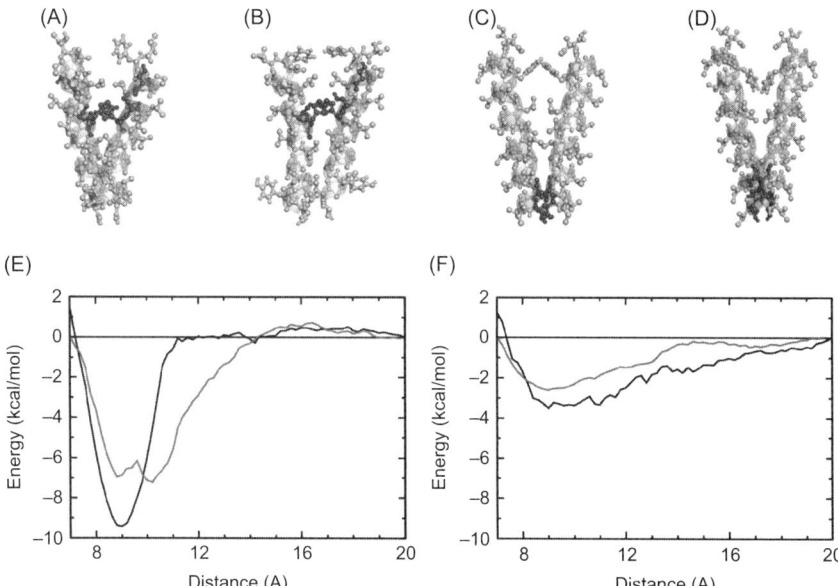

FIG. 2. Computational assessment of the free energy of dimerization of transmembrane fragments of Bnip3 and ErbB2. NMR-derived (A, C) and -predicted (B, D) structures of Bnip3 and ErbB2, respectively. Peptides are given in CPK and ribbon representations. Polar residues are shown in dark-gray. Potential of mean force (PMF) plots calculated for NMR (*black line*) and predicted (*red line*) models of the Bnip3 (E) and ErbB2 (F) dimers. Values on the *X*-axis of the plots correspond to the distance between centers of mass of the helices. (For interpretation of the references to color in this figure legend, the reader is referred to the Web version of this chapter.)

TM helix–helix complex was predicted via tandem MC/MD technique as described elsewhere (Vereshaga et al., 2007). For prediction of the ErbB2 dimer, we used the PREDDIMER approach (see above). In both cases, experimental (NMR) and predicted structures displayed similar packing geometry and interfacial residues (Fig. 2). Comparison of the helix association energies computed from PMFs (Fig. 2E, F) showed that in accordance with the experimental data, association of Bnip3 is more efficient. In the latter case, association free energies are $-9.2$ and $-7.3$ kcal/mol for NMR and predicted models, respectively. At the same time, the energy gain upon ErbB2 dimerization was estimated as twice smaller: $-4.0$ and $-3.1$ kcal/mol, respectively.

PFM-based technique was also used to study of the role of hydrogen bonding and helix–lipid interactions in TM helix association in a lipid bilayer (Lee and Im, 2008). Two model TM peptides were constructed based on GCN4 leucine zipper and studied using MD in DMPC bilayer: in the first peptide, asparagine residue was placed on the dimerization interface (pVNVV), while the second peptide did not have any polar residues (pVVVV). Simulations showed that hydrogen bonding significantly contributed to dimerization: Asn-to-Val mutant reveals a dramatic free energy change upon the residue substitution. This change was enough for mutant does not form a stable dimer below a certain peptide concentration.

### C.  Application of CG Models

CG representation of peptides and membranes allows free energy estimation without preliminary knowledge of the spatial structure of a dimer. At the initial stage of the simulation, the system exists as two monomers. Then, long-term MD is performed and association energy is derived based on the found populations of monomeric and dimeric states. Such an approach was applied for determination of association energy and its difference for GpA and several its mutants (Psachoulia et al., 2008). Conformational space was explored using several long-term CG MD runs in dipalmitoyl-phosphatidylcholine (DPPC) bilayer. It was found that GpA prefers to form right-handed dimers with the interface close to one determined by NMR spectroscopy. In spite of the simplified system representation, these structures displayed quite a good agreement with the experimental ones (mean RMSD $= 3.6 \pm 1.3$ Å, similar packing geometry and dimerization interface). Analogous computational protocol was applied to analyze dimerization efficiency for a set of critical and noncritical mutants of GpA. It was shown that the peptides with noncritical mutations form dimers in the similar way with the wild-type peptide, while critical mutations significantly distort the helix packing, although do not block dimerization during simualtions. The population rates of dimeric and monomeric states for different peptides were used to estimate the values of $\Delta\Delta G_{ass.}$. The results were in quite a good agreement with available experimental data. The main advantage of this approach is the absence of any initial guess about the dimeric structure. On the other hand, as mentioned by the authors, the free energy values were modulated by a systematic error caused by the coarse-grain approximation. Therefore, this approach seems to be useful mainly for $\Delta\Delta G_{ass.}$-based ranking of different HH-models proposed for the same pairs of helices.

Effect of lipid composition of a bilayer on the dimerization energy of GpA was also addressed using CG Monte Carlo conformational search (Janosi et al., 2010). Several lipid bilayers with different physicochemical properties were used (DLPC, DPPC, DOPC). The conformational search was performed using $(MW)_2$-XDOS—algorithm. It was shown that association is assisted by the lipid-induced interactions with the most favorable contributions rising from the bilayer revealing the highest structural order. However, at intermediate distances, repulsive lipid-induced interactions appear. They become especially noticeable in the thickest membranes. In addition, TM α-helices tilt to a different extent, depending both on membrane properties and amino acid sequence. This leads to formation of multiple favorable conformational states of the dimers along the same interface, which equally contribute to dimerization.

Recently, CG MD simulations were applied to determination of association energy of the wild-type GpA and its disruptive mutants (Sengupta and Marrink, 2010) using PMF calculations. The proposed approach was tested in studies of association of considered peptides from monomeric state in the DPPC bilayer. After several microseconds of diffusion, all the peptides formed stable dimers, which existed till the end of MD simulations. The most frequent structure of GpA was in a good agreement with the experimental one. Then, the PMF profile was calculated using umbrella sampling. The set of distance probability histograms obtained from MD trajectories were analyzed using WHAM approach. It was shown that the shape of PMF curves significantly depended on the simulation time (sampling rate). When a time of simulations was sufficiently long (∼8 μs) one additional minimum at the energy profile appeared at the distance ~1.0 nm. Analysis of MD data showed existence of a large variety of structures at this distance. In addition, some of them demonstrated packing significantly different from that in the NMR-like structure. It was proposed that this minimum has an entropic nature. Comparison of PMF for GpA and its mutant forms gave some principal reasons for the association loss such as less-favorable protein–protein assembling and disruption of lipid packing around mutant dimers. It was shown that the role of the nonspecific "lipid–phobic" contribution appears to be as important as the specific "helix–helix" contribution. Although the differences between the wild type and the mutants were subtle, the described simulations correctly predicted a dimerization state not only for the wild-type GpA but also for these "disruptive" mutants.

To summarize this section, we should outline that various approaches of assessment of dimerization free energy of TM helices were developed last decade. Each of them has some reasonable advantages and difficulties. For instance, CG simulations allow efficient sampling of the conformational space that is good enough to explore fine details of PMF (Sengupta and Marrink, 2010). On the other hand, simple molecular description used in this case and predetermined secondary structure of peptides leads to problems with realistic modeling of protein dynamics and evaluation of the conformational entropy. Implicit membrane (Zhang and Lazaridis, 2006, 2009) model also provide good sampling but lack effect of the membrane environment (solvent entropy). Full-atomic simulations maintain accurate consideration of all energy terms but are limited in conformational sampling. Anyway, contribution of all of these factors in the dimerization energy can be estimated using combination of different methods, which in each particular case provide results in rather good agreement with experimental data.

## IV. Combination of Modeling and Experimental Techniques

Along with predictive power of modeling approaches, they have several important applications in combination with experimental techniques. A particular example represents relaxation of NMR-derived structures. Often, a protein molecule contains regions lacking direct NMR information about interatomic contacts. In this case, conformation of these regions is usually determined based on indirect data (other contacts, the most probable side chain rotamers, etc.). At the same time, influence of the environment is a very important factor, especially in the case of the MPs, consideration of which can significantly help in deciphering of conformation in the "problematic" regions (Fig. 3A). Another NMR problem, which can be solved using computer simulations, corresponds to atoms' assignment. For instance, in

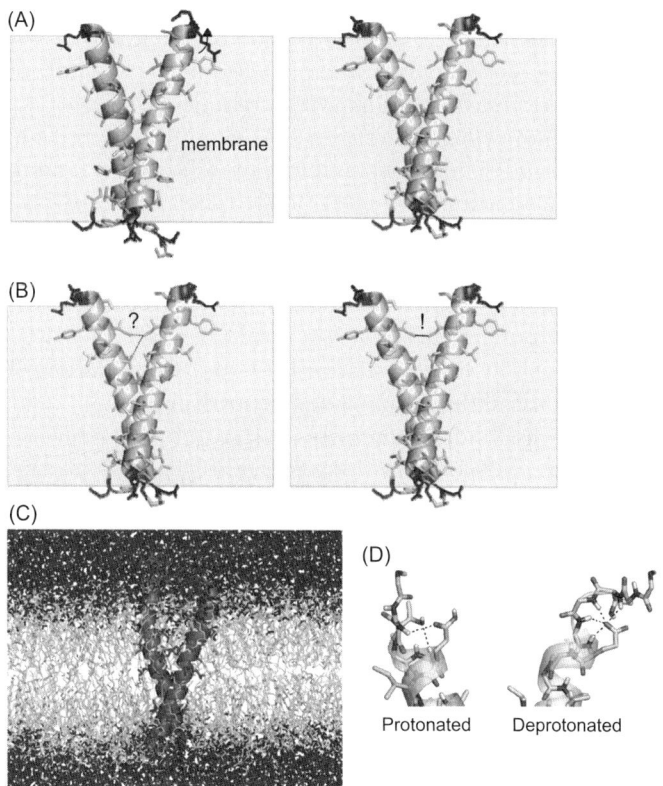

FIG. 3. Application of MD simulations in hydrated lipid bilayers to refinement of NMR-derived structures of transmembrane helix–helix dimers. (A) Refinement of the poor-contact regions between the monomers. (B) Assignment of new contacts (unavailable from NMR data). (C) Analysis of the mutual protein-membrane influence and relaxation of the system. (D) Simulations of the effects caused by changes of protonation/deprotonation state of particular residues. The nonpolar layer of membrane is shown in gray.

case of strong sequence degeneracy several atoms exhibit overlapping signals in spectra, thus making difficult assignments for individual residues. This problem can also be solved via MD simulation in the membrane-like environment with applied constrains based on the well-resolved signals (Fig. 3B). Moreover, molecular modeling can provide information about mutual influence of a dimer and a membrane (Fig. 3C). For example, MD relaxation of the NMR-derived models of the Bnip3 dimer in explicit DMPC bilayer clearly showed permeability of a membrane for water molecules in the vicinity of the N-termini of the peptides (Bocharov et al., 2007). As a result, possible molecular mechanism of Bnip3 activity was suggested. Another important issue, which can be addressed in simulations, is exploration of the phenomena, which resist easy characterization in experiments, for example, effect of a ionization state of particular protein residues (Fig. 3D). Thus, we demonstrated that protonation of His in TM segments of Bnip3 is accompanied with increasing of the membrane permeability for water molecules and formation of a network of hydrogen bonds involving several water molecules and imidazole rings (Bocharov et al., 2007). Also, we found that protonation of Glu residue in the interfacial membrane region of RTK EphA1 seriously affects conformation of the dimer (Bocharov et al., 2008b). MD simulations with this residue in protonated/deprotonated forms revealed prominent rearrangements of the hydrogen bonding patterns and even changes in the monomers secondary structure.

Modeling techniques can also help in rationalization of experiments. For instance, we used analysis of structural models of the EphA1 dimer to design mutants, which were later efficiently applied to study protein dimerization in ToxR system (Volynsky et al., 2010). Such a computational approach allowed selection of only structurally important residues that facilitated fast realization of the ToxR assays. Taking together, theoretical and experimental approaches were successful in providing an unambiguous and self-consistent model of the 3D structure of a homodimer formed by TM helices of the EphA1 receptor.

## V. Limitations and Shortcomings of the Computational Methods

One of severe limitations of the computational and experimental biophysical methods used to study association of TM helices in model systems is missing from the calculations the extramembrane parts of the corresponding proteins. Obviously, this is done in order to make the system as simple as possible and to prevent side effects related to improper folding of TM segments in artificial membranes, overlapping and broadening of signals in NMR spectra, tremendous enlargement of the conformational space to be sampled in computation, and so forth. In most of these studies, it is therefore assumed that the water-soluble protein parts do not affect self-assembling of the helices. Although there is a body of experimental and computational evidence supporting such an assumption, the question requires special consideration for each particular system.

Another "thin point" in the computational methods is limited (and therefore incomplete) sampling of the conformational space available for a pair of TM helices in the presence of heterogeneous membrane environment. Obviously, large-scale exploration of the configurational space can be done using implicit or CG membrane models. On the other hand, for the computational efficiency these techniques have to pay by ignoring (lacking) some details critical for proper packing of the peptides. Among them are local distortions of α-helical structures (both, on the termini and in the middle part), water rearrangement induced by microscopic interactions with lipids, protein, ions, etc. The latter effects can be assessed via much more demanding all-atom simulations. Meanwhile, the "rule of thumb" is a combined application of methods based on different levels of approximation. Thus, initial rough but extensive conformational search is performed with full-atom or CG peptide models in implicit membrane. Then, the resulting low-energy states are taken as starting point for simulations in CG lipid environment. Finally, the most energetically favorable configurations are converted into full-atom representation and subjected to computations with the most detailed description of TM helices in hydrated bilayer membranes. It is important that all these models and protocols are used in a self-consistent manner—each stage must provide maximum information about putative low-energy states of the system.

Further, the following important aspects mediating interactions of TM helices are not taken into account in computer simulations. First, it is known that properties of lipid bilayer in the vicinity of many bitopic MPs (including RTKs, BNIP3, and others) can drastically differ from those in a "bulk" bilayer. For example, proper functioning of RTKs requires a raft-like environment, where composition of lipids and their packing density are quite specific (e.g., Cymer and Schneider, 2010). In addition, real membranes have asymmetric distribution of physicochemical properties along the normal to the membrane plane, whereas theoretical models of membranes are usually symmetrical. Because association of TM helices is sensitive to the lipid neighborhood, neglecting of such effects may distort the results. Second, ionization states of residues in TM segments are normally not exactly known—they are taken based on "realistic assumptions" formulated by the authors (e.g., local pH in the cellular compartment containing a given protein, etc.).

Finally, atomistic simulations of proteins in hydrated lipid membranes face the convergence problems and lack of precision in the description of hydrophobic interactions between the protein and its surrounding. As pointed out by Edholm and Jähnig (1988), the environmental effects result from small differences between strong interactions of the protein with water and lipid molecules. The correct treatment of such interactions therefore demands precise determination of parameters in van der Waals and electrostatic energy terms.

## VI. Comparison with NMR Models: How Reliable Is The Reference?

Most often, correctness of the computational approaches is validated by comparison of the predicted spatial models of TM helix–helix complexes with those obtained in experiments (see above). Usually, the latter ones are solved by NMR in various membrane mimics, like detergent micelles and lipid bicelles. Of course, this is the first reasonable check to be done. If all the calculated structures fall far away from the experimental ones, it seems that the computational approach does not work and should be improved. On the other hand, modeling often results in a number of possible states (families of conformers), some of which are similar to NMR-derived models, while others can be quite different. In our opinion, the presence of the former states proves overall correctness of the method, while occurrence of the "decoys" can be explained either by computational artifacts or by existence of potential alternative conformations. Thus, taking into account that NMR structures are obtained in a very artificial environment (bicelles and, especially, micelles are not equivalent to multicomponent lipid bilayers forming cell membranes), as well as the influence of medium effects on association of TM helices (see Section III), it is reasonable to suggest that at least some of such "misfolded" dimers represent the conformations that can be realized in particular membrane environments different from those employed in NMR experiments. This makes computational experiments very promising for exploration of a hypothetical conformational lability of TM helical oligomers, which often is functionally important. Attribution of different conformations to their suitable environments is not a trivial task because it demands large-scale computations mainly with explicit membrane models. In addition, still there are no experimental structures of non-covalent TM helix dimers obtained in membranes with drastically different properties. Recently, it was shown that the NMR structures of GpA dimers in DPC micelles (MacKenzie et al., 1997) and DMPC/DHPC bicelles are very close to each other (E. V. Bocharov, unpublished data). This can be explained by stability of the dimer, by quite similar physicochemical properties of the two membrane mimics, or by both of these factors.

## VII. Is an Accurate Structure Prediction Always Required?

It should be emphasized that in many practical tasks, accurate atomic-scale prediction of the spatial structure of TM helix–helix complexes is not absolutely required. Moreover, taking into consideration approximations inherent in the methods—simplified definition of the system and reliability of the computational models (see above)—this would overestimate possibilities of the technique. Instead, the problem often is formulated as follows: whether given TM helices are capable of forming stable dimers in a particular membrane environment? If so, what are the most probable conformations of TM helix–helix complexes (geometry of packing) and to what membrane media they are well-adapted? Knowledge of the "most probable conformations" generally means that the integral parameters of helix packing—tilt angle/distance between the helix axes, residues on the contact interface, strongest intermolecular interactions (e.g., H-bonds, salt bridges, etc.) are delineated. This data are invaluable for understanding the driving forces of TM helix association, planning of new experiments destined to better understanding of biological action of MPs and subsequent rational design of molecules targeting TM segments of these proteins with high affinity and selectivity.

## VIII. What Makes TM Helices Suitable Pharmacological Targets?

Actually, why does oligomerization of TM helices provoke so strong interest in the recent years? Apart from the obvious fundamental impact (deciphering of basic physical principles driving folding, membrane insertion, and functioning of a wide class of MPs), this topic is extremely "hot" from the pharmacological point of view (see Section I). This is because protein TM domains involved in functionally important dimerization/oligomerization processes represent a new class of biological targets with quite specific features. For instance, in contrast to extracellular and cytoplasmic parts of RTKs (which have been very extensively used in drug development since many years; Zwick et al., 2002; Johnson, 2009), their membrane-spanning segments potentially can provide unique specificity to drugs, especially to medium- and large-molecular weight compounds like peptides, peptidomimetics, peptide-containing chimeric constructs, etc. This is because the target TM segments of RTKs "make high demands" to the partner molecule to associate with. First, the latter one should also have a membrane-exposed region for recognition of the target. Second, such intermolecular interactions require extended contact surface and usually involve at least 10 amino acid residues, that is much higher than binding of a typical low-molecular-weight ligand (e.g., drug) in extramembrane domains of RTKs. Hopefully, this can help to solve the well-known problem of low specificity of small ligands designed for binding in water-soluble parts of kinases. Also, the peptide-like molecules targeting TM helices will probably have another advantage over traditional drugs: hopefully, they can assist to overcome the problem of drug resistance, which became very serious in the last years (see, e.g., Martínez, 2008).

## IX. Modeling of TM Helical Dimers as a First Step Toward 3D Structure of Polytopic MPs

As shown above, even for the simplest systems—two individual TM helices in membrane-mimicking medium—*in silico* prediction of structural and energetic properties of dimers still represents a big challenge. At the same time, solving of such tasks is only the first stage toward much more biologically relevant problem—namely, understanding of TM helix packing, stabilization, and functioning of homo- and hetero-oligomers with more than two protein segments. The multi-helix TM associates are formed both—within the same protein and/or by different proteins. Here, a question arises: whether pairs of neighboring TM helices in large helix bundles are packed like in the simplest two-helix models or their interaction is strongly affected by other TM segments? Preliminary MC calculations in implicit membrane performed for individual TM helix–helix hairpins of bacteroirhodopsin (BRh) (initially, the helices were taken in random orientation with respect to each other and to the membrane) have shown (manuscript in preparation) that only in few cases the found low-energy states of dimers (LESDs) correspond to those realized in the X-ray structure of BRh (Luecke et al., 1999). It should be also noted that the latter ones were present among the sets of LESDs obtained for each helix–helix pair, although not with the minimal energics. This clearly demonstrates that individual TM helices have only a limited number of possibilities to associate with other partners—upon assembling of three and more helices their packing modes are just "selected" from these predefined sets of conformations, or LESDs.

## X. From Structure and Thermodynamics to Function and Design

The fact that most of the TM helix–helix complexes reveal a certain degree of conformational heterogeneity may have important biological impact, especially in the case of weak association of the protein segments, like in RTKs. Knowledge of the predicted possible dimeric structures and corresponding free energies of binding helps in elucidation of molecular details of their functioning. Thus, occurrence of multiple putative dimerization interfaces in TM helices of RTKs may shed light on conformational rearrangements accompanying ligand binding and signal transduction across the membrane (Moore et al., 2008; Cymer and Schneider, 2010). Further, some of these states can be assigned to active or inactive states of the dimers. Hypothetical mechanisms of the related activation processes have been proposed for ErbB2 (Fleishman et al., 2002; Bocharov et al., 2008a) and EphA1 (Bocharov et al., 2010b) receptors based on the 3D structures of their TM helix–helix complexes.

Although in this review the main accent is done on interactions of TM α-helical peptides, the problem seems to be much wider. Understanding on molecular level the forces driving dimerization of TM helices represents just a first stage because this information can be further used to design not only "peptide-interceptors" (or "computed helical anti-MPs," CHAMPs; Yin et al., 2007), but creates a basis for development of other nonpeptidic molecules—various peptidomimetics, chimeric constructs (with and without peptide parts), low-molecular-weight compounds, and so forth. Here, one of the main difficulties is caused by the necessity of addressed delivery of such molecules to the corresponding TM target. Usually, these agents possess low solubility in water, thus demanding addition of hydrophilic moieties, which can affect intermolecular interactions with other protein segments, lipids, and water. For example, in a case of peptides these polar groups are often introduced near the termini. In TM helix–helix complexes, geometry of packing depends on the sequence context, especially in the interfacial region. Therefore, the designed peptide or mimetic should be properly oriented (head-to-head, head-to-tail) in the membrane and have a suitable length of the hydrophobic region to satisfy the well-known hydrophobic match effect (Marsh, 2008).

## XI. Conclusion

The problem of TM helix–helix interactions in proteins already comes of age. Recently, the wide arsenal of powerful physicochemical techniques traditionally used to characterize these systems was enlarged by modern NMR methods of structure determination in membrane-mimicking medium. Altogether, such approaches provided a wealth of valuable information about the fundamental principles of TM helix association. In particular, several high-resolution 3D structures of dimers formed by membrane-spanning helices were solved. Apart from the fundamental importance—understanding on a molecular level of the main factors driving MP folding and oligomerization—the problem of helices assembling in cellular membranes looks very promising for biomedicine and biotechnology. Perspectives in the latter areas are primarily related to molecular design of a new class of pharmaceutical compounds that can modulate in a predetermined way specific helix–helix association in cell membrane, providing a novel form of therapy of many human diseases related with abnormal activity of the MPs involved. This is especially actual for oncologic and neurodegenerative disorders.

Among the modern methods employed in these studies, computational molecular modeling is one of the key players. Although direct *in silico* observation of association of TM helices in full-atomic membrane still represents a challenge, a number of state-of-the-art modeling techniques have been proved to be very useful. As it was demonstrated in this review, they are used to support interpretation of the experimental data, predict the spatial structure of TM helix–helix complexes, and assess dynamic/energetic aspects of their association, evaluate the consequences of point mutations and/or chemical modifications as well as the role of medium effects in such processes, and design new molecules targeting TM domains of a wide class of receptors, channels, etc. On the other hand, it became clear now that detailed description of a large variety of intra- and intermolecular interactions in TM domains of proteins and elucidation of the roles of the TM domains in normal and abnormal functioning of these proteins along with their proper localization in cell membranes can only be achieved using a combination of experimental and theoretical approaches, which complement each other. It seems that no one of the individual methods is capable of giving complete and adequate picture of the complex events occurring in such "simple" systems like a pair of TM helices in a membrane environment.

To conclude, we would like to note that modeling of oligomerization of TM helices is a necessary step toward our knowledge of detailed aspects of folding of MPs and subsequent *ab initio* prediction of their complete (with extramembrane parts) spatial organization, which is still far from being a routine.

### Acknowledgments

This work was supported by the Russian Foundation for Basic Research, by the RAS Programs (MCB and "Basic fundamental research for nanotechnologies and nanomaterials"), and by the Russian Ministry of Science and Education (MK-8439.2010.4). Access to computational facilities of the Joint Supercomputer Center (Moscow) and Computer Center of M.V. Lomonosov Moscow State University is gratefully acknowledged. We thank Prof. A.S. Arseniev and Dr. E.V. Bocharov (both from Shemyakin-Ovchinnikov Institute of Bioorganic Chemistry RAS, Moscow) for providing us with the NMR-models of helical dimers prior to publication.

### References

Apgar, J. R., Gutwin, K. N., Keating, A. E. (2008). Predicting helix orientation for coiled-coil dimers. *Proteins* **72**, 1048–1065.

Barth, P., Schonbrun, J., Baker, D. (2007). Toward high-resolution prediction and design of transmembrane helical protein structures. *Proc. Natl. Acad. Sci. USA* **104**, 15682–15687.

Bennasroune, A., Fickova, M., Gardin, A., Dirrig-Grosch, S., Aunis, D., Cremel, G., et al. (2004). Transmembrane peptides as inhibitors of ErbB receptor signaling. *Mol. Biol. Cell.* **15**, 3464–3474.

Bennasroune, A., Gardin, A., Auzan, C., Clauser, E., Dirrig-Grosch, S., Meira, M., et al. (2005). Inhibition by transmembrane peptides of chimeric insulin receptors. *Cell. Mol. Life Sci.* **62**, 2124–2131.

Bocharov, E. V., Mayzel, M. L., Volynsky, P. E., Goncharuk, M. V., Ermolyuk, Y. S., Schulga, A. A., et al. (2008). Spatial structure and pH-dependent conformational diversity of dimeric transmembrane domain of the receptor tyrosine kinase EphA1. *J. Biol. Chem.* **283**, 29385–29395.

Bocharov, E. V., Mayzel, M. L., Volynsky, P. E., Mineev, K. S., Tkach, E. N., Ermolyuk, Y. S., et al. (2010). Left-handed dimer of EphA2 transmembrane domain: helix packing diversity among receptor tyrosine kinases. *Biophys. J.* **98**, 881–889.

Bocharov, E. V., Mineev, K. S., Volynsky, P. E., Ermolyuk, Ya.S., Tkach, E. N., Sobol, A. G., et al. (2008). Spatial structure of dimeric transmembrane domain of growth factor receptor ErbB2. *J. Biol. Chem.* **283**, 6950–6956.

Bocharov, E. V., Pustovalova, Y. E., Pavlov, K. V., Volynsky, P. E., Goncharuk, M. V., Ermolyuk, Y. S., et al. (2007). Unique dimeric structure of BNip3 transmembrane domain suggests membrane permeabilization as a cell death trigger. *J. Biol. Chem.* **282**, 16256–16266.

Bocharov, E. V., Volynsky, P. E., Pavlov, K. V., Efremov, R. G., Arseniev, A. S. (2010). Structure elucidation of dimeric transmembrane domains of bitopic proteins. *Cell Adh. Migr.* **4**, 284–298.

Bond, P. J., Sansom, M. S. (2006). Insertion and assembly of membrane proteins via simulation. *J. Am. Chem. Soc.* **128**, 2697–2704.

Bowie, J. U. (2005). Solving the membrane protein folding problem. *Nature* **438**, 581–589.

Braun, R., Engelman, D. M., Schulten, K. (2004). Molecular dynamics simulations of micelle formation around dimeric glycophorin a transmembrane helices. *Biophys. J.* **87**, 754–763.

Bu, L., Im, W., Brooks, C. L., 3rd. (2007). Membrane assembly of simple helix homo-oligomers studied via molecular dynamics simulations. *Biophys. J.* **92**, 854–863.

Cherezov, V., Rosenbaum, D. M., Hanson, M. A., Rasmussen, S. G., Thian, F. S., Kobilka, T. S., et al. (2007). High-resolution crystal structure of an engineered human beta2-adrenergic G protein-coupled receptor. *Science* **318**, 1258–1265.

Chugunov, A. O., Novoseletsky, V. N., Nolde, D. E., Arseniev, A. S., Efremov, R. G. (2007a). Method to assess packing quality of transmembrane alpha-helices in proteins. 2. Validation by "correct vs misleading" test. *J. Chem. Inf. Model.* **47**, 1163–1170.

Chugunov, A. O., Novoseletsky, V. N., Nolde, D. E., Arseniev, A. S., Efremov, R. G. (2007b). Method to assess packing quality of transmembrane alpha-helices in proteins. 1. Parametrization using structural data. *J. Chem. Inf. Model.* **47**, 1150–1162.

Cox, D., Brennan, M., Moran, N. (2010). Integrins as therapeutic targets: lessons and opportunities. *Nat. Rev. Drug Discov.* **9**, 804–820.

Crick, F. H. C. (1953). The packing of alpha-helices—simple coiled-coils. *Acta Crystallogr.* **6**, 689–697.

Cymer, F., Schneider, D. (2010). Transmembrane helix-helix interactions involved in ErbB receptor signaling. *Cell Adh. Migr.* **4**, 299–312.

DeGrado, W. F., Gratkowski, H., Lear, J. D. (2003). How do helix-helix interactions help determine the folds of membrane proteins? Perspectives from the study of homo-oligomeric helical bundles. *Protein Sci.* **12**, 647–665.

Ducarme, P., Thomas, A., Brasseur, R. (2000). The optimisation of the helix/helix interaction of a transmembrane dimer is improved by the IMPALA restraint field. *Biochim. Biophys. Acta* **1509**, 148–154.

Edholm, O., Jähnig, F. (1988). The structure of a membrane-spanning polypeptide studied by molecular dynamics. *Biophys. Chem.* **30**, 279–292.

Efremov, R. G., Chugunov, A. O., Pyrkov, T. V., Priestle, J. P., Arseniev, A. S., Jacoby, E. (2007). Molecular lipophilicity in protein modeling and drug design. *Curr. Med. Chem.* **14**, 393–415.

Efremov, R. G., Gulyaev, D. I., Modyanov, N. N. (1992). Application of 3D molecular hydrophobicity potential to the analysis of spatial organization of membrane domains in proteins. II. Optimization of hydrophobic contacts in transmembrane hairpin structures of Na,K-ATPase. *J. Protein Chem.* **11**, 699–708.

Efremov, R. G., Gulyaev, D. I., Vergoten, G., Modyanov, N. N. (1992). Application of 3D molecular hydrophobicity potential to the analysis of spatial organization of membrane domains in proteins. I. Hydrophobic properties of transmembrane segments of Na, K-ATPase. *J. Protein Chem.* **11**, 665–675.

Efremov, R. G., Nolde, D. E., Konshina, A. G., Syrtcev, N. P., Arseniev, A. S. (2004). Peptides and proteins in membranes: what can we learn via computer simulations? *Curr. Med. Chem.* **11**, 2421–2442.

Efremov, R. G., Vereshaga, Y. A., Volynsky, P. E., Nolde, D. E., Arseniev, A. S. (2006). Association of transmembrane helices: what determines assembling of a dimer? *J. Comput. Aided Mol. Des.* **20**, 27–45.

Efremov, R. G., Volynsky, P. E., Nolde, D. E., Arseniev, A. S. (2001). Implicit two-phase solvation model as a tool to assess conformation and energetics of proteins in membrane-mimic media. *Theor. Chem. Acc.* **106**, 48–54.

Eilers, M., Patel, A. B., Liu, W., Smith, S. O. (2002). Comparison of helix interactions in membrane and soluble alpha-bundle proteins. *Biophys. J.* **82**, 2720–2736.

Eilers, M., Shekar, S. C., Shieh, T., Smith, S. O., Fleming, P. J. (2000). Internal packing of helical membrane proteins. *Proc. Natl. Acad. Sci. USA* **97**, 5796–5801.

Fleishman, S. J., Ben-Tal, N. (2002). A novel scoring function for predicting the conformations of tightly packed pairs of transmembrane alpha-helices. *J. Mol. Biol.* **321**, 363–378.

Fleishman, S. J., Schlessinger, J., Ben-Tal, N. (2002). A putative molecular-activation switch in the transmembrane domain of erbB2. *Proc. Natl. Acad. Sci. USA* **99**, 15937–15940.

Fleming, K. G., Ackerman, A. L., Engelman, D. M. (1997). The effect of point mutations on the free energy of transmembrane alpha-helix dimerization. *J. Mol. Biol.* **272**, 266–275.

Forrest, L. R., Sansom, M. S. (2000). Membrane simulations: bigger and better? *Curr. Opin. Struct. Biol.* **10**, 174–181.

Gerber, D., Sal-Man, N., Shai, Y. (2004). Two motifs within a transmembrane domain, one for homodimerization and the other for heterodimerization. *J. Biol. Chem.* **279**, 21177–21182.

Gervais, C., Wust, T., Landau, D. P., Xu, Y. (2009). Application of the Wang-Landau algorithm to the dimerization of glycophorin A. *J. Chem. Phys.* **130**, 215106-1–215106-7.

Green, W. N., Millar, N. S. (1995). Ion-channel assembly. *Trends Neurosci.* **18**, 280–287.

Gurezka, R., Laage, R., Brosig, B., Langosch, D. (1999). A heptad motif of leucine residues found in membrane proteins can drive self-assembly of artificial transmembrane segments. *J. Biol. Chem.* **274**, 9265–9270.

Henin, J., Pohorille, A., Chipot, C. (2005). Insights into the recognition and association of transmembrane alpha-helices. The free energy of alpha-helix dimerization in glycophorin A. *J. Am. Chem. Soc.* **127**, 8478–8484.

Im, W., Feig, M., Brooks, C. L., III. (2003). An implicit membrane generalized Born theory for the study of structure, stability and interactions of membrane proteins. *Biophys. J.* **85**, 2900–2918.

Janosi, L., Prakash, A., Doxastakis, M. (2010). Lipid-modulated sequence-specific association of glycophorin A in membranes. *Biophys. J.* **99**, 284–292.

Johnson, L. N. (2009). Protein kinase inhibitors: contributions from structure to clinical compounds. *Q. Rev. Biophys.* **42**, 1–40.

Johnson, R. M., Hecht, K., Deber, C. M. (2007). Aromatic and cation-pi interactions enhance helix-helix association in a membrane environment. *Biochemistry* **46**, 9208–9214.

Kim, S., Chamberlain, A. K., Bowie, J. U. (2003). A simple method for modeling transmembrane helix oligomers. *J. Mol. Biol.* **329**, 831–840.

Kim, S., Jeon, T.-J., Yang, D., Schmidt, J. J., Bowie, J. U. (2005). Transmembrane glycine zippers: physiological and pathological roles in membrane proteins. *Proc. Natl. Acad. Sci. USA* **102**, 14278–14283.

King, G., Dixon, A. M. (2010). Evidence for role of transmembrane helix-helix interactions in the assembly of the Class II major histocompatibility complex. *Mol. Biosyst.* **6**, 1650–1661.

Kokubo, H., Okamoto, Y. (2004). Classification and prediction of low-energy membrane protein helix configuration by replica-exchange Monte Carlo method. *J. Phys. Soc. Jpn.* **73**, 2571–2585.

Kolibaba, K. S., Druker, B. J. (1997). Protein tyrosine kinases and cancer. *Biochim. Biophys. Acta* **1333**, F217–F248.

Kovacs, J. A., Yeager, M., Abagyan, R. (2007). Computational prediction of atomic structures of helical membrane proteins aided by EM maps. *Biophys. J.* **93**, 1950–1959.

Kumar, S., Bouzida, D., Swendsen, R. H., Kollman, P. A., Rosenberg, J. M. (1992). The Weighted Histogram Analysis Method for Free-Energy Calculations on Biomolecules.1. The Method. *J. Comput. Chem.* **13**, 1011–1021.

Langosch, D., Heringa, J. (1998). Interaction of transmembrane helices by a knobs-into-holes packing characteristic of soluble coiled coils. *Proteins* **31**, 150–159.

Lee, J., Im, W. (2008). Role of hydrogen bonding and helix-lipid interactions in transmembrane helix association. *J. Am. Chem. Soc.* **130**, 6456–6462.

Lemmon, M. A., Flanagan, J. M., Treutlein, H. R., Zhang, J., Engelman, D. M. (1992). Sequence specificity in the dimerization of transmembrane alpha-helices. *Biochemistry* **31**, 12719–12725.

Li, E., Hristova, K. (2006). Role of receptor tyrosine kinase transmembrane domains in cell signaling and human pathologies. *Biochemistry* **45**, 6241–6251.

Li, E., Hristova, K. (2010). Receptor tyrosine kinase transmembrane domains: function, dimer structure and dimerization energetics. *Cell Adh. Migr.* **4**, 249–254.

Liang, J. (2002). Experimental and computational studies of determinants of membrane-protein folding. *Curr. Opin. Chem. Biol.* **6**, 878–884.

Lohse, M. J. (2010). Dimerization in GPCR mobility and signaling. *Curr. Opin. Pharmacol.* **10**, 53–58.

Luecke, H., Schobert, B., Richter, H.-T., Cartailler, J.-P.h., Lanyi, J. K. (1999). Structure of bacteriorhodopsin at 1.55Å resolution. *J. Mol. Biol.* **291**, 899–911.

MacKenzie, K. R. (2006). Folding and stability of alpha-helical integral membrane proteins. *Chem. Rev.* **106**, 1931–1977.

MacKenzie, K. R., Prestegard, J. H., Engelman, D. M. (1997). A transmembrane helix dimer: structure and implications. *Science* **276**, 131–133.

Marsh, D. (2008). Protein modulation of lipids, and vice-versa, in membranes. *Biochim. Biophys. Acta* **1778**, 1545–1575.

Martínez, J. L. (2008). Antibiotics and antibiotic resistance genes in natural environments. *Science* **321**, 365–367.

Matthews, E. E., Zoonens, M., Engelman, D. M. (2006). Dynamic helix interactions in transmembrane signaling. *Cell* **127**, 447–450.

McDonnell, A. V., Jiang, T., Keating, A. E., Berger, B. (2006). Paircoil2: improved prediction of coiled coils from sequence. *Bioinformatics* **22**, 356–358.

Mineev, K. S., Bocharov, E. V., Pustovalova, Y. E., Bocharova, O. V., Chupin, V. V., Arseniev, A. S. (2010). Spatial structure of the transmembrane domain heterodimer of ErbB1 and ErbB2 receptor tyrosine kinases. *J. Mol. Biol.* **400**, 231–243.

Mitsutake, A., Sugita, Y., Okamoto, Y. (2001). Generalized-ensemble algorithms for molecular simulations of biopolymers. *Biopolymers* **60**, 96–123.

Miyashita, N., Straub, J. E., Thirumalai, D., Sugita, Y. (2009). Transmembrane structures of amyloid precursor protein dimer predicted by replica-exchange molecular dynamics simulations. *J. Am. Chem. Soc.* **131**, 3438–3439.

Moore, D. T., Berger, B. W., DeGrado, W. F. (2008). Protein-protein interactions in the membrane: sequence, structural, and biological motifs. *Structure* **16**, 991–1001.

Nolde, D. E., Arseniev, A. S., Vergoten, G., Efremov, R. G. (1997). Atomic solvation parameters for protein in a membrane environment. Application to transmembrane α-helices. *J. Biomol. Struct. Dyn.* **15**, 1–18.

Nugent, T., Jones, D. T. (2010). Predicting transmembrane helix packing arrangements using residue contacts and a force-directed algorithm. *PLoS Comput. Biol.* **6**, e1000714.

Park, Y., Elsner, M., Staritzbichler, R., Helms, V. (2004). Novel scoring function for modeling structures of oligomers of transmembrane alpha-helices. *Proteins* **57**, 577–585.

Psachoulia, E., Fowler, P. W., Bond, P. J., Sansom, M. S. (2008). Helix-helix interactions in membrane proteins: coarse-grained simulations of glycophorin a helix dimerization. *Biochemistry* **47**, 10503–10512.

Psachoulia, E., Marshall, D. P., Sansom, M. S. (2010). Molecular dynamics simulations of the dimerization of transmembrane alpha-helices. *Acc. Chem. Res.* **43**, 388–396.

Rath, A., Johnson, R. M., Deber, Ch. M. (2007). Peptides as transmembrane segments: determinants for helix-helix interactions in membrane proteins. *Peptide Science* **88**, 217–231.

Richter, L., Munter, L. M., Ness, J., Hildebrand, P. W., Dasari, M., Unterreitmeier, S., et al. (2010). Amyloid beta 42 peptide (Abeta42)-lowering compounds directly bind to Abeta and interfere with amyloid precursor protein (APP) transmembrane dimerization. *Proc. Natl. Acad. Sci. USA* **107**, 14597–14602.

Roskoski, R., Jr. (2004). The ErbB/HER receptor protein-tyrosine kinases and cancer. *Biochem. Biophys. Res. Commun.* **319**, 1–11.

Sal-Man, N., Gerber, D., Bloch, I., Shai, Y. (2007). Specificity in transmembrane helix-helix interactions mediated by aromatic residues. *J. Biol. Chem.* **282**, 19753–19761.

Sal-Man, N., Gerber, D., Shai, Y. (2005). The identification of a minimal dimerization motif QXXS that enables homo- and hetero-association of transmembrane helices in vivo. *J. Biol. Chem.* **280**, 27449–27457.

Samna Soumana, O., Garnier, N., Genest, M. (2007). Molecular dynamics simulation approach for the prediction of transmembrane helix-helix heterodimers assembly. *Eur. Biophys. J.* **36**, 1071–1082.

Schneider, D., Finger, C., Prodohl, A., Volkmer, T. (2007). From interactions of single transmembrane helices to folding of alpha-helical membrane proteins: analyzing transmembrane helix-helix interactions in bacteria. *Curr. Protein Pept. Sci.* **8**, 45–61.

Sengupta, D., Marrink, S. J. (2010). Lipid-mediated interactions tune the association of glycophorin A helix and its disruptive mutants in membranes. *Phys. Chem. Chem. Phys.* **12**, 12987–12996.

Sodt, A. J., Head-Gordon, T. (2010). Driving forces for transmembrane alpha-helix oligomerization. *Biophys. J.* **99**, 227–237.

Souaille, M., Roux, B. (2001). Extension to the weighted histogram analysis method: combining umbrella sampling with free energy calculations. *Comput. Phys. Commun.* **135**, 40–57.

Stockner, T., Ash, W. L., MacCallum, J. L., Tieleman, D. P. (2004). Direct simulation of transmembrane helix association: role of asparagines. *Biophys. J.* **87**, 1650–1656.

Sugita, Y. (2009). Free-energy landscapes of proteins in solution by generalized-ensemble simulations. *Front. Biosci.* **14**, 1292–1303.

Tarasova, N. I., Rice, W. G., Michejda, C. J. (1999). Inhibition of G-protein-coupled receptor function by disruption of transmembrane domain interactions. *J. Biol. Chem.* **274**, 34911–34915.

Tarasova, N. I., Seth, R., Tarasov, S. G., Kosakowska-Cholody, T., Hrycyna, C. A., Gottesman, M. M., et al. (2005). Transmembrane inhibitors of P-glycoprotein, an ABC transporter. *J. Med. Chem.* **48**, 3768–3775.

Ubarretxena-Belandia, I., Engelman, D. M. (2001). Helical membrane proteins: diversity of functions in the context of simple architecture. *Curr. Opin. Struct. Biol.* **11**, 370–376.

Vereshaga, Y. A., Volynsky, P. E., Nolde, D. E., Arseniev, A. S., Efremov, R. G. (2005). Helix interactions in membranes: lessons from unrestrained Monte Carlo simulations. *J. Chem. Theory Comput.* **1**, 1252–1264.

Vereshaga, Y. A., Volynsky, P. E., Pustovalova, J. E., Nolde, D. E., Arseniev, A. S., Efremov, R. G. (2007). Specificity of helix packing in transmembrane dimer of the cell death factor BNIP3: a molecular modeling study. *Proteins* **69**, 309–325.

Volynsky, P. E., Mineeva, E. A., Goncharuk, M. V., Ermolyuk, Y. S., Arseniev, A. S., Efremov, R. G. (2010). Computer simulations and modeling-assisted ToxR screening in deciphering 3D structures of transmembrane alpha-helical dimers: ephrin receptor A1. *Phys. Biol.* **7**, 16014.

Wallin, E., von Heijne, G. (1998). Genome-wide analysis of integral membrane proteins from eubacterial, archaean, and eukaryotic organisms. *Protein Sci.* **7**, 1029–1038.

Walshaw, J., Woolfson, D. N. (2001). Socket: a program for identifying and analysing coiled-coil motifs within protein structures. *J. Mol. Biol.* **307**, 1427–1450.

Walters, R. F. S., DeGrado, W. F. (2006). Helix-packing motifs in membrane proteins. *Proc. Natl. Acad. Sci. USA* **103**, 13658–13663.

Wang, F., Landau, D. (2001). Efficient, multiple-range random walk algorithm to calculate the density of states. *Phys. Rev. Lett.* **86**, 2050–2053.

Yin, H., Slusky, J. S., Berger, B. W., Walters, R. S., Vilaire, G., Litvinov, R. I., et al. (2007). Computational design of peptides that target transmembrane helices. *Science* **315**, 1817–1822.

Zhang, J., Lazaridis, T. (2006). Calculating the free energy of association of transmembrane helices. *Biophys. J.* **91**, 1710–1723.

Zhang, J., Lazaridis, T. (2009). Transmembrane helix association affinity can be modulated by flanking and noninterfacial residues. *Biophys. J.* **96**, 4418–4427.

Zwick, E., Bange, J., Ullrich, A. (2002). Receptor tyrosine kinases as targets for anticancer drugs. *Trends Mol. Med.* **8**, 17–23.

# PROTEINS MOVE! PROTEIN DYNAMICS AND LONG-RANGE ALLOSTERY IN CELL SIGNALING

### By ZIMEI BU* AND DAVID J. E. CALLAWAY*,†

*Department of Chemistry, The City College of New York, New York, New York, USA
†New York University School of Medicine, New York, New York, USA

| | | |
|---|---|---|
| I. | Introduction .................................................................. | 164 |
| II. | NHERF1 Modulates the Macromolecular Assembly, Cell Surface Retention, and Subcellular Localization of Membrane Proteins ................................ | 167 |
| | A. CFTR ................................................................ | 167 |
| | B. NaPiT2a ............................................................ | 168 |
| | C. Podocalyxin Complexes ............................................ | 169 |
| | D. Tyrosine Kinase Receptor Complexes .............................. | 170 |
| III. | Signal Transduction by Allosteric Scaffolding Protein Interactions............ | 171 |
| | A. Allosteric Modulation of NHERF1 to Assemble Membrane Complexes..... | 173 |
| | B. Negative Cooperativity and Feedback Loop ................................ | 174 |
| IV. | Structural Basis of Autoinhibition and Long-Range Allostery in NHERF1 ...... | 178 |
| | A. NHERF1 Has a Highly Elongated Shape, Allowing Long-Range Interdomain Allosteric Communications.................................... | 179 |
| | B. Allosteric Control in NHERF1 Originates from Domain–Domain Interactions................................................ | 181 |
| | C. Redefinition of the Structural Boundary of PDZ Domains ................. | 183 |
| | D. Autoinhibition of PDZ2 by the C-Terminal Domain of NHERF1............ | 185 |
| | E. Disease-Associated NHERF1 Mutations Affect the Structure, Stability, and Binding Capability of PDZ Domains ........................... | 185 |
| V. | Dynamic Propagation of Allosteric Signals by Nanoscale Protein Motion ....... | 186 |
| | A. The Physical Concepts Behind Protein Dynamics.......................... | 187 |
| | B. Nanoscale Protein Dynamics: The Emergence of a New Frontier........... | 189 |
| | C. Dynamic Propagation of Allosteric Signals by Nanoscale Protein Motion... | 202 |
| | D. A Simple Four-Point Model Describes Domain Motion..................... | 207 |
| | E. The Importance of NSE and a Plan for the Future ......................... | 211 |
| VI. | Summary and Perspective ...................................................... | 213 |
| | References.................................................................... | 214 |

### Abstract

An emerging point of view in protein chemistry is that proteins are not the static objects that are displayed in textbooks but are instead dynamic actors. Protein dynamics plays a fundamental role in many diseases, and spans a large hierarchy of timescales, from picoseconds to milliseconds or even longer. Nanoscale protein domain motion on length scales comparable to protein dimensions is key to understanding how signals are relayed through multiple protein–protein interactions. A canonical example is how the scaffolding proteins NHERF1 and ezrin work in coordination to assemble crucial membrane complexes. As membrane–cytoskeleton scaffolding proteins, these provide excellent prototypes for understanding how regulatory signals are relayed through protein–protein interactions between the membrane and the cytoskeleton. Here, we review recent progress in understanding the structure and dynamics of the interaction. We describe recent novel applications of neutron spin echo spectroscopy to reveal the dynamic propagation of allosteric signals by nanoscale protein motion, and present a guide to the future study of dynamics and its application to the cure of disease.

## I. Introduction

*Proteins move!* This simple statement encapsulates a wide variety of phenomena that are central for the understanding of life and for the cure of disease. Proteins are composed of multiple domains, whose flexibility and mobility lead to a great deal of versatility in their function. Protein dynamics (particularly at the domain level) is a controlling influence in the allosteric formation of protein complexes, in catalysis, in cell signaling and regulation, in metabolic transport, and in cellular locomotion. Yet, despite the importance of protein domain dynamics, the study of this field is in its infancy, largely because of the paucity of biophysical methods that are able to probe this regime. We here provide a review of some of the field, and propose a roadmap for future exploration. In order to understand what is known about protein dynamics and the significance of the challenges ahead, it is essential to review some of the relevant concepts. In order to motivate this discussion, we will first present a review of the biological relevance of long-range allosteric effects that couple the dynamics of protein domains.

In cells, membrane channels and receptors are often assembled into macromolecular complexes in specialized subcellular domains for the dynamic control of diverse cellular events. For instance, forming "quarternary" complexes of receptors, such as the EGF receptor or the PDGF receptor is necessary for initializing cascades of signaling events for cell growth and proliferation (Schlessinger, 1988). The function of ion transport proteins, such as the cystic fibrosis transmembrane conductance regulator (CFTR) or sodium–phosphate cotransporter 2a (NaPiT2a), is regulated by a network of interactions with other membrane proteins, such as the G-protein coupled receptors and other ion channels by forming membrane oligomers either directly, or via cytosolic proteins as adapters or scaffolds. Forming large adherence membrane complexes at the cell–cell junctions is essential to maintain tissue integrity and to suppress tumor cell invasion (Yap et al., 1997; Perez-Moreno et al., 2003; Pujuguet et al., 2003). Understanding how transmembrane protein complexes are regulated and disregulated in disease state can help to identify elements as target to treat various diseases.

The mammalian $Na^+/H^+$ exchange regulatory factor (NHERF) family proteins are scaffolding proteins that assemble macromolecular complexes of transmembrane proteins, and regulate receptor signaling and ion transport (Shenolikar et al., 2004; Lamprecht and Seidler, 2006; Weinman et al., 2006). Members of this protein family, which contain two or more copies of modular PDZ (PSD-95/Discs-large/ZO-1) domains (Fig. 1), localize in the apical membrane region of polarized epithelial cells (Donowitz et al., 2005; Thelin et al., 2005). The PDZ domains are protein–protein interaction modules that are capable of binding to specific PDZ-binding motifs residing in the cytoplasmic portion of a large number of transmembrane proteins (Harris and Lim, 2001). Scaffolding proteins containing multiple PDZ domains and/or other protein–protein interaction modular domains can assemble membrane complexes as well as bring the membrane complexes into proximity with other cytosolic signaling molecules to assemble highly regulated signaling complexes (Sheng and Sala, 2001).

As the first member of the NHERF family, NHERF1 consists of two PDZ domains, PDZ1 and PDZ2 that bind to membrane proteins. NHERF1 also contains carboxy-terminal domain that binds to the membrane–cytoskeleton linker protein ezrin (EB) (Reczek and Bretscher, 1998) (Fig. 1). NHERF1 was shown to interact with ezrin, and therefore is also called ezrin-binding protein 50 or EBP50 (Reczek et al., 1997). The PDZ domains of NHERF1 interact with a number of transmembrane proteins. Some of the NHERF target proteins are implicated in human diseases (Takahashi et al., 2006; Kwon et al., 2007; Sizemore et al., 2007). The well-known functions of NHERF1 include assembling signaling complexes and regulating the endocytic recycling of the CFTR, cell surface adhesion and

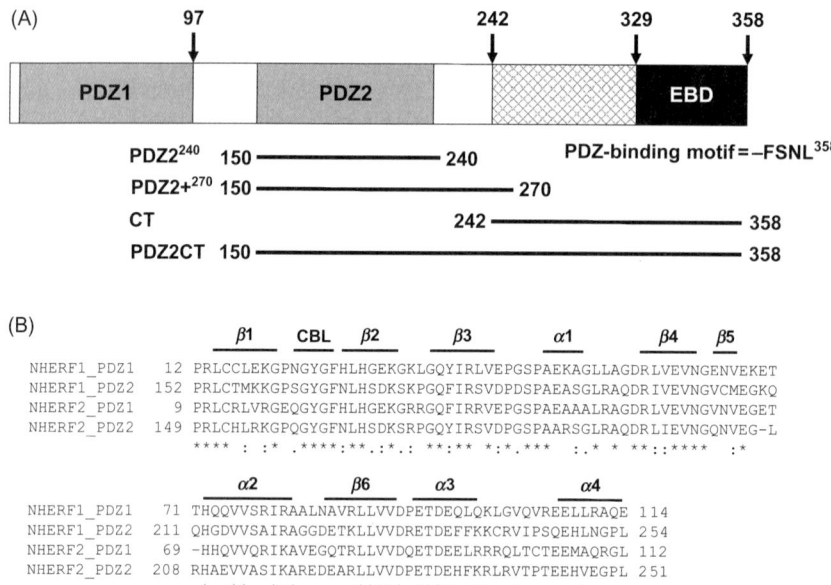

FIG. 1. (A) Schematic representation of the domain structure of human NHERF1. The carboxy-terminal end of the EB domain contains a canonical PDZ-binding motif. The amino acid positions of the differently truncated domains, including the putative PDZ2$^{240}$ and that of PDZ2+$^{270}$ with the extra carboxy-terminal helical subdomain, are shown. (B) Sequence alignment of the PDZ domains of human NHERF1 and NHERF2 proteins, annotated with secondary structure elements (CBL, carboxylate-binding loop). The residues involved in ligand binding are shown in bold. The alignment indicates that the sequence in the extra subdomain formed by α3 an α4 is conserved.

antiadhesion proteins such as podocalyxin, G-protein coupled receptors, and tyrosine kinase receptors PDGFR and EGFR (Hall et al., 1998a; Cao et al., 1999; Ko et al., 2004; Schmieder et al., 2004; Weinman et al., 2006). NHERF1 mutations, which affect its ability to assemble the transmembrane NaPiT2a, are correlated with impaired renal phosphate reabsorption in patients with chronicle kidney disease (Karim et al., 2008). Altered subcellular localization of NHERF1 is associated with breast cancer progression (Mangia et al., 2009).

The function of NHERF1 and the interactions of NHERF1 with membrane transport proteins and receptors have been reviewed earlier (Shenolikar et al., 2004; Weinman et al., 2006). Recent progress in understanding the structure and function of ezrin and other ezrin–moesin–radixin (ERM) proteins has been reviewed (Bretscher et al., 2002; Fievet et al., 2007; McClatchey and Fehon, 2009; Fehon et al., 2010). Here, we first summarize several representative cases that identify NHERF1 as an important factor to assemble complexes of transmembrane receptors or transport proteins. We review the findings that the assembling of signaling complexes by NHERF1 is allosterically regulated. These biochemical studies illustrate that scaffolding or adapter proteins not only function as scaffolds to dock signaling partners but are also regulated transistors and switches that control the effective propagation of signals from a remote site to a specific location over a long distance. We then review the studies that aimed at understanding the structural and dynamic mechanisms of NHERF1 and its interactions with the membrane–cytoskeleton linker protein ezrin in the allosteric regulation of the assembly of membrane complexes. We show that the long-range allosteric-binding behavior is communicated through interdomain conformational and dynamic changes (Li et al., 2009; Bhattacharya et al., 2010; Farago et al., 2010). A recent study using a novel neutron spin echo (NSE) spectroscopy reveals the activation of long-range interdomain motions in NHERF1 on nanometer length scales and on submicrosecond timescale upon binding to ezrin (Farago et al., 2010). Protein domain motion on these timescales and on length scales comparable to protein dimensions can thus propagate allosteric binding signals dynamically. The long-range conformational changes and nanoscale dynamics during the interactions of NHERF1 and ezrin provide a paradigm for studying how cellular signals are transmitted allosterically over a long distance in the cellular signaling network.

## II. NHERF1 Modulates the Macromolecular Assembly, Cell Surface Retention, and Subcellular Localization of Membrane Proteins

### A. CFTR

The PDZ domains of NHERF1 interact with the C-terminal tail of CFTR (Hall et al., 1998a). Interaction of NHERF1 with CFTR increases the polarized expression of CFTR in the apical plasma membrane, as well as enhances the vectorial transport of chloride ions (Moyer et al., 2000;

Raghuram et al., 2001). Moreover, NHERF1 overexpression increases the cell surface expression of a disease-causing mutant of CFTR with a deletion at amino acid Phe508 (ΔF508) (Guerra et al., 2005; Bossard et al., 2007; Favia et al., 2009). The ΔF508 mutant, responsible for 80% of the cases of the genetic disease cystic fibrosis, is trapped in the endoplasmic reticulum after biosynthesis and fails to reach the cell membrane to perform its normal functions as a chloride ion channel. NHERF1 interacts with both CFTR and the G-protein coupled beta 2 adrenergic receptor ($\beta_2$-AR), and assembles a signaling complex comprised of CFTR and $\beta_2$-AR (Naren et al., 2003). This complex mediates the stimulation of the CFTR ion channel by the $\beta_2$-AR receptor (Taouil et al., 2003; Singh et al., 2009).

Using fluorescence photobleaching recovery and single particle tracking, the effects of NHERF1 on the lateral mobility of CFTR in living cells (Haggie et al., 2004, 2006; Bates et al., 2006). CFTR has a significant immobile population (50%), but adding 10 histidine residues at the C-terminus of CFTR to mask the PDZ-binding motif abolished its association with NHERF1, reduced the immobile fraction, and increased mobility. The effects of CFTR interactions with the F-actin cytoskeleton via NHERF1 and ezrin were studied previously with similar methods with N-terminal GFP tag (Haggie et al., 2006). Although the immobile population of CFTR with N-terminal GFP tag is significantly smaller than that measured by Bates et al., the GFP labeled CFTR became mobile after truncation or blocking of the C-terminal PDZ-binding motif, disrupting CFTR association with actin by expressing a mutant NHERF1 lacking the ezrin-binding domain (EBD), or disrupting the F-actin cytoskeleton by latrunculin. The association of the CFTR C-terminus with NHERF1 and ezrin has been proposed to physically tether CFTR to the actin cytoskeleton (Short et al., 1998). Together these studies suggest that NHERF1 and ezrin function as adapters between CFTR and the F-actin cytoskeletal network to stabilize CFTR at the cell membrane and impedes endocytosis retrieval.

### B. NaPiT2a

NHERF1 regulates the reabsorption of phosphate ions in the kidney (Hernando et al., 2002). In the proximal tubule of kidneys, the transmembrane transporter NaPiT2a is responsible for the reabsorption of phosphate from urine. Impaired renal phosphate reabsorption leads to kidney stone formation and bone demineralization. The ability of NaPiT2a to transport phosphate ions depends on the correct localization of NaPiT2a at the apical membrane of polarized epithelial cells, which is modulated by the parathyroid hormone (PTH). PTH binding to receptor PTH1R triggers a cascade of cellular signaling events, including the activation of protein kinase C, which regulates NaPiT2a endocytosis and thus the capacity of NaPiT2a to uptake phosphate. NHERF1 interacts with both NaPiT2a and PTH1R by binding to the PDZ motifs in their respective cytoplasmic tails (Gisler et al., 2001; Mahon et al., 2002). However, it is not clear if NHERF1 assembles NaPiT2a and PTH1R into the same complex. NHERF1 is required for correct apical localization of NaPiT2a (Hernando et al., 2002). Ezrin and NHERF1 assemble the PTH1R and NPT2a complexes localized in the actin-containing microvilli in the apical domains of these cells. Upon PTH treatment, the PTH1R, NaPiT2a, NHERF1, and ezrin colocalize to endocytic vesicles and NaPiT2a-dependent phosphate uptake is markedly inhibited (Mahon, 2008).

A recent study correlates NHERF1 mutations with impaired renal phosphate reabsorption in patients (Karim et al., 2008). This study finds that three NHERF1 mutations, L110V, R153Q, and E225K are found in patients of chronic kidney disease with impaired renal phosphate reabsorption (Karim et al., 2008). The disease mutations of NHERF1, L110V, R153Q, and E225K inhibit phosphate transport by NPT2a, in a similar fashion as in NHERF1$^{-/-}$ kidney cells (Cunningham et al., 2005).

### C. Podocalyxin Complexes

The cell surface antiadhesion molecule podocalyxin is connected to the F-actin cytoskeleton via NHERF1 or NHERF2 and ezrin (Takeda et al., 2001; Schmieder et al., 2004). Podocalyxin is the main molecular component of the apical plasma membrane of podocyte foot processes (Kerjaschki et al., 1984). Disruption of the interactions of podocalyxin/NHERF/actin cytoskeleton results in loss of glomerular foot processes and the glomerular disease proteinuria in animal models (Takeda et al., 2001). Podocalyxin activates *Rho*A through NHERF and ezrin, leading to redistribution of actin filaments (Schmieder et al., 2004). Moreover, podocalyxin contributes to the progression of breast cancer by perturbing tumor cell adhesion (Somasiri et al., 2004). The interaction of podocalyxin with NHERF1 and/or ezrin increases the aggressiveness of breast and prostate cancer cells (Sizemore et al., 2007).

## D. Tyrosine Kinase Receptor Complexes

NHERF1 interacts with a number of important cell surface receptors and organizes multiple signal transduction pathways between cell membranes and the cytoskeletal network. NHERF1 interacts with cell growth factor tyrosine kinase receptors such as the platelet-derived growth factor receptor (PDGFR; Maudsley et al., 2000) and the epidermal growth factor receptor (Lazar et al., 2004), and several G protein-coupled receptors that include the $\beta_2$-AR (Hall et al., 1998b), the parathyroid hormone receptor (PTH1R) (Mahon and Segre, 2004), and the kappa-opioid receptor (Liu-Chen, 2004). NHERF1 and ezrin are responsible for organizing the trafficking, localization, and membrane targeting of these receptors (Hall et al., 1998b; Maudsley et al., 2000; Mahon et al., 2002; Sneddon et al., 2003). NHERF1 and ezrin work together to link PDGFR to the actin cytoskeletal network, and are responsible for transmitting signals from PDGFR to the cytoskeletal networks so as to influence a cell's ability to spread and to migrate (James et al., 2004).

The PDZ1 domain of NHERF interacts with an internal peptide motif located within the C-terminal regulatory domain of EGFR (Lazar et al., 2004). This interaction slows the rate of EGF-induced receptor degradation, and stabilizes EGFR at the cell surface. Recent evidence shows that increased cytoplasmic expression of NHERF1 is correlated with tumor progresses in breast cancer (Mangia et al., 2009). In metastatic breast tumors, the localization of NHERF1 is mainly cytoplasmic. This same study finds that NHERF1 colocalizes with the oncogenic receptor HER2/neu in invasive breast cancer cells, although it is not known if NHERF1 directly binds to the cytoplasmic domain of HER2/neu directly or through other proteins. Because NHERF1 and ezrin can anchor the receptors to F-actin, they could play the dual roles of forming transient receptor dimers and localizing the receptors in the right location in response to stimulation. Besides ligand-induced dimerization of the receptors, preformed receptor dimers are thought to be primed for ligand binding and signaling and may enable cells to respond in a polarized fashion to growth factor stimulation (Chung et al., 2010), especially during cell migration when a cell's leading edge is typically enriched in actin filaments.

### III. Signal Transduction by Allosteric Scaffolding Protein Interactions

The important feature of NHERF1 is its binding to ezrin and other members of the ERM proteins. The interaction of NHERF1 with F-actin is due to the NHERF1 C-terminal EBD (Reczek and Bretscher, 1998). Ezrin and other ERM proteins are the membrane–cytoskeleton adapter or scaffolding proteins that link cell membrane to the F-actin cytoskeleton. The ERM proteins contain an N-terminal FERM (*4.1 e*zrin–*r*adixin–*m*oesin) domain of about 300 amino acid residues, and a C-terminal actin-binding domain that is connected to the FERM domain by a coiled-coil helical domain (Gary and Bretscher, 1995; Li et al., 2007b).

In the dormant form, ezrin and other ERM proteins are negatively regulated by head-to-tail intramolecular interactions between the FERM domain and the C-terminal domain, reviewed in Fehon et al. (2010). Although the activation mechanisms are not fully understood, ezrin and other ERM proteins become activated when the autoinhibition interactions are thought to be disrupted upon phosphorylation and/or phospholipid PIP2 binding (Matsui et al., 1999; Fievet et al., 2004; Roch et al., 2010). The activated ERM proteins are proposed to undergo large conformational changes (Matsui et al., 1998; Bretscher et al., 2000; Yonemura et al., 2002). The unmasked FERM domain in the activated ezrin binds to target membrane proteins either directly or indirectly through NHERF1 or 2, while the carboxy-terminal domain of about 32 amino acid residues of ezrin binds to cytoskeletal actin. The interaction of NHERF1 and ezrin is thus regulated by the autoinhibition and activation of ezrin.

Because ezrin and other ERM proteins bind to both cell membrane and the F-acitn cytoskeleton, they contribute to membrane–cytoskeleton interface that influences a range of cellular functions, such as cell–cell adhesion, cell morphology, cell surface tension, and lateral mobility and exocytosis/endocytosis of the assembled transmembrane protein complexes. The manner that ezrin binds to the cytoskeletal actin could also control the spacing and movement of the membrane protein complexes that ezrin binds directly or indirectly via NHERF1 or NHERF2. The ERM proteins are responsible for generating specialized membrane domains and structures such as membrane ruffles and microvilli, which host a large variety of transmembrane channels, transporters, and receptors (Bretscher et al., 2002; Gautreau et al., 2002; Fehon et al., 2010). Ezrin is involved in the formation of the immunological synapse and in T cell activation (Roumier et al., 2001). The interactions of the ezrin/radixin/moesin family proteins with the cell membranes and with the cytoskeletal actin facilitate the transmission of human immunodeficiency virus (HIV) into uninfected cells (Liu et al., 2009; Wong and Gough, 2009). There is also increasing evidence that ezrin promotes cancer metastasis and progression (McClatchey, 2003; Curto and McClatchey, 2004; Khanna et al., 2004; Yu et al., 2004; Elliott et al., 2005).

The interaction of NHERF1 and ezrin is essential for the cell surface assembly and the normal function of membrane proteins. NHERF1 and ezrin interact and cooperate in regulating ion transport in epithelial cells. Disruption of this link diminishes cell surface expression of ion transport proteins or receptors, which destroys their ability to transport ions across the cell membrane or to transduce signals. Expressing NHERF1 without the carboxy-terminal EB domain in cells leads to the internalization of $Na^+/H^+$ exchanger 3 (NHE3) and abolishes ion transport activities of NHE3 (Weinman et al., 2003). Truncation of the EBD results in loss of functional expression of the sodium–potassium-ATPase transporter at the cell membrane (Lederer et al., 2003). Similarly, expressing NHERF1 in cells impedes antagonist-induced endocytosis of PTH1R, but deleting the EBD of NHERF1 results in otherwise inactive ligands to internalize PTH1R (Sneddon et al., 2003). NHERF and ezrin are thus key elements to transmit the regulation of receptor signaling and ion transport functions by the actin cytoskeletal network.

NHERF and ezrin are necessary to anchor CFTR to the cytoskeleton for the proper function of CFTR (Short et al., 1998). An organized actin cytoskeleton is necessary to retain CFTR in the cell membrane so that CFTR can function properly (Prat et al., 1999). In turn, NHERF1 overexpression-dependent increase of cytoskeleton organization is necessary for rescuing the ΔF508 mutant of CFTR (Favia et al., 2009). Taken together, these studies suggest the important roles of NHERF1 and ezrin in the bidirectional communication between CFTR and the F-actin cytoskeleton.

In summary, NHERF1 is a scaffolding protein that binds to transmembrane proteins and assembles membrane protein complexes. The important feature of NHERF1 is its binding to ezrin and to other (ERM) proteins, thus integrating cell surface membrane protein complexes into the cytoskeletal F-actin network. Such interactions stabilize the

NHERF1-assembled complexes of receptors and/or ion transport proteins at the cell membrane, and thus influence the cellular trafficking of the membrane proteins that NHERF1 interacts with. As a result, the cytoskeletal actin network controls the surface expression and the assembly of membrane proteins that have crucial impacts on the cellular or physiological functions. Moreover, the transmission of signals by NHERF1·ezrin complexes could be bidirectional between the cell membrane and the cytoskeletal F-actin network. The NHERF1·ezrin-assembled membrane complexes could influence the cellular distribution, the assembly, and the dynamics of the F-actin network. The NHERF1·ezrin complexes are thus important connectors that transduce signals between the cell membrane and the cytoskeleton.

### A. Allosteric Modulation of NHERF1 to Assemble Membrane Complexes

The integration of the NHERF1·ezrin-assembled membrane complexes into the F-actin filament fit into the emerging picture that the F-actin cytoskeleton and actin-associated adapter proteins underlying the cell membrane regulate the function of membrane complexes. The cytoskeletal actin filament has long been recognized to provide a network of barriers that hampers the diffusion of membrane complexes inside cell membranes (Lee et al., 2007). However, F-actin can also exert active control on regulating the assembly of membrane protein complexes. During such processes, the F-actin-associated adapter proteins and scaffolding proteins function as signal transducers. The interactions among these adapter proteins are regulated by the dynamic conformational changes upon posttranslational modification or binding to signaling lipids and to other proteins. The interactions among these adapter proteins, such as NHERF1 and ezrin, provide a means to relay allosteric signals from the F-actin cytoskeleton to the membrane for the effective control of the membrane assembly. Moreover, it is becoming increasingly evident that NHERF1 and ezrin are signal tranducers through which the F-actin cytoskeleton organizes the assembly of membrane receptor and ion transport protein complexes allosterically.

Recent biochemical and biophysical experiments have provided detailed molecular mechanisms that ezrin positively modulates the interactions of the PDZ domains of NHERF1 to assemble multiprotein complexes in a cooperative fashion. Biophysical methods have been applied to determine the stoichiometry and affinity of NHERF1 binding to the C-terminal domain of CFTR (Li et al., 2005). Both static light scattering and analytical ultracentrifugation experiments show that NHERF1 exists as monomer in solution. The PDZ1 domain of NHERF1 was found to have high binding affinity, while PDZ2 plus the C-terminal domain have a lower binding affinity for C-CFTR. However, when the FERM domain of ezrin binds to the carboxy-terminal EBD of NHERF1 with high affinity (with $K_d = 19$ nM), the binding affinity of PDZ2 for the last 70 amino acid residue CT domain of CFTR (C-CFTR), which contains a type I C-terminal PDZ-binding motif—DTRL, increases by 26-fold (see Fig. 2). As a result of EB, the stoichiometry of the full-length NHERF1 binding to C-CFTR is increased from 1:1 to 1:2. Moreover, the binding affinity of PDZ1 for the target protein also increases significantly upon FERM binding. A thermodynamic cycle analysis indicates that ezrin positively modulates the intramolecular domain–domain interactions in NHERF1 and controls NHERF1 to assemble membrane signaling complexes allosterically (Li et al., 2009) (see Fig. 3).

EB to NHERF1 also positively regulates the interactions of NHERF1 with other signaling proteins, such as PDZK1, which is a four PDZ domain scaffolding protein that belongs to the NHERF family of proteins (LaLonde et al., 2010). The binding of the PDZ domains of NHERF1 to PDZK1 in turn disrupts the autoinhibition-like interactions in PDZK1 to enable it to assemble larger protein complexes that contribute to cellular microvillar organization (LaLonde and Bretscher, 2009). In addition, EB can also activate NHERF1 to assemble a heterogeneous complex, PTEN at PDZ1 and β-catenin at PDZ2 (Morales et al., 2007).

## B. Negative Cooperativity and Feedback Loop

NHERF1 can also send negative feedback to the network through domain–domain interactions. In the presence of C-CFTR in the PDZ1 domain of NHERF reduces the affinity of the NHERF C-terminus for FERM by sixfold (Li et al., 2005). The reduction in binding affinity of NHERF1 for ezrin suggests that binding of C-CFTR to the PDZ1 domain of NHERF can negatively regulate the interaction of NHERF with ezrin. The weakened NHERF–ezrin interaction, due to the binding of CFTR to the PDZ1 domain of NHERF, can also serve as a feedback loop to weaken the interaction of the PDZ2 domain with CFTR, by uncoupling EB

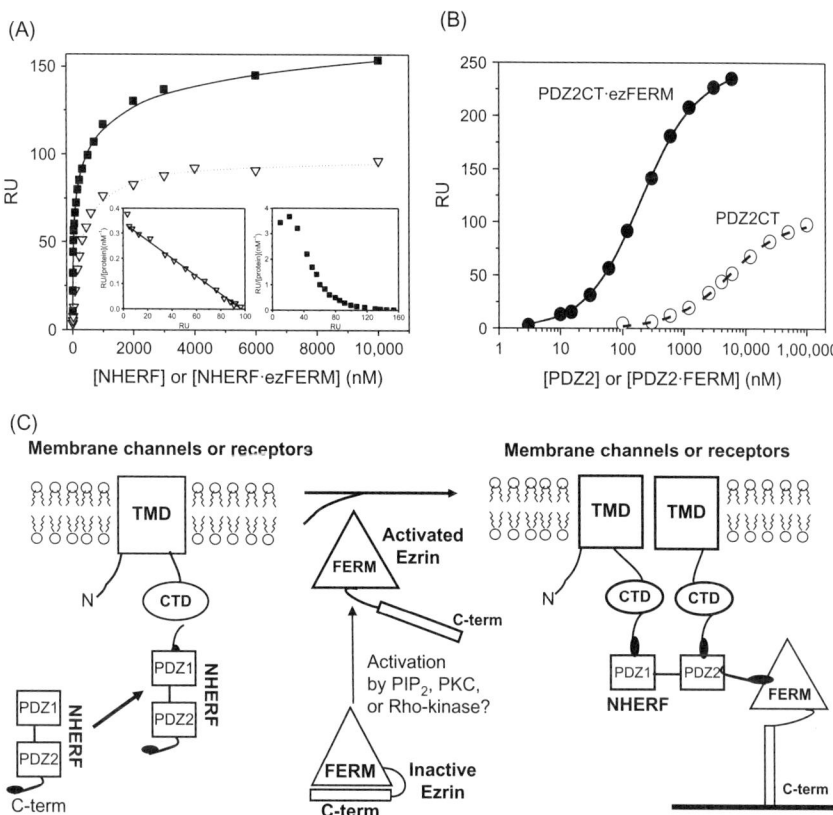

FIG. 2. (A) SPR analysis indicates increased affinity and stoichiometry of C-CFTR to NHERF1 binding when the FERM of ezrin is bound to NHERF1. (▽) NHERF1 alone binding to C-CFTR is monovalent as shown by the linear Scatchard plot (inset). (■) NHERF·FERM binding to C-CFTR is bivalent as also shown by the nonlinear Scatchard plot (inset). (B) Increased binding affinity of C-CFTR for PDZ2, when FERM is bound to the PDZ2CT construct. Note that in (B), the bindings of both PDZ2CT and PDZ2CT·FERM to C-CFTR have reached saturation. The $X$-axis is in logarithmic scale in order to show both binding curves in the same plot. (C) Our hypothesis about the regulation of NHERF1 by ezrin, which changes the stoichiometry of NHERF interaction with membrane channels or receptors. *TMD*, transmembrane domain; *CTD*, the cytoplasmic domain. Ezrin binding to the C-terminus of NHERF activates PDZ2 to interact tightly with the cytoplasmic domain of channels or receptors. The schematic also shows that the C-terminus of ezrin binds to the filamentous actin.

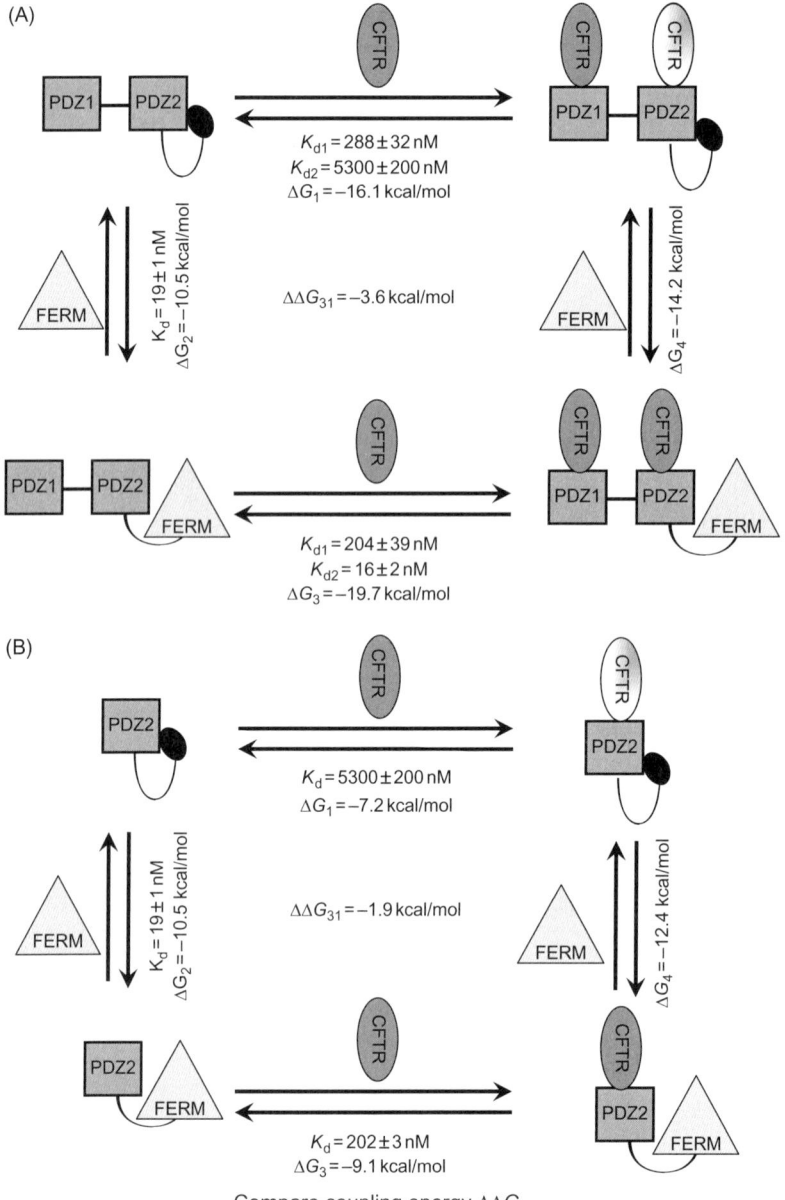

FIG. 3. Thermodynamic cycle analysis reveals FERM-induced long-range inter-domain allostery in the scaffolding protein NHERF1. FERM binding to NHERF1 increases the binding affinities of both PDZ1 and PDZ2 for C-CFTR. (A) The ''coupling''

to NHERF. As a result, the stimulation of CFTR channel due to the activation of NHERF by ezrin is weakened.

NHERF1 also displays both positive and negative long-range allosteric communications between its two PDZ domains. A study by LaLonde et al. (2010) shows that mutations in PDZ1 reduce the binding affinity of PDZ2, or occupancy of a ligand in PDZ1 enhances the binding capability of PDZ2. Intriguingly, a recent study by Garbett et al. (2010) shows that serine to aspartic acid mutations in NHERF1, which mimic phosphorylation at S280 and 302 by cdc42 or phosphorylation at S162, 339 and 340 by PKC inhibits ligand accessibility to PDZ1 when PDZ2 is occupied, suggesting that NHERF1 phosphorylation by either PKC or Cdc2 inhibits simultaneous binding to both PDZ domains.

These studies indicate that there are long-range domain–domain allosteric communications in NHERF1. Adapter and scaffolding proteins, NHERF1 and ezrin are thus not just passive adapters to dock the binding partners. Instead, ezrin and NHERF function as signal transducers between cell membranes and the actin cytoskeleton. The cortical actin cytoskeleton is thus not only a passive support of cell membranes but also actively controls the assembly of membrane proteins. F-actin and associated membrane–cytoskeleton adapter or scaffolding proteins thus integrate diverse signals in space and time to influence all aspects of biology from receptor signaling to cellular homeostasis (Pollard and Cooper, 2009).

The allosteric regulation of membrane protein–NHERF1–ezrin–actin cytoskeleton is bidirectional. The above cases show inside-out types of regulation that transmits the control signal of membrane assembly by the cytoskeletal F-actin. The NHERF1·ezrin can also transmit outside-in type signals. The NHEF1·ezrin complex allosterically assembles the CFTR homo and/or hetero complexes and anchors the assembled complexes to the F-action cytoskeleton, which promote the surface expression and cross

---

energy (Fersht, 1998) for FERM binding to the NHERF1 carboxy-terminus and for NHERF binding to C-CFTR is $\Delta\Delta G = \Delta G_3 - \Delta G_1 = \Delta G_4 - \Delta G_2 = -3.64$ kcal/mol. $K_d$ values, taken from our previous publication, are measured by surface plasmon resonance (Li et al., 2005). (B) The coupling energy for FERM binding to PDZ2CT and for PDZ2CT binding to C-CFTR is $\Delta\Delta G = -1.93$ kcal/mol. Comparing the coupling energy of C-CFTR binding to the full-length NHERF1 with that of binding to PDZ2CT suggests that FERM binding induces long-range allosteric-binding behavior in NHERF.

talk signaling of CFTR. Alternatively, anchoring ion transport proteins to the cytoskeleton alters the organization and assembly of cytoskeletal actin, and is believed to influence cell shapes and the ability of a cell to migrate (Denker and Barber, 2002). Similarly, disruption of the interactions of podocalyxin/NHERF2/actin cytoskeleton results in loss of glomerular foot processes and the glomerular disease proteinuria in animal models (Takeda et al., 2001). For instance, podocalyxin activates *RhoA* and induces actin reorganization through NHERF1 and ezrin (Schmieder et al., 2004). The NHERF1·ezrin complex promotes βPDGFR signaling that results in actin cytoskeletal rearrangements to the cytoskeleton and promote cell spreading and migration (James et al., 2004).

## IV. Structural Basis of Autoinhibition and Long-Range Allostery in NHERF1

The above examples show that the assembly of membrane complexes by NHERF1 is allosterically regulated. The allosteric regulation is not confined to a single protein, but rather it is an allosteric relay of signals through a chain of multiple protein–protein interactions. Such a relay of signals, in a Rube Goldberg device style, can also be found in almost all other signaling pathways (Ma and Nussinov, 2009; Scott and Pawson, 2009). The relay of signals, using multiple proteins as transistors, has the advantage of being much more efficient and specific than thermal diffusion in transmitting regulatory signals from one location to a reach a target. Because of the multiple domains of NHERF1, EB activation of PDZ domains enable NHERF1 to assemble homo and/or hetero membrane complexes, thus integrating signals to promote cross talks among membrane proteins. In return, binding to NHERF1 could enhance the binding of ezrin to the F-actin filament, thus integrating the membrane signals to the cytoskeleton network. The regulated complex formation between NHERF1 and ezrin in the membrane–cytoskeleton thus provide an excellent prototype to understand how multiple proteins interact to form complexes, transmit and relay signals.

We next review structural and dynamic studies of NHERF1 and its interactions with ezrin with an emphasis on understanding the long-range allosteric binding communicated through interdomain conformational changes (Li et al., 2009; Bhattacharya et al., 2010). Our recent study using NSE spectroscopy reveals that interdomain motions among PDZ1,

PDZ2, and CT in NHERF1 on nanometer length scales and on submicrosecond timescale can propagate allosteric-binding signals dynamically (Farago et al., 2010). The long-range conformational changes and nanoscale dynamics during the interactions of NHERF1 and ezrin provide a first glimpse for understanding the mechanisms of how cellular signals are transmitted allosterically over a long distance in the signaling network.

### A. NHERF1 Has a Highly Elongated Shape, Allowing Long-Range Interdomain Allosteric Communications

Understanding the scaffolding function of a multidomain protein requires the structure of the full-length protein. Because of the quite large size and the dynamic nature, the structure of the full-length NHERF1 has eluted high-resolution structural studies. Combining solution small angle X-ray scattering (SAXS) with high-resolution structure data on the fragment can provide information about the 3D shape, domain–domain distances, and domain orientation of such heterogeneous multidomain proteins (Bu et al., 1998; Bu and Engelman, 1999). The structural information from SAXS on the full-length scaffolding protein NHERF1, although at low resolution, is accurate and necessary to understand how NHERF1 and other scaffolding proteins assemble complexes.

NHERF1 is monomeric in solution at concentrations as high as 2 mg/ml (corresponding to 50 μM) as determined by a combination of gel filtration, static and dynamic light scattering, and analytical ultracentrifugation experiments (Li et al., 2005, 2007a, 2009; Garbett et al., 2010). Above 3 mg/ml, NHERF1 starts to show a small fraction of weak association in light-scattering experiments, but analytical ultracentrifugation suggests that NHERF1 is a monomer even at higher concentrations (Li and Bu, unpublished data). A recent study shows that NHERF1 is an elongated monomer regardless of the phosphorylation states (Garbett et al., 2010).

SAXS provides the 3D shape of NHERF1 and the spatial arrangement of the different domains in NHERF1 (Li et al., 2005, 2007a, 2009). The 3D molecular shape of NHERF1 shows that the full-length NHERF1 monomer is elongated. The elongated shape of NHERF1 is also manifested by the very asymmetric shape of the length distribution function $P(r)$ obtained from SAXS experiments with a maximum dimension of 140 Å (Fig. 4). The radius of gyration ($R_g$) of NHERF1 is $40.9 \pm 0.6$ Å.

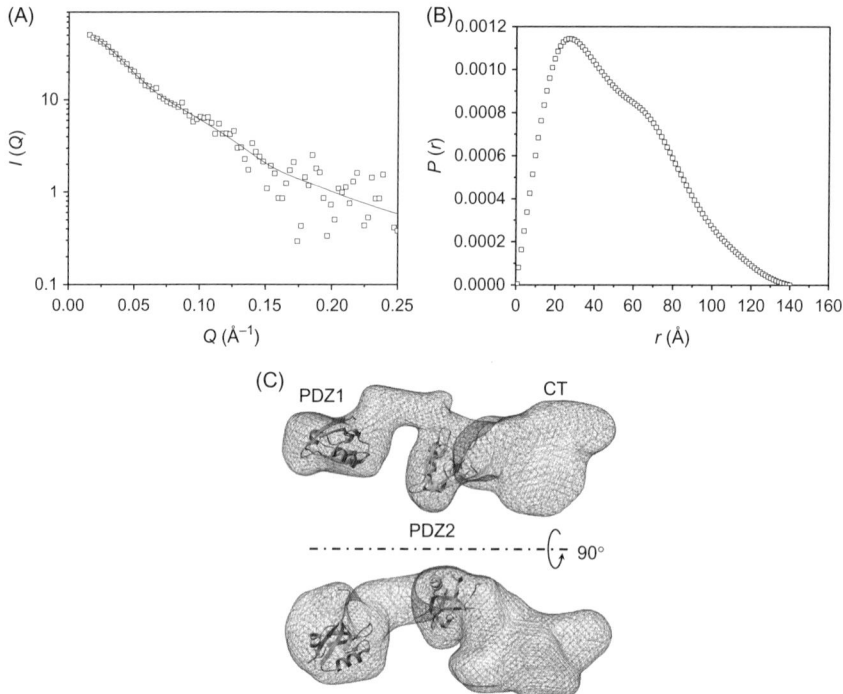

FIG. 4. Solution small angle X-ray reveals the 3D shape of full-length NHERF1. (A) SAXS data of the scattered intensity $I(Q)$ versus $Q$. (B) Length distribution function $P(r)$. (C) 3D shape of NHERF1 reconstructed from SAXS data using the program DAMMIN (Svergun, 1999).

There are three lobes in the 3D map of NHERF1 reconstructed from SAXS. The lobe representing the PDZ1 domain is well separated from the other two lobes. The two lobes representing PDZ2 and CT domains are in close contact with each other, indicating that there are domain–domain interactions between PDZ2 and CT (Li et al., 2005, 2007a). The center-of-mass distance between PDZ2 and CT is 45.8 Å. The distance between PDZ1 and PDZ2 is 57.1 Å, and the distance between PDZ1 and CT is about 110 Å.

In the 3D map of NHERF1, the lobe representing the CT domain is compact with a clearly defined boundary. Our NMR studies find that the CT domain is largely unstructured (Bhattacharya et al., 2010). Combining the SAXS and NMR results suggests that the CT domain adopts a compact but disordered conformation in NHERF1.

Because of the elongated shape of NHERF1, the PDZ1 domain is separated about 110 Å away from the CT domain (Li et al., 2007a, 2009). PDZ1 and CT are unlikely to form head-to-tail like interactions as previously thought. Rather, the modulation of the binding capability of PDZ1 by FERM binding at a remote site is through a long-range allosteric behavior.

### B. Allosteric Control in NHERF1 Originates from Domain–Domain Interactions

Small angle neutron scattering (SANS) has the unique advantage of allowing the structure of multicomponent complexes by contrast variation and deuterium labeling. The technique of contrast variation neutron scattering relies on the tremendous scattering difference between hydrogen and deuterium. By selectively labeling portions of a protein or a complex with deuterium, and changing the $D_2O$ composition of buffer, the subunit can be made essentially invisible to neutrons at the contrast match point. Thus, one can highlight the component of interest for study. This feature of neutron scattering makes it particularly useful for studying the structure of biological macromolecular complexes. SANS thus has the advantage of studying the structure of multicomponent complexes by contrast variation and deuterium labeling. This critical feature of neutron scattering makes it particularly useful for studying biological macromolecular complexes.

We have used SANS to determine the conformational changes of NHERF1 upon forming a complex with ezrin, using deuterated NHERF1 and hydrogenated FERM domain (Li et al., 2009). Contrast variation SANS reveals that when FERM binds to the C-terminus of NHERF1, FERM induces large conformational changes in NHERF1. The shape of the length distribution function $P(r)$ of NHERF1 in the complex is significantly different from that in solution (Fig. 5). The radius of gyration changes $40.9 \pm 0.6$ Å in solution to $45.8 \pm 0.8$ Å in the complex, but the change in $D_{max}$ is less dramatic from 140 to 145 Å. This result suggests a significant change in the geometry and size of NHERF1.

Comparing the 3D shape of $^d$NHERF1 in solution and in the complex shows that the region linking PDZ2 and CT becomes more extended in NHERF1 (Fig. 5). An angle of about 120° is formed between PDZ1–PDZ2 and PDZ2–CT at the location of PDZ2 in $^d$NHERF1 in the complex, which

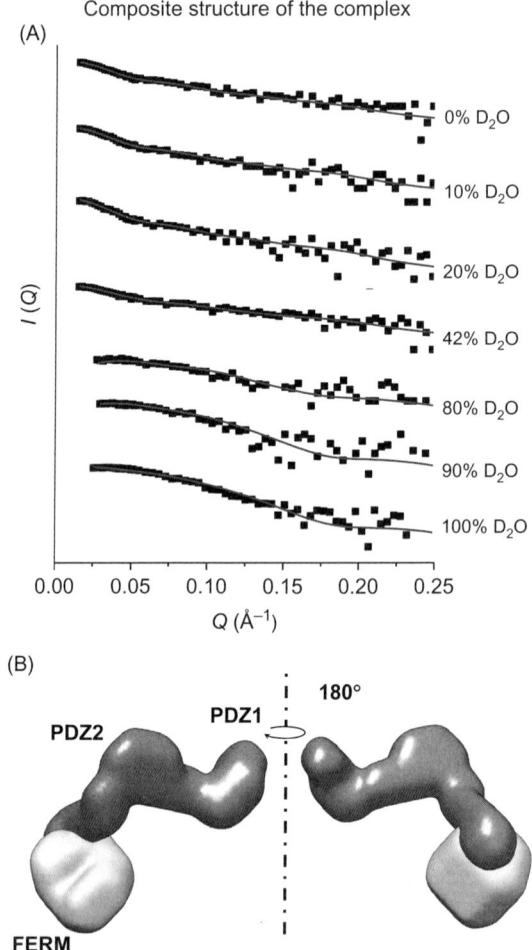

FIG. 5. Contrast variation SANS reveals the composite structure of the deuterated NHERF in complex with unlabeled FERM ($^{d}$NHERF1·FERM) complex. (A) SANS data of $^{d}$NHERF1·FERM at various contrasts. The lines are fits to the scattering curves when reconstructing the 3D composite structure of the complex. The $\chi^2$ values of fitting are shown in the plots. (B) The 3D models of the $^{d}$NHERF1·FERM reconstructed *ab initio* from contrast variation SANS (Petoukhov and Svergun, 2006).

is consistent with the $P(r)$ shape changes of $^d$NHERF1 in the FERM-bound complex. The distance between the centers of PDZ2 and the EBD changes from 45.8 Å in solution to 63 Å in the complex, suggesting that domain–domain contacts between PDZ2 and CT are disrupted.

Significant conformational changes are also apparent in the region that links PDZ1 and PDZ2. The distance between PDZ1 and PDZ2 changes from 57.1 Å in solution to 67.0 Å in the complex. In addition, the SANS results also suggest that the FERM domain does not have global conformational changes when it is bound to NHERF1. The SANS results thus provide a structural explanation of the binding and thermodynamic analyses, which demonstrate positive allosteric regulation of NHERF1 by ezrin as it assembles membrane protein complexes.

### C. Redefinition of the Structural Boundary of PDZ Domains

The structures of the isolated PDZ1, PDZ2 of domains of NHERF1, and the structures of the NHERF1 PDZ1 domain in complex with the carboxy-terminal peptides of membrane receptors and channels have been determined by X-ray crystallography (Karthikeyan et al., 2001, 2002). The putative NHERF1 PDZ structures adopt similar fold as in other PDZ proteins. The PDZ domains have a characteristic β-sandwich structure that is composed of six strands stacked in an antiparallel fashion into two β-sheets, which are flanked by two α1 and α2 helices (Fig. 6). The partially open hydrophobic cavity enclosed by the β-sandwich (Fig. 6) serves as a robust scaffold to recruit peptide based ligands. The conformation of the carboxylate binding (CB) loop (–GYGF–) is capable of forming H-bond pairs with the peptide ligand (Doyle et al., 1996). Primary sequence analysis predicts the "conventional" PDZ1 domain of NHERF1 starts amino acid L14 and ends at D92, and the PDZ2 domain starts at L154 and ends at V231.

However, our NMR structural studies on a larger PDZ2 plus the CT domain (PDZ2CT) find that, besides the putative PDZ2 structure, the structured region of NHERF1 PDZ2 extends to N252, well beyond V231 at the predicted boundary for a putative PDZ domain (Fig. 6) (Bhattacharya et al., 2010). We find that a C-terminal extension (R233-N252) consists of two helices α3 and α4, forming a closed hydrophobic cluster at the C-terminal end of the putative PDZ fold. Thus, the C-terminal extension is an integral part of the PDZ2 domain. The extended

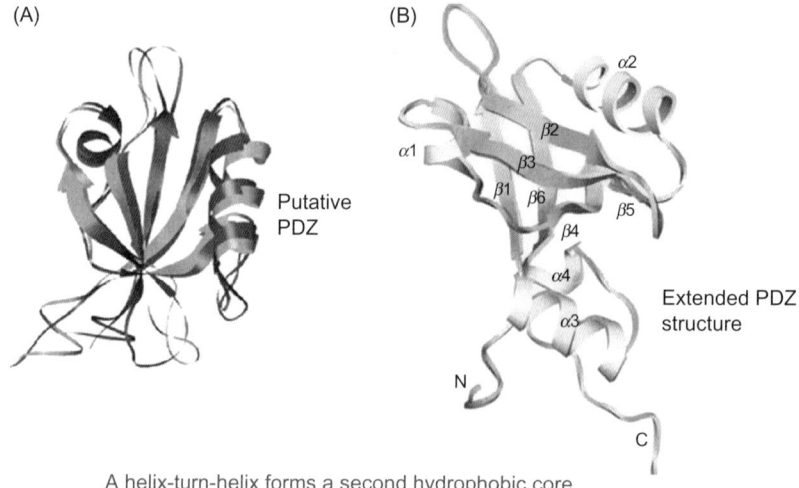

A helix-turn-helix forms a second hydrophobic core

Fig. 6. The extended structure of PDZ2 of NHERF1 with a novel helix-turn-helix subdomain at the carboxy-terminal end of the putative PDZ fold. (A) The structure of a putative PDZ domain (PDB code: 2JXO). (B) The extended structure of PDZ2+$^{270}$ with a C-terminal helix-turn-helix motif (PDB code: 2KJD).

PDZ structure is thermodynamically more stable, and has higher affinity for ligand than the putative PDZ2$^{240}$. Because the helical extension does not interact with the ligand, this extension augments the PDZ2 stability and affinity for ligands by an allosteric mechanism instead of direct engagement.

The C-terminal helical extension of a PDZ fold appears to be a feature that is frequently shared by many PDZ domains (Bhattacharya et al., 2010). Based on multiple sequence alignment alone, the carboxy-terminal hydrophobic residues are conserved across all the PDZ domains in the NHERF family of proteins, suggesting similar functional roles for this extended helical fold (Bhattacharya et al., 2010). Moreover, multiple alignment and secondary structure analysis predict α-helical propensity at the C-terminal end in a majority of the PDZ domains that we have analyzed even though the amino acid sequence at the C-terminal extension is not well conserved. The C-terminal helical extensions have been found in many of the PDZ domains that are important in cell signaling, such as those of the human harmonin, protein tyrosine phosphatase, tamalin and PARR3, and PDZ1 of PSD-95, as well as the PDZ domains of *Drosophila melanogaster* INAD (Bhattacharya et al., 2010). In addition to regulating the target affinity, the C-terminal extensions to PDZ domains are also known to mediate multimerization in some examples such as harmonin (Verpy et al., 2000; Wang et al., 2010). Thus, amino acid sequence variation within a PDZ domain or in the PDZ-binding motif is not the only means to render specificity for PDZ/target protein interactions. The structure flanking the core PDZ fold can also influence target peptide binding. The roles of the extended structure(s) may include modification of "dynamic allostery."

### D. Autoinhibition of PDZ2 by the C-Terminal Domain of NHERF1

The NMR structure of the PDZ2CT structure reveals that a PDZ-binding motif (-SNL) located at the C-terminus of NHERF1 binds to the ligand-interaction pocket of PDZ2 (Bhattacharya et al., 2010). The NMR structure thus corroborates the biochemical data that there are autoinhibitory interactions between PDZ2 and the CT domain of NHERF1 (Morales et al., 2007; Li et al., 2009). While NMR and circular dichroism indicate that the CT domain of more than 100 amino acid residues adopts a largely disordered structure, SAXS shows that this CT domain is a collapsed domain. When binding to the FERM domain of ezrin, the EBD, which resides in the CT domain and overlaps with the C-terminal PDZ-binding motif, adopts a distinct helical structure (Terawaki et al., 2006), but the linker region between PDZ2 and EBD becomes totally unfolded (Bhattacharya et al., 2010). The binding affinity between FERM and EBD is strong with $K_d = 10$ nM. Binding to FERM domain thus disrupts the autoinhibition between PDZ2 and CT domain, and the disordered linker region acts as a flexible spacer between PDZ2 and the FERM domain.

### E. Disease-Associated NHERF1 Mutations Affect the Structure, Stability, and Binding Capability of PDZ Domains

Using NMR and biophysical experiments, we have analyzed the effects of the disease-associated R153Q and E225K mutations on protein structure and stability (Bhattacharya et al., 2010). These mutations were identified in patients with impaired renal phosphate reabsorption (Karim et al., 2008). The R153Q and E225K mutations are located in

PDZ2 outside the ligand-binding sites (Bhattacharya et al., 2010). The R153Q mutant is considerably less stable (with an unfolding transition temperature $T_m = 37\,°C$) than the wild-type protein ($T_m = 55\,°C$) (Bhattacharya et al., 2010), due to the loss of H-bond/salt bridge between the positively charged bidentate R153 $N^\eta$ group in strand β1 and the negative charge of the COO$^-$ groups of D197 (β4) and D232 at the N-terminus of α2 helix. The binding affinity of R153Q for the peptide ligand has also decreased.

The E225K mutation has a dramatic effect on the conformational stability of PDZ2 domain and this mutant fails to express as an intact protein in *Escherichia coli* at either 37 or 20 °C. In the wild-type protein, the negatively charged E225 is complemented by surrounding positive charge of lysine side-chains (K158 and K227) on the exposed surface of the β-sheet (Bhattacharya et al., 2010). The unfavorable electrostatic energy of the E225K mutant would destabilize the protein.

Thus, although these mutations are located outside the ligand-binding site of the PDZ2 domain, the mutations R153Q and E225K evidently destabilize the native state and would therefore accelerate degradation in a cellular context. Reduced protein stability could translate into the loss of functional NHERF1 expressed in cells, and diminish the ability of NHERF1 to assemble transmembrane protein complexes of NPT2a at the cell membrane.

## V. Dynamic Propagation of Allosteric Signals by Nanoscale Protein Motion

The above examples demonstrate that allostery of signals are transmitted over a long distance within a single protein, as well as in multiple protein–protein interactions. There is increasing evidence that the transmission of allosteric-binding signals requires both conformational changes and protein motion (Kern and Zuiderweg, 2003). Protein dynamics can initiate and control protein function. Protein motion regulates the transition state dynamics of enzyme catalysis (Benkovic and Hammes-Schiffer, 2003; Eisenmesser et al., 2005), and protein motion has been proposed to contribute significantly to the propagation of allosteric signals (Cooper and Dryden, 1984). Information arising from ligand binding can be communicated to a distal site in a protein by altering internal dynamic modes (Hawkins and McLeish, 2004).

Most studies have been focused on how allosteric signals are transmitted within a single protein domain, such as in the PDZ domains. However, the overall dynamic architectures, and in particular the long-range motion properties of scaffolding proteins that control the function of assembled macromolecular signaling complexes, remain a largely unexplored territory. Protein motions on nanoscales are indispensible for relaying signals allosterically in the cellular networks (Ma and Nussinov, 2009). This concept is emerging as a powerful theme in cell signaling. To understand the mechanism of how allosteric signals are propagated in multidomain proteins and over a cascade of multiple protein–protein interactions requires the study of structure and dynamics on nanolength scales that are comparable to the dimension of multidomain signaling proteins such as in NHERF1 and ezrin.

In the following, we first summarize our new applications of NSE spectroscopy to studying nanoscale long-range domain motions in NHERF1 upon forming a complex with ezrin. We expect that NSE can be extended to determine how signals are propagated in multiprotein signaling complexes.

### A. The Physical Concepts Behind Protein Dynamics

Many experimental studies show that it is sensible to attribute the dynamic properties of bulk materials to proteins (Howard, 2001). For example, the Young's moduli of proteins are typically found to be mechanically isotropic, independent of the direction of the applied force (Howard, 2001, chapter 8). Moreover, mechanical measurements show that the Young's moduli of very different proteins are fairly similar (Howard, 2001, Table 3.2, page 31), supporting the notion that proteins can often be thought of as being largely composed of fairly uniform soft matter. According to the domain concept of structural biology (Creighton, 1993), multidomain proteins can be considered as being comprised of somewhat rigid domains connected by soft spring linkers (Gerstein et al., 1994; Zaccai, 2000; Farago et al., 2010). The conceptual virtue of attributing materials properties to proteins can easily be seen by comparing the difficulty of retaining an atomic-level description, whereby one must perform difficult molecular dynamics simulations to achieve an understanding of protein motion that may provide only a limited improvement over the materials point of view.

## 1. Protein Motions Are Overdamped, Creeping Movements Rather than Underdamped Oscillations

The environment in which proteins act is one of low Reynolds number (usually abbreviated $Re$). The Reynolds number (which was actually introduced by Stokes!) is the ratio of the magnitude of inertial forces to that of the forces that arise from the viscous drag that opposes motion. If inertial forces are more important, Reynolds number is large, and forces are proportional to mass times acceleration. If viscous drag is more important, the Reynolds number is small, and mechanical forces are proportional to the velocity of the protein (more generally incorporating a concept known as the mobility tensor, which we will discuss in detail below). Reynolds number is actually an imprecise concept, usually used as a way to argue that certain terms in the Navier–Stokes equations of fluid dynamics can safely be neglected. Reynolds number can be *estimated* by the simple formula $Re = Lv/\nu$, where $L$ is a characteristic length scale of the protein, $v$ is a characteristic velocity, and $\nu$ is the kinetic viscosity of the solvent (for water, $\nu = 10^5 \text{ Å}^2/\text{ns}$). At large $Re$, one has oscillatory (underdamped) phenomena such as the ringing of a bell, while at small $Re$ (which occurs when $Re$ is less than about 1000), dynamics involves slow, creeping *overdamped* motion.

The Reynolds number for typical proteins is less than 0.1, indicating that they are well within the low $Re$ regime. Simple calculations show that even a multiprotein complex as large as a ribosome is still in the overdamped regime (Howard, 2001, page 43). The environment of a protein thus has more in common with playing badminton at the bottom of a swimming pool full of molasses (low $Re$) than in crossing the Atlantic in an ocean liner (high $Re$), see reference Howard (2001, chapter 3 and Table 3.4). Since for proteins, inertial forces are less important than diffusive, viscous effects, protein dynamics should be largely independent of the mass of the protein (or of the relative masses of internal domains). For example, the diffusion constant of a deuterated protein should be almost the same as a hydrogenated protein, even though deuterium has twice the mass of hydrogen. This effect will be seen to be important for our later discussion of deuterium contrast matching.

*Proteins obey Brownian dynamics.* Protein dynamics arises as a result of interplay between the mechanical forces mentioned above, and the thermal forces that arise from the collision of the protein with solvent molecules. These thermal forces are random in magnitude and direction, and lead to the protein undergoing a process known as *diffusion*. A freely diffusing object displays what is called *Brownian motion*, with frequent changes in the direction and speed of its movement. The world in which proteins operate is therefore characterized by the presence of a significant amount of noise and the resultant diffusion of protein subunits arising from thermal motion. This thermal motion is essential for the protein to reach its equilibrium state.

*The limiting rate for the propagation of conformational changes is the speed of sound in proteins.* This is true even in the case of overdamped motion, see reference Howard (2001, page 307). We can estimate the speed of sound by $c \approx \sqrt{(E/\rho)}$, where $E \approx 1$ GPa is the Young's modulus of a protein, and $\rho \approx 10^3$ kg/m$^3$ is the density of protein. A crude estimate of the limiting speed of conformational change is thus $10^4$ Å/ns. Since most proteins are of the order of a few hundred Angstroms in size, we see that long-range coupled motion between separate protein domains is physically possible on nanosecond timescales.

*Protein dynamics involves a hierarchy of length scales and timescales.* Biologically relevant protein motions occur on timescales ranging from femtoseconds to hundreds of seconds and from sub-Angstrom length scales up to hundreds of Angstroms. Small-amplitude local conformational changes (partially directed by mechanical forces) drive a protein to explore its energy landscape and find the saddle-like passes between energy valleys. These semideterministic *local* motions inspire larger scale stochastic *global* conformational changes between these valleys. The small-scale local dynamic motions are typically on the picosecond-to-nanosecond timescale, while the global kinetic motions are generally of the order of microseconds to milliseconds or longer (Lindorff-Larsen et al., 2005; Fehon, 2006). The existence and nature of the motions in this hierarchy are summarized in Table I.

### B. Nanoscale Protein Dynamics: The Emergence of a New Frontier

Although the importance of protein domain dynamics has long been recognized, it has only been in recent years that the field has burgeoned. This situation arises in part from two causes. First, there has been a shortage of methods that are capable of addressing such a complicated issue. Second, the theoretical methods needed to understand protein

TABLE I
The Hierarchy of Protein Dynamics (Gerstein et al., 1994; Palmer, 2004)

| Types of motion | Function | Timescale | Amplitude |
| --- | --- | --- | --- |
| *Local motions* | | | |
| Atomic fluctuation | Ligand docking flexibility | Femtoseconds (fs) to picoseconds (ps) | <1 Å |
| Side-chain motion | Thermal diffusion | $10^{-15}$–$10^{-12}$ s | |
| *Medium-scale motions* | | | |
| Loop motion | Active site conformation adaptation. | Nanoseconds (ns) to microseconds (μs) | 1–5 Å |
| Rigid-body motion | Binding Specificity | $10^{-9}$–$10^{-6}$ s | |
| *Large-scale motions* | Hinge-bending motion | Microseconds (μs) to milliseconds (ms) | 5–10 Å |
| Domain motion | Allosteric transitions | $10^{-6}$–$10^{-3}$ s | |
| Subunit motion | | | |
| *Folding–unfolding transition* | Ligand-binding, protein–protein interaction | Milliseconds (ms) to hours $10^{-3}$–$10^{4}$ s | >10 Å |

dynamics (which have their origins in the study of polymers) have only been developed recently.

*Biophysical techniques: the promise of neutron scattering.* The observation of protein motion on nanometer length scales and on nanosecond to microsecond timescales has remained an elusive goal because this is a spatial–temporal regime that has not been reached by existing biophysical techniques. Thus there is an important information gap, on nanoscales, between the structural dynamics occurring at atomic resolution and the cellular organization and dynamics on the much larger micron length scales and slower timescales. This information gap is both spatial and temporal. One technique of great promise for addressing this knowledge gap is neutron scattering (Higgins and Benoit, 1994), which is emerging as a powerful way to address protein dynamics. The neutron appears as a natural candidate for the study of proteins, as thermal neutrons possess a wavelength that is of the order of Angstroms, making it an appropriate probe for protein structure. Moreover, the existence of such tools as contrast matching is unique to this field. We illustrate this concept with an example shown in Fig. 7.

In a multicomponent contrast-matching experiment, one can selectively deuterate a particular protein domain or a subunit in a complex (Fig. 7). In $D_2O$ buffer solution in which the NSE experiments are conducted, the deuterated component becomes invisible to neutron. Since deuteration should not significantly affect protein dynamics, despite the fact that deuterium has about twice the mass of hydrogen.

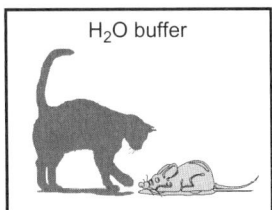
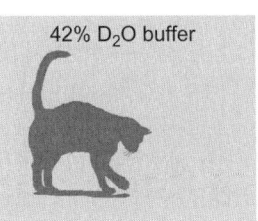
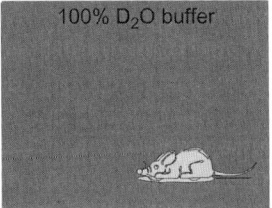

FIG. 7. Contrast variation in neutron-scattering experiments from a protein complex with selective deuteration of one subunit. (A) In $H_2O$ buffer solution, both the deuterated cat and hydrogenated mouse scatter neutrons. (B) In 42% $D_2O$ at the contrast match point of hydrogenated component (hydrogenated mouse), only the deuterated component (cat) scatters neutron. (C) In 100% $D_2O$ buffer and at the match point deuterated component (cat), only the hydrogenated component (mouse) scatter neutron. The deuterated component (cat) is invisible to neutrons.

So why were neutrons not used more extensively? The answer is largely historical (Higgins and Benoit, 1994). The neutron was first discovered by Chadwick in 1932, and the first nuclear reactors were built during the 1940s and 1950s. However, the first scientists to use these reactors for research were solid state physicists employed at the reactor centers themselves. The high-flux reactors needed for polymer science were designed and built in the USA and France during the late 1960s, and slowly became available (and of interest) to polymer scientists. Today, there are a number of research reactors and neutron scattering centers worldwide. There are a number of spectrometric techniques, such as time-of-flight and backscattering, in addition to NSE spectroscopy. NSE is possibly the most powerful technique available today for the study of protein dynamics. (It is also the rarest, as there are only a handful of NSE spectrometers in the world!).

*The development of theoretical techniques for protein dynamics.* Concomitant with the rise of experimental techniques for the study of polymer dynamics, the theoretical technology need to understand the experimental results began to emerge. As will be explicated below, the theory behind protein dynamics emerges as a natural (if significantly more complex) extension of the theoretical framework developed for understanding polymer dynamics. Readers interested in a detailed look at this fascinating field may find (Berne and Pecora, 1976; Doi and Edwards, 1986) useful for further study. We present here only a summary of the basic ideas. We begin by considering the interpretation of experimental results from NSE spectroscopy. Although it is perhaps the most difficult experimental technique to understand, the results lend themselves to easy interpretation.

NSE measures the intermediate scattering function $I(Q,t)$, which is the spatial Fourier transformation of the space–time van Hove correlation function $G(r,t)$ (Mezei, 1980), $I(Q,t) = \int_V G(r,t)\exp(-iQ \cdot r)dr$, where $Q = (4\pi \sin(\theta/2))/\lambda$ is the magnitude of the scattering vector with $\theta$ the scattering angle, $\lambda$ the wavelength of the neutron, $t$ is the time, and $r$ is the position of a scattering center. (The designation "intermediate" arises precisely because only one of the variables of $G(r,t)$ is Fourier transformed.) Like the static SAXS, in the low $Q$ region, $I(Q,t)$ is dominated by coherent scattering that yields the cross-correlation $G(r,t)$, that is, the probability of finding a nucleus at position $r_i$ at time $t=0$ and finding another nucleus at position $r_j$ at time $t$.

For a given $Q$, $I(Q,t)$ typically can be fit to a single exponential in time (and is difficult to fit to more exponentials). A natural way to interpret neutron-scattering data is therefore to examine the effective diffusion constant $D_{\text{eff}}(Q)$ as a function of $Q$, which is determined by the normalized intermediate scattering function $I(Q,t)/I(Q,0)$:

$$\Gamma(Q) = -\lim_{t \to 0} \frac{\partial}{\partial t} \ln[I(Q,t)/I(Q,0)]$$

$$D_{\text{eff}}(Q) = \frac{\Gamma(Q)}{Q^2} \tag{1}$$

where $I(Q,0)$ is the static form factor. As $I(Q,t)/I(Q,0)$ is generally amenable to a single-exponential fit in time (see Fig. 1), $D_{\text{eff}}(Q)$ can be accurately estimated by the first cumulant expression (Bu et al., 2005):

$$D_{\text{eff}}(Q) = \frac{k_B T}{Q^2} \frac{\sum_{jl}\left\langle b_j b_l \left(Q \cdot H_{jl}^T \cdot Q + L_j \cdot H_{jl}^R \cdot L_l\right) e^{iQ\cdot(r_j - r_l)}\right\rangle}{\sum_{jl}\left\langle b_j b_l e^{iQ\cdot(r_j - r_l)}\right\rangle} \tag{2}$$

which is a generalization of the remarkable Akcasu–Gurol (AG) formula (Akcasu and Gurol, 1976) to rotational motion (Bu et al., 2005). Here, $b_j$ is the scattering length of a subunit $j$, $H^T$ is the translational mobility tensor, and $H^R$ is the rotational mobility tensor. The coordinates of the various subunits ("subunits" can be atoms, beads, or domains), taken relative to the center of *friction* of the protein, are given by $\mathbf{r}_j$ (note that $\Sigma \mathbf{r}_j = 0$); $k_B T$ is the usual temperature factor; and $\mathbf{L}_j = \mathbf{r}_j \times \mathbf{Q}$ is the angular momentum vector for each coordinate. The brackets <> denote an orientational average over the vector $\mathbf{Q}$, so that $\langle Q_a Q_b \exp(iQr)\rangle Q^{-2} = (1/3)\delta_{ab} j_0(Qr) + [(1/3)\delta_{ab} - (r_a r_b/r^2)] j_2(Qr)$ can be expressed in terms of the spherical Bessel functions $j$. The translational mobility tensor is illustrated in Fig. 3.

The AG approach described in Eq. (2) is valid for either rigid-bodies or rigid-body subunits connected by soft spring linkers (Bu et al., 2005). The translational mobility tensor $H^T$ is defined by the velocity response $\mathbf{v} = H^T \mathbf{F}$ to an applied force $\mathbf{F}$. The rotational mobility tensor $H^R$ is defined by the angular velocity response $\mathbf{\omega} = H^R \mathbf{\tau}$ to an applied torque $\mathbf{\tau}$. In practice, the structural coordinates of a protein may be obtained from high-resolution crystallography or NMR or from low-resolution EM, SAXS, or SANS. Comparison of the calculations Eq. (2) to experimental $D_{\text{eff}}(Q)$ thus allows one to test models of the mobility tensors. For example, with a completely flexible body, the rotational diffusion term (involving $H^R$) is absent.

(Generally speaking, the rotational mobility tensor arises from the consideration of rigid-body constraints, introduced via Lagrange multipliers or by generalized coordinates (Doi and Edwards, 1986)).

Note that the mobility tensors implicitly have two sets of indices. First, there is the index that indicates the specific subunit under consideration, which we denote with Latin letters ($j, l, \ldots$). There is also a second set of indices which indicate the spatial orientation ($x,y,z$). We will omit this second set of indices for clarity, and use a bold font to indicate vectors. For a rigid-body composed of $N$ identical subunits, the translational mobility tensor $H^T$ is a matrix with $N^2$ identical $3 \times 3$ elements This must be so, since $H^T$ yields the velocity response of for example, subunits B and C to a force applied to subunit A. If the mobility tensor components $H_{AB}$ and $H_{AC}$ are unequal, the velocity response of B and C will be different, B and C will move apart, and the body will no longer remain rigid. Thus, the mobility tensor provides a direct indication of the existence of internal degrees of freedom (Fig. 8).

It is important to stress a key point in Eq. (2). This formula shows that the effective diffusion constant $D_{\text{eff}}(Q)$ can be calculated if we know the structure of the protein and have a model of the mobility tensor.

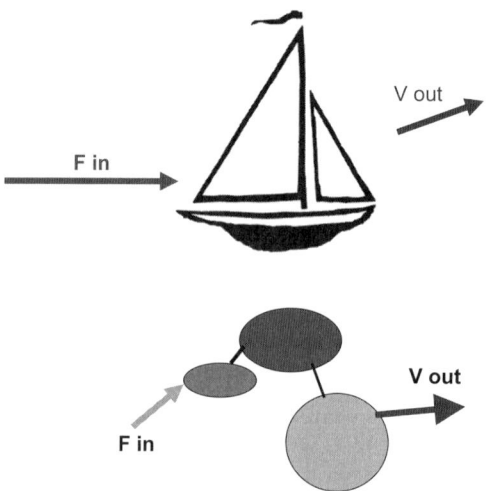

FIG. 8. Translational mobility tensor. The translational mobility tensor gives the velocity response (speed and direction) of a given protein domain to a force applied to itself or to another domain.

In particular, there is no explicit dependence upon force fields or potentials between components of the protein. This is a significant improvement in reliability over complex molecular simulations, which often depend upon numerous parameters that cannot be determined directly. If a complex molecular dynamics simulation yields the wrong answer, it is not immediately obvious how to improve the simulation, as there is no direct connection between experimentally determined quantities and the input parameters.

Our strategy of reducing the NSE data in terms of the AG formula allows a wealth of information to be easily extracted. As another example, note that the diffusion constant for an object is normally defined by the long-time limit of $\langle [r(t) - r(0)]^2 \rangle / 6t \equiv D_\infty$, with $\mathbf{r}(t)$ denoting any vector subunit coordinate of the molecule. Yet, a second definition is commonly used in neutron scattering, that of taking the $Q \rightarrow 0$ limit of the effective diffusion constant, so that $D_{\text{eff}}(Q=0) \equiv D_0$. From the AG result for the first cumulant Eq. (2), we see that for an object composed of $N$ subunits $D_0 = (k_B T \sum_{mn} H_{mn})/N^2$ (note that the contribution of rotational motion to $D_{\text{eff}}$ is zero at $Q=0$ since $\sum_n r_n = 0$). By contrast, an easy result that follows from an elegant paper of Fixman (1983) is that $D_\infty = k_B T/(\sum_{mn} H_{mn}^{-1})$, where $H_{mn}^{-1}$ is the matrix inverse of $H_{mn}$. Generally $D_\infty \leq D_0$, so that the rate of diffusion of a protein will tend to decrease over time as internal modes damp out. The two diffusion constants are equal if and only if a vector whose components are all equal is an eigenvector of $H$—in other words, if an identical force applied to all subunits results in an identical velocity response in each subunit. This is of course simply the statement that the object undergoes exact rigid-body motion, typically requiring permutation symmetry among subunits (like the vertices of a Platonic solid). *Thus, the decrease of D with time can provide another signature of internal motion.*

In contrast to NSE spectroscopy, other neutron-scattering techniques (e.g., time-of-flight or backscattering) have significantly less precision in determining energy transfer (Higgins and Benoit, 1994), and so typically measure only the dynamic form factor $S(Q,\omega) = \int e^{i\omega t} I(Q,t) dt$, which then must be fit to mathematical functions or Fourier transformed numerically. However, in all cases, the important quantity to determine is the effective diffusion constant $D_{\text{eff}}(Q)$ [or equivalently $\Gamma(Q) \equiv Q^2 D(Q)$] as a function of $Q$. This directly yields the desired information about the degree and nature of internal protein dynamics.

These concepts can be illustrated by applying them to study bulk properties of a simple protein. Earlier, we reported a neutron-scattering study of the nanosecond and picosecond dynamics of native and denatured alpha-lactalbumin (Bu et al., 2001). The quasielastic-scattering intensity shows that there are alpha-helical structure and tertiary-like side-chain interactions fluctuating on subnanosecond timescales under extremely denaturing conditions and in the absence of disulfide bonds. The nanosecond dynamics of the native and the denatured proteins was found to have three dynamic regimes (Fig. 9). When $0.05 < Q < 0.5$ Å$^{-1}$ (where the scattering vector, $Q$, is inversely proportional to the length scale), the decay rate, $\Gamma(Q)$, shows a power law relationship, with $\Gamma(Q)$ proportional to $Q^{(2.42\pm0.08)}$, that is analogous to the dynamic behavior of a random coil. However, when $0.5 < Q < 1.0$ Å$^{-1}$, the decay rate exhibits a $\Gamma(Q)$ proportional to $Q^{(1.0\pm0.2)}$ relationship. The effective diffusion constant of the protein therefore decreases with increasing $Q$, a striking dynamic behavior that is not found in any chain-like macromolecule.

These results stand in dramatic contrast to the results obtained in the canonical models of polymer dynamics, which we review here (Doi and Edwards, 1986). The main models of polymer dynamics can be summarized by the Rouse model (Ferrand et al., 1993) and the Zimm model (Fixman, 1983). In the Rouse model, one considers a polymer whose dynamics are given by Brownian motion. The polymer is simply a chain of beads that interact via harmonic oscillator springs, with each bead on the chain interacting only with the beads behind and ahead of it. Hydrodynamic interactions and repulsive interactions between the beads are ignored. The Rouse model can be shown to yield a result for the decay rate $\Gamma(Q) \approx Q^4$ when $Q$ is large enough that internal motion can be seen. The Zimm model is mathematically more elaborate, and also includes a crude approximation to the hydrodynamic contributions to the mobility tensor, as estimated via the Navier–Stokes equation. In the Zimm model, one arrives at the large $Q$ result $\Gamma(Q) \approx Q^3$. (Both of these results can also be easily derived by using scaling arguments originally due to De Gennes (Mi et al., 1994.) Finally, we note that for small $Q$, $\Gamma(Q) \approx Q^2$ (the limit of overall rigid-body diffusion).

We suggested that this unusual internal protein dynamics is due to the presence of a strongly attractive force and collective conformational fluctuations in both the native and the denatured states of the protein. Above $Q > 1.0$ Å$^{-1}$ is a regime that displays the local dynamic behavior of individual

# STRUCTURE AND DYNAMICS OF MEMBRANE-CYTOSKELETON PROTEINS 197

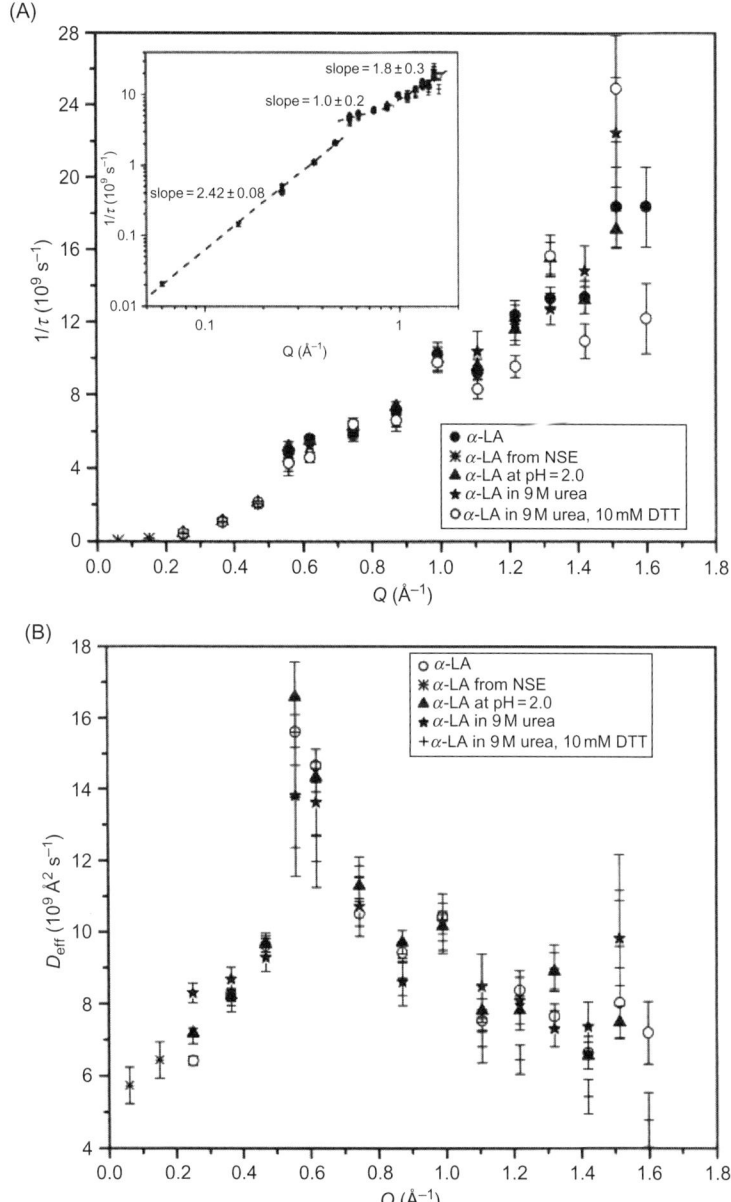

FIG. 9. (A) The nanosecond decay rate (or $1/\tau$) as a function of $Q$. (B) The effective diffusion constant $D_{\text{eff}} = \Gamma/\hbar Q^2$ of the native protein and the denatured proteins as a function of the scattering vector.

residues, with $\Gamma(Q)$ proportional to $Q^{(1.8\pm 0.3)}$ indicating free-body diffusion. Our results provide a dynamic view of the native-like topology established in the early stages of protein folding, and yield a sense of the unusual bulk dynamics properties of proteins. In particular, this result suggests that the conformational entropy in protein matter is significantly lower in an unfolded protein than in a random coil. This dynamic behavior is reminiscent of collective density fluctuations found in fluids, suggesting the existence of strongly nonlocal, attractive forces within proteins.

We now return to the general problem of understanding long-range protein dynamics via Eq. (2). The rotational mobility tensor $H^R$ for the entire protein is derived by evaluating the torque by summing over all subunits $m$:

$$\boldsymbol{\tau} = (H^R)^{-1}\boldsymbol{\omega} = \sum_m (\mathbf{r}_m \times \mathbf{F}_m)$$
$$\mathbf{F}_m = \sum_n (N^2 H^T)^{-1}_{mn} \mathbf{v}_n = \sum_n (N^2 H^T)^{-1}_{mn} (\boldsymbol{\omega} \times \mathbf{r}_n) \quad (3)$$

Equation (3) generally yields the simple estimate $D_{\text{eff}}(Q \to \infty) = 2D_{\text{eff}}(Q=0)$ for *rigid-bodies composed of identical (e.g., nondeuterated) point subunits*. We adopt the simplifying assumption that the three principal spatial components of the translational mobility tensor for each subunit are equal to $ND_0/(k_B T) = 1/\zeta$ with $\zeta$ the friction constant of a subunit and $D_0$ the measured diffusion constant of the protein. Then $\mathbf{F}_n = \zeta \mathbf{v}_n = \zeta(\boldsymbol{\omega} \times \mathbf{r}_n)$. Thus, Eq. (3) yields the rotational mobility tensor $H^R$ via a straightforward inversion of a $3 \times 3$ matrix. A fair estimate of the rigid-body $D_{\text{eff}}(Q)$ measured by NSE can therefore be made using only the coordinates and diffusion constant $D_0$ of the system. We now explain our reasoning.

We take the rigid-body as consisting of a collection of $N$ identical beads. The case of a continuous solid can be reached by an appropriate limit if desired. First, we note that for a rigid-body, the rotational and translational mobility tensors are simply $3 \times 3$ matrices, identical for each bead. This must be so, since otherwise a force applied to a given bead would result in different resultant velocities for other beads, and the body would not remain rigid. Thus, we see that the mobility tensor defines and characterizes internal motion for a body. We also make the simple observation that $\exp(i\mathbf{Q}\mathbf{r})$ is one for all values of the vector $\mathbf{r}$ at zero $\mathbf{Q}$, and that, as $Q$ increases without bound, this quantity is one when $r$ is zero and equals zero otherwise. The remainder of our analysis begins with the Akcasu–Gurol (AG) formula Eq. (2). We shall adopt the convention that the indices that identify a given bead will be labeled with Latin subscripts ($m, n, \ldots$) while spatial indices

($x,y,z$) will be indicated with Greek symbols ($\alpha$, $\beta$, $\gamma$, ...), and are typically omitted for clarity, with vector quantities indicated by bold script. We note that at $Q=0$, the contribution to $D_{\text{eff}}(Q)$ from translational diffusion is $D_{\text{eff}}^{\text{T}}(Q=0) = k_B T(\sum_{mn} H_{mn}^{\text{T}})/N^2$, while at infinite $Q$ this contribution becomes $D_{\text{eff}}^{\text{T}}(Q\to\infty) = k_B T(\sum_{nn} H_{nn}^{\text{T}})/N = k_B T Tr(H^{\text{T}})/N$. We point out for later use that for a rigid-body, $D_{\text{eff}}^{\text{T}}(Q=0) = D_{\text{eff}}^{\text{T}}(Q\to\infty)$ since $H^{\text{T}}$ is independent of $n$. By contrast, $D_{\text{eff}}^{R}(Q)$, the contribution to $D_{\text{eff}}(Q)$ from rotational diffusion, is zero at $Q=0$, since $\sum_n \mathbf{r}_n = 0$ by definition of our coordinate system. Of course, $D_{\text{eff}}(Q) = D_{\text{eff}}^{\text{T}}(Q) + D_{\text{eff}}^{R}(Q)$. The calculation of the contribution of rotational diffusion at large $Q$ is slightly more involved, and will now be considered in detail.

The contributions to rotational diffusion are normally evaluated using the usual methods for systems involving rigid constraints, such as Lagrange multipliers or generalized coordinates (see Doi and Edwards, 1986, Section 3.8). The following course is simpler and suffices here. The angular velocity vector of the rigid object is given by $\boldsymbol{\omega} = H^R \boldsymbol{\tau}$, with the torque $\boldsymbol{\tau} = \sum_n \mathbf{r}_n \times \mathbf{F}_n$. The vector force $\mathbf{F}_n$ on bead $n$ is given in terms of the velocity $\mathbf{v}_n$ and overall angular velocity $\boldsymbol{\omega}$ by Eq. (3). Thus for an arbitrary three-component vector $\boldsymbol{\omega}$ following Eq. (3)

$$\omega = H^R \sum_{mn} r_m \times (N^2 H^{\text{T}})^{-1}_{mn} (\omega \times \mathbf{r}_n) \tag{4}$$

We note that Eq. (4) is of the form $\boldsymbol{\omega} = M\boldsymbol{\omega}$ for an arbitrary vector $\boldsymbol{\omega}$, implying that $M$ is the identity matrix. It immediately follows that the $3 \times 3$ matrix $H^R$ can be evaluated by an inversion of a $3 \times 3$ matrix calculated by summing over bead coordinates $n$. *The rotational mobility tensor is thus entirely determined by the translational mobility tensor.* In the simplified case (adopted in this work), where the $x$, $y$, and $z$ principal components of the translational mobility tensor $H^{\text{T}}$ are all set equal to a friction constant $1/\zeta$ (see discussion surrounding Eq. (3), a compact formula arises in terms of a $3 \times 3$ matrix inverse, arising from the protein diffusion constant $D_0$ (measured by NMR) and the $N$ structural coordinates of the protein (defined so that $\sum_n \mathbf{r}_n = 0$):

$$H^R_{\alpha\beta} = N(D_0/k_B T) \left[ \sum_n \left( \delta_{\alpha\beta} r^2_n - r_{n\alpha} r_{n\beta} \right) \right]^{-1} \tag{5}$$

To complete the proof that $D_{\text{eff}}(Q\to\infty) = 2D_{\text{eff}}(Q=0)$, we now use the fact that $\boldsymbol{\omega}$ is arbitrary, set $\boldsymbol{\omega} = H^{\text{T}} \mathbf{Q}$ in Eq. (4), contract the remaining vector index on the left-hand side with $\mathbf{Q}$, and perform an average $\langle\ldots\rangle$ over the orientation of $\mathbf{Q}$:

$$\begin{aligned}D_{\text{eff}}^{T}(Q\to\infty) &= k_{B}T\sum_{n}\langle\mathbf{Q}H^{T}\mathbf{Q}\rangle/(NQ^{2}) \\ &= k_{B}T\sum_{n}\langle(\mathbf{Q}\times\mathbf{r}_{n})H^{R}(\mathbf{Q}\times\mathbf{r}_{n})\rangle/(NQ^{2}) \\ &= D_{\text{eff}}^{R}(Q\to\infty)\end{aligned} \quad (6)$$

from which we see that at large $Q$ the translational and rotational contributions to the AG formula an identical, so $D_{\text{eff}}(Q=0) = D_{\text{eff}}^{T}(Q=0) = D_{\text{eff}}^{T}(Q\to\infty) = D_{\text{eff}}^{R}(Q\to\infty)$. Since $D_{\text{eff}}(Q) = D_{\text{eff}}^{T}(Q) + D_{\text{eff}}^{R}(Q)$ this completes the proof.

We will also demonstrate that in general one *only* has the bound $D_{\text{eff}}(Q\to\infty) \geq D_{\text{eff}}(Q=0)$. We consider a system with no rigid constraints, so there is only one mobility tensor $H$. We also omit spatial indices in the interests of clarity.

Note that the second law of thermodynamics assures us that the power dissipated by a system is generally nonnegative (Doi and Edwards, 1986, Eq. (3.18)), therefore

$$\sum_{n}\mathbf{v}_{n}\mathbf{F}_{n} = \sum_{mn}\mathbf{F}_{m}H_{mn}\mathbf{F}_{n} \geq 0 \quad (7)$$

for any set of applied forces $\mathbf{F}$, and therefore the mobility tensor is positive semidefinite (has no negative eigenvalues). If we choose $\mathbf{F}$ to have only two nonzero components, $\mathbf{F}_{m}=1$ and $\mathbf{F}_{n}=-1$, we see that Eq. (7) implies that $H$ is dominated by its diagonal elements:

$$H_{mm} + H_{nn} \geq H_{mn} + H_{nm} \quad (8)$$

We now sum over all indices $m$ and $n$, and note that Eq. (8) implies

$$D_{\text{eff}}(Q\to\infty) = k_{B}TTr(H)/N \geq k_{B}T\sum_{mn}H_{mn}/N^{2} = D_{\text{eff}}(Q=0) \quad (9)$$

The inequality approaches the equality $D_{\text{eff}}(Q\to\infty) = D_{\text{eff}}(Q=0)$ when all elements of $H$ are equal (the delicate singular limit of a stiff but still flexible body, discussed in Bu et al. (2005)). In the other extreme limit, when the mobility tensor is entirely diagonal (the limit of noninteracting beads in a flexible system) we have

$$D_{\text{eff}}(Q\to\infty) = ND_{\text{eff}}(Q=0) \quad (10)$$

For a system with an infinite number $N$ of subunits, $D_{\text{eff}}(Q\to\infty)$ thus increases without bound in the case of a diagonal mobility tensor (cf. the

Rouse model of polymers). Thus we see that the uniform rigid body result $D_{\text{eff}}(Q\to\infty) = 2\ D_{\text{eff}}(Q=0)$ is quite unusual. In the above calculation of $H^R$, as well as below, we assume that the $x$, $y$, and $z$ diagonal components of $H^T$ are equal for each submit and are the only nonzero components. In general, of course, both the rotational and translational mobility tensors have different values for each of the three principal axes, so that there are six independent qualities for each domain. In a multidomain complex like NHERF1 bound to the FERM domain of ezrin, there are generally at least 24 independent quantities for the mobility tensor (three translational plus three rotational for each of the four subunits). These quantities are difficult to evaluate to the precision required to compare with NSE data. Programs such as HYDROPRO utilize continuum Navier–Stokes equations to estimate the mobility tensor components from structural coordinates (Garcia De La Torre et al., 2000), but a continuum approximation is insufficiently accurate for our purposes because many structural features of a protein are of the same size as water molecules. By contrast, our simple approach requires neither complicated molecular dynamics simulations nor Navier–Stokes hydrodynamics. The effects of scattering length inhomogeneity can also be neglected as neutron scattering occurs mostly from hydrogen atoms.

For an object with internal domain motion, comparing the calculated $D_{\text{eff}}(Q)$ with data allows one to extract the relative degree of dynamic coupling between the various components of the system, for this dynamic coupling is defined by the mobility tensor. For example, a rigid two-domain system will be described by a mobility tensor

$$H = H_0 \begin{pmatrix} 1 & 1 \\ 1 & 1 \end{pmatrix} \tag{11a}$$

with all elements of the tensor equal, and yields [via Eq. (2)] the simple result that the translational contribution to the effective diffusion constant is given by $D_{\text{eff}}^T(Q) = k_B T H_0$, independent of $Q$. By contrast, a two-domain system with internal motion will possess a mobility tensor

$$H = \begin{pmatrix} H_1 & 0 \\ 0 & H_2 \end{pmatrix} \tag{11b}$$

in principal coordinates. Thus, the application of equal forces to the two domains will result in their having different velocities, revealing internal motion. For the case where there is one internal translational mode between subunits 1 and 2 with $D_1 = k_B T H_1$ and $D_2 = k_B T H_2$ (Bu et al., 2005), the translational contribution to the effective diffusion constant is

$$D_{\text{eff}}^T(Q) = \frac{D_1 S_1(Q) + D_2 S_2(Q)}{S(Q)} \quad (11c)$$

Here, $S_1(Q)$ and $S_2(Q)$ are the form factors of the separate individual protein domains, while $S(Q)$ is the form factor of the entire protein. Orientational averages are performed, so that for example, $S(Q) = \sum j_0(Qr)$; and $S(Q)$ is normalized so that $S(0) = N^2$.

Rotational diffusion will introduce additional contributions to the numerator, as per Eq. (2). The calculations we perform here consist of rigid-body motion (including both translational and rotational motion), and an internal translational mode as per Eqs. (11b) and (11c). We stress that, in principle, it is possible to include the effects of arbitrary translational and rotational internal motion in the calculation (Bu et al., 2005). Therefore, the combination of NSE and first cumulant analysis allows one to test complex models of the mobility tensors of the system, and extract dynamical information about the internal motion of the protein.

### C. Dynamic Propagation of Allosteric Signals by Nanoscale Protein Motion

The virtue of the above simple approach can be seen by comparing the NSE data $D_{\text{eff}}(Q)$ for unbound NHERF1 with the rigid-body calculation using Eqs. (2), (3), and (11a) (Farago et al., 2010). The rigid-body calculation uses as input only the translational diffusion coefficient $D_0$ of NHERF1 obtained from pulsed-field gradient (PFG) NMR, and the "dummy atom" structural coordinates (Svergun, 1999) reconstructed from SAXS (Li et al., 2007a, 2009). Figure 10B shows that the calculated rigid-body $D_{\text{eff}}(Q)$ fits the NSE experimental data quite well, except at high $Q$ where the experimental $D_{\text{eff}}(Q)$ is slightly smaller than the computed values (Farago et al., 2010). This is possibly due to fine structural differences from that represented by the coordinates reconstructed from SAXS. Thus, NHERF1 behaves essentially as a rigid-body on the time and length scales detected by our NSE experiment. The rotational diffusion relaxation time $1/H_0^R$ can be estimated (via Eq. (3)) to be about 1000 ns. The Fourier time-window employed in our NSE experiments is between 0.3 and 200 ns. Thus rotational diffusion is present in the time-window of the NSE experiments.

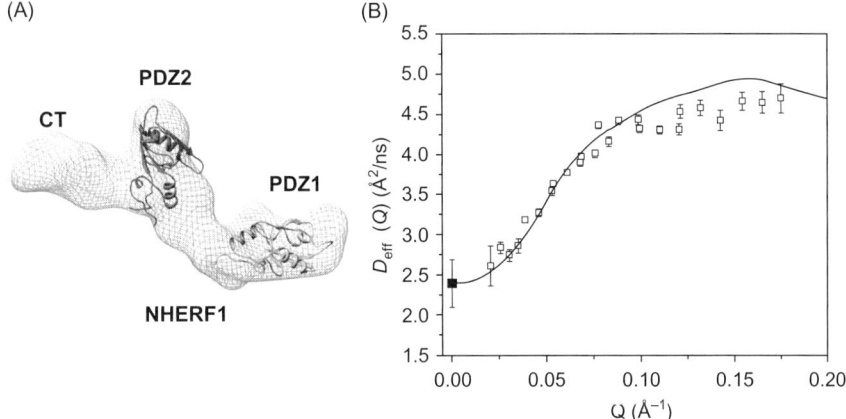

FIG. 10. NHERF1 alone can be described by a rigid-body model. (A) The 3D shape of NHERF1 reconstructed from SAXS (Li et al., 2009) using the *ab initio* program DAMMIN (Svergun, 1999). The known high-resolution structures of the PDZ1 (PDB code: 1I92) and PDZ2 (PDB code: 2KJD) domains are docked into the 3D shape, using UCSF chimera (Pettersen et al., 2004). EBD, which overlaps with the last 13 amino acid residues that interact with PDZ2 is not marked in the graph. (B) Comparing the experimental NSE $D_{eff}(Q)$ of NHERF1 (black open square) with the rigid-body calculation (black solid line). The overall translational diffusion constant $D_0$ (filled black square) at $Q=0$ Å$^{-1}$ is $D_0 = 2.4$ Å$^2$/ns from pulsed-field gradient (PFG) NMR measurements.

As one might expect from the *unbound* NHERF1, the salient features of protein domain motion, as viewed by NSE, can be understood in terms of simple models. We next use this direct approach and construct models of increasing sophistication to demonstrate domain motion in the complex of NHERF1 bound to the FERM domain of ezrin. Because our approach depends on few assumptions, it is subject to less unquantifiable uncertainty than a large scale multiparameter fit or molecular dynamics simulation.

We have performed NSE experiments on two types of complexes of NHERF1 bound to the FERM domain of ezrin. One is the hydrogenated NHERF1 bound to hydrogenated FERM (NHERF1·$^h$FERM), and the other is hydrogenated NHERF1 bound to deuterium labeled FERM (NHERF1·$^d$FERM). We then performed a series of computations of $D_{eff}(Q)$ for both the deuterated and hydrogenated complexes. When calculating $D_{eff}(Q)$ for the NHERF1·$^d$FERM complex, the scattering from the deuterated component is treated as "invisible" in Eq. (2) because the neutron-scattering length density of the deuterated component contrast matches that of the D$_2$O buffer background.

We have previously shown that deuteration does not cause aggregation or conformational changes in the NHERF1·FERM complex (Li et al., 2009). At low Reynolds number, the dynamics of a protein should not depend upon its mass, but rather upon its size (Howard, 2001). Thus, the dynamics of the deuterated complex can be treated as similar to that of the hydrogenated complex. The PFG NMR results show that NHERF1·$^h$FERM and NHERF1·$^d$FERM complexes have very similar translational diffusion constants, in support of this assertion (Fig. 11B). In our calculations, we thus always

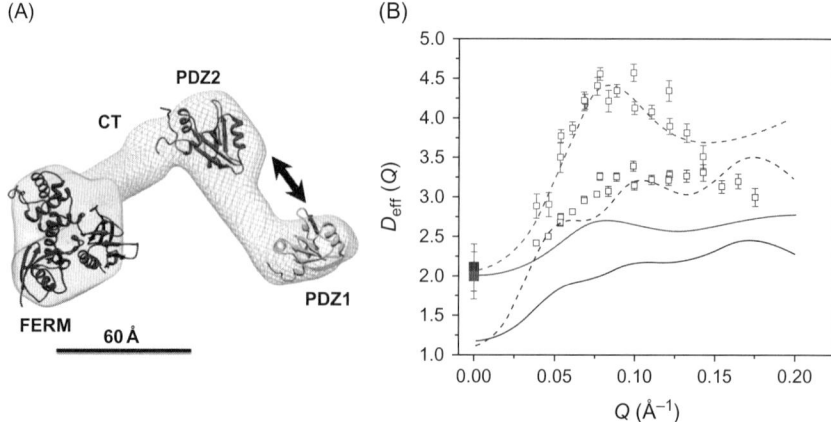

FIG. 11. (A) A model representing domain motion between PDZ1 and PDZ2 in the complex. The 3D shape of the complex is reconstructed from SANS (Li et al., 2009). The known high-resolution structure fragments of PDZ1, PDZ2, and ezrin FERM domain (PDB code: 1NI2) are docked into the envelope using UCSF chimera (Pettersen et al., 2004). The arrows represent translational motion between PDZ1 and PDZ2. A length scale bar of 60 Å is shown. (B) Comparing experimental $D_{eff}(Q)$ of NHERF1·FERM with calculations incorporating interdomain motion between PDZ1 and PDZ2. The calculations were performed using the coordinates of the docked high-resolution structures of different domains. The calculations were performed with only one adjustable parameter $H^T$ for the domain motion model of both the deuterated and the hydrogenated complexes. Open blue square is the NSE data from hydrogenated complex. Open red square is NSE data from hydrogenated NHERF1 in complex with deuterated FERM domain. Solid blue square and solid red square are the self-diffusion constants $D_0$ of the hydrogenated and the deuterated complexes, respectively, obtained from PFG NMR. The calculated curves are shown for NHERF1·$^d$FERM (solid red line for rigid-body model and dashed red line for model incorporating domain motion) and NHERF1·$^h$FERM (solid blue line for rigid-body model and dashed blue line for incorporating domain motion). The comparisons suggest that deuteration of the FERM domain amplifies the effects of protein internal motions detected by NSE. (See color plate 4).

impose the constraint that the dynamics (and therefore the mobility tensors) of the hydrogenated and deuterated components is the same. As we show below, this provides a significant and essential constraint.

Using the same approach applied to NHERF1, we have first performed a rigid-body calculation of $D_{\text{eff}}(Q)$ for the deuterated NHERF1·$^{\text{d}}$FERM complex. The structural coordinates of the complex are the "dummy atoms" (Petoukhov and Svergun, 2006) reconstructed from SANS data (Li et al., 2009) (shown in Fig. 11A), and the one constraint parameter $D_{\text{eff}}(Q=0)$ is the self-diffusion constant $D_0$ for this deuterated complex, taken from PFG NMR. We then use the same approach and parameter to compute the $D_{\text{eff}}(Q)$ of the hydrogenated NHERF1·$^{\text{h}}$FERM complex. As shown in Fig. 11B, the agreement between the experimental data and *rigid-body* calculations is poor for both the NHERF1·$^{\text{d}}$FERM and the NHERF1·$^{\text{h}}$FERM complex.

We argue that the difference between the experimental NSE results and dummy atom rigid-body calculations arises because of the internal motion of the protein. This internal motion produces various effects: the first way in which internal motion manifests itself is through the fact that the mobility tensor associated with a protein with internal motion is different than the mobility tensor for a rigid-body. This was discussed above (see Eqs. (11a)–(11c)).

Second, it is essential to note that the evaluation of Eq. (2) implicitly requires an average over a distribution of particle densities. For the purposes of this calculation, the SAXS/SANS dummy atom structural data may be an accurate representation for a rigid-body, but will be inaccurate for a protein with a significant degree of internal dynamics. This is because *ab initio* programs utilized for shape reconstruction from SANS or SAXS data typically produce an envelope of the calculated structure, in which the density inside the envelope is assumed constant. In a highly mobile object, the reconstructed shape may thus be a poor representation of the fluctuating structures. Thus, for example, if the linker regions between domains are highly mobile, the size of the linker regions may be overestimated, and the dummy atom shape reconstruction will not be a good representation of the entire protein. We speculate that the disagreement between the experimental NSE data and that computed from the SANS reconstructed shape model (Fig. 11B) is partly due to this variation of density within the reconstructed shape.

Thus, in the following, we construct two models incorporating these internal motion effects in order to understand the discrepancy between the calculations and experimental data. We first present a more detailed rigid-body model, in which the known high-resolution structural fragments of the PDZ1, PDZ2CT, and the FERM domains are docked into the 3D shapes reconstructed from SANS (Fig. 11A) using the software package UCSF Chimera (Pettersen et al., 2004). This "docked" model therefore incorporates a crude form of density variation within the complex, by ignoring the density of the linker regions. The mobility tensor for this first model is taken to be that for a rigid-body. It will be seen that this density variation alone does not yield a good comparison with the NSE data. We therefore construct a second "docked" model in which the mobility tensor used is that for interdomain motion between the two PDZ domains. It will be seen that the second model produces a sizeable improvement in explaining the data.

When calculating the $D_{eff}(Q)$ using the docked coordinates, the rigid-body mobility tensor docked calculations again provide poor fits to the NSE data for both the hydrogenated and deuterated complexes. The comparison thus suggests that NHERF1·$^h$FERM and NHERF1·$^d$FERM do not behave as rigid-bodies on the length scales and timescales of the NSE experiments. This observation is supported by our previous SANS and NMR structural studies that find large conformational changes in NHERF1 upon binding to FERM (Li et al., 2009; Bhattacharya et al., 2010). In particular, the region that links PDZ1 and PDZ2 becomes more extended, and the CT region of NHERF1 becomes largely unfolded upon binding to FERM. Thus, structural fluctuations in the complexes can become significant in the complex on the length scales and timescales of the NSE experiments.

We next incorporate interdomain motion in the mobility tensor for the NHERF1·$^d$FERM and NHERF1·$^h$FERM complexes in our calculation. To compute $D_{eff}(Q)$ with domain motion using Eqs. (2)–(4), we use the coordinates of the docked model (Fig. 11A), and assume translational interdomain motion between PDZ1 and PDZ2 (Eq. (11b)). Here we perform the calculation, as always, with only one adjustable parameter, the translational diffusion constant for each atom. Again, this parameter is adjusted so that $D_{eff}(Q)$ for the deuterated complex agrees with the value measured by PFG NMR. As with the above calculations, we use this same parameter for both the deuterated and hydrogenated complexes. When performing the calculations, the rotational contributions to diffusion (Eq. (3)) are taken to be the same as for the rigid-body docked model calculation. The difference with the previous docked calculation is thus solely that in this second docked model calculation we employ the mobility tensor for a protein with an internal mode between the PDZ1 and PDZ2 domains, rather than a rigid-body mobility tensor. After incorporating interdomain motion between PDZ1 and PDZ2 in the NHERF1·$^d$FERM complex, the calculated $D_{eff}(Q)$ with internal motion agrees well the NSE results (Fig. 11B). In particular, the docked calculation with the internal mode mobility tensor generates a peak at a $Q$ value of 0.07 Å$^{-1}$ in Fig. 3D, which agrees well with the NSE results. For the NHERF1·$^h$FERM complex, there is also better agreement between the experimental data and the calculation after incorporating interdomain motion between PDZ1 and PDZ2 (Fig. 11B). Nevertheless, we note that, for the docked NHERF1·$^h$FERM complex, the computed $D_0$ at $Q=0$ is not close to the experimental values from PFG NMR measurement. As pointed out above, we attribute this discrepancy to large conformational fluctuations in the CT–FERM region caused by the unfolding of the CT domain upon binding to FERM (Bhattacharya et al., 2010), which cannot be represented by a single reconstructed SANS structure shown in Fig. 3A. Such complications are minimal in the NHERF1·$^d$FERM complex because the deuterated $^d$FERM is "invisible" to neutrons. Future experiments could use selective deuteration of other portions of the complex in order to highlight the motions of PDZ2–CT–FERM domains for NSE study.

### D. A Simple Four-Point Model Describes Domain Motion

The simple calculations we presented above require only the structural coordinates and a single constraint (the diffusion constant at $Q=0$ Å$^{-1}$ for the deuterated complex, measured independently by PFG NMR) to generate the computed $D_{eff}(Q)$. It is possible to argue however that the structural coordinates are insufficiently accurate to explain the NSE data, or that some coincidental artifact produces the peak at 0.07 Å$^{-1}$ that implies internal motion. We therefore introduce an even more simplified model that yields the same effect, and serves to explain its origin. The simplified model is taken by extracting four points that represent the coordinates of the center-of-mass of domains obtained from the SANS data of the NHERF1·FERM complex. These points form a triangle model as shown in Fig. 12A with the distances FERM–PDZ2 = 80 Å, PDZ2–PDZ1 = 59 Å, and FERM–PDZ1 = 110 Å. The CT domain is taken as being halfway between the FERM and PDZ2 domains (Fig. 12A). We include the point representing the FERM domain with a weight factor of 3 to account for its larger size relative to the other

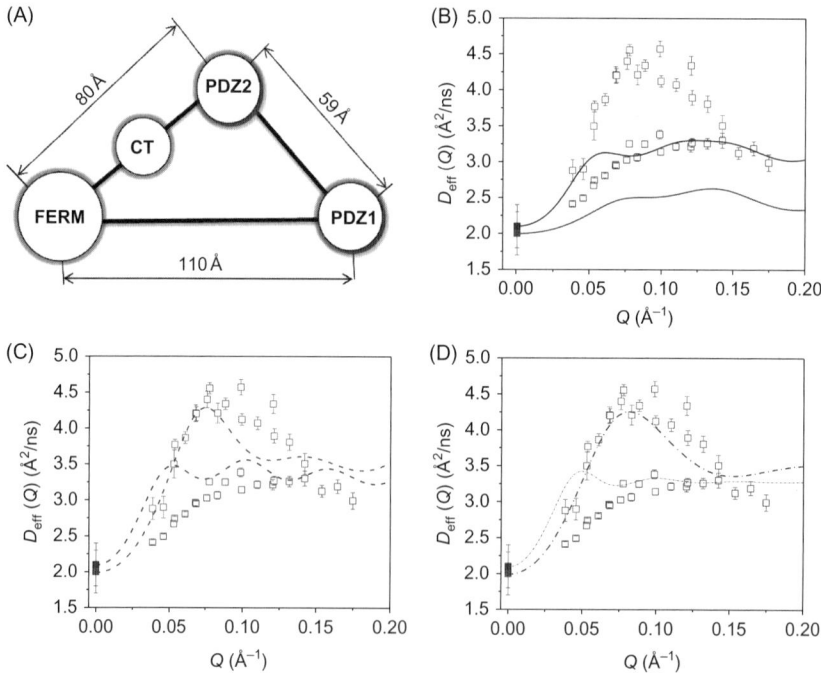

FIG. 12. A simple four-point model can well describe domain motion in the complex. (A) The four-point model represents the NHERF1·FERM complex, with the centers of PDZ1, PDZ2, CT, and FERM domain taken from Fig. 3A. (B) Comparing the experimental NSE data with the four-point rigid-body calculations for NHERF1·$^h$FERM (blue open squares are the experimental data and blue solid line is the calculated data) and for NHERF1·$^d$FERM (red open squares are experimental data and red solid line is the calculated data). $D_0$ of NHERF1·$^d$FERM (solid red squares) and NHERF1·$^d$FERM (solid blue squares) from PFG NMR are shown. (C) Comparing the experimental data with calculations assuming interdomain motion between PDZ1 and PDZ2 in NHERF1·$^d$FERM (red dash line) and NHERF1·$^h$FERM (blue dash line). The experimental symbols are the same as in (B). (D) Comparing the experimental data with calculations incorporating interdomain motion between PDZ1 and PDZ2, as well as assuming finite size form factor of spheres of 20 Å radius for the FERM domain and for both PDZ domains in NHERF1·$^d$FERM (red dash dot line) and in NHERF1·$^h$FERM (blue dash dot line). (See color plate 5).

domains. Because it is possible to obtain the center-of-mass distances between the domains with confidence even with low resolution SAXS or SANS data, this model possesses fewer uncertainties than a model based upon the molecular shape.

We first present the calculation of the four-point model representing the deuterated NHERF1·$^{\text{d}}$FERM and hydrogenated NHERF1·$^{\text{h}}$FERM complexes. The calculation has one adjustable parameter, the domain translational diffusion constant $D_{\text{domain}}^{\text{T}}$, which is chosen to yield the correct value for the diffusion constant $D_0$ of the deuterated complex (as measured by PFG NMR). For the model with internal motion, the domain diffusion constant is taken the same for the FERM, the CT, and the two PDZ domains. We use $D_{\text{domain}}^{\text{T}} = 2.9 \times 10^{-7}$ cm$^2$/s for the domain diffusion constant for both the deuterated and hydrogenated systems. Note that the diffusion constant $D^{\text{T}}$ for the individual domains is larger than that for the complex, as expected. Interestingly enough, the diffusion constant for the hydrogenated complex ($2.1 \times 10^{-7}$ cm$^2$/s) is estimated correctly by this procedure, and is thus an output. The rotational diffusion constant is then estimated using the Stokes formula for a sphere $D_{\text{domain}}^{\text{R}} = (3/4) D_{\text{domain}}^{\text{T}} / R_{\text{S}}^2$, with the Stokes–Einstein radius $R_{\text{S}} = k_{\text{B}} T / (6 \pi \eta D_{\text{domain}}^{\text{T}})$ and is taken to be identical for all domains. Such an estimate has been shown to be valid for a number of proteins (Yao et al., 2008).

In our four-point calculation, we assume that the PDZ1 domain is a separate subunit, and so there is a degree of internal motion in the protein, appearing as *translational* mode between PDZ1 and PDZ2. The translational mobility tensor for the PDZ1 domain is thus a simple constant, while the FERM, CT, and PDZ2 domains are treated as rigid subunits and thus their translational mobility tensor is a $3 \times 3$ matrix (whose *xyz* elements are all equal). The rotational mobility tensor is a $4 \times 4$ matrix, taken as the same as a rigid-body. Thus, we see that the topological dynamic connectivity of the mobility tensors defines the $Q$ dependence of the effective diffusion constant, while their numerical values largely determine only its overall scale. It is, of course, the connectivity that defines the degree and nature of protein internal motion.

Figure 4B compares the experimental NSE data with the calculated $D_{\text{eff}}(Q)$ from the rigid four-point model for the hydrogenated and the partially deuterated complexes. Figure 12C is the $D_{\text{eff}}(Q)$ of the four-point model incorporating internal domain motion between PDZ1 and the rest of the complex. After incorporating internal motion, the overall $D_{\text{eff}}(Q)$ from the four-point model agrees well with the experimental data for both the partially deuterated and the hydrogenated complexes. There are however some oscillations remaining, for we have approximated the domains as point objects.

The comparison between calculation and experimental data improves considerably after including the form factor of a 20 Å radius sphere for the FERM domain and both PDZ domains in the calculation (Fig. 12D). Thus, the NSE data is better represented by the four-point model that includes PDZ1–PDZ2 oscillatory motion than by a model that assumes the complex as a rigid-body. Further improvement likely requires the use of methods of evaluating the mobility tensors for proteins with high accuracy.

Moreover, from the four-point model calculations, we note that $D_{\text{eff}}(Q)$ for the hydrogenated rigid complex and the hydrogenated complex with internal motion are nearly indistinguishable (Fig. 13A). For the deuterated complex, $D_{\text{eff}}(Q)$ obtained from the interdomain motion model is

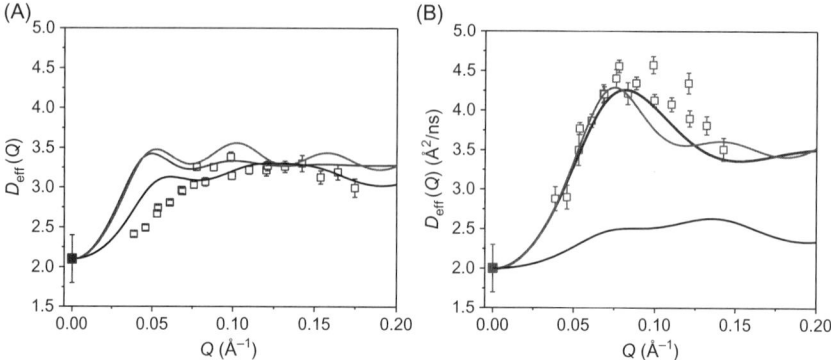

FIG. 13. For the hydrogenated NHERF1·$^{\text{h}}$FERM complex, the difference in $D_{\text{eff}}(Q)$ between the rigid-body model and domain-motion models is very small, but is significantly increased in the deuterated complex. (A) Comparing the rigid-body calculation with the domain-motion calculation in the four-point model in the hydrogenated NHERF1·$^{\text{h}}$FERM complex. NSE data from the NHERF1·$^{\text{h}}$FERM (blue open squares), the four-point rigid-body model (black line), four-point model incorporating domain motion between PDZ1 and PDZ2 (red line), four-point model incorporating domain motion between PDZ1 and PDZ2 and finite size form factor of 20 Å radius for the FERM domain, PDZ1 and PDZ2 (blue line). $D_0$ at $Q = 0$ Å$^{-1}$ as measured from PFG NMR is shown in blue solid square. (B) Comparing the rigid-body calculation with the domain-motion calculation in the four-point model in the deuterated NHERF1·$^{\text{d}}$FERM complex. NSE data from the NHERF1·$^{\text{d}}$FERM (red open squares), the four-point rigid-body model (black line), four-point model incorporating domain motion between PDZ1 and PDZ2 (red line), four-point model incorporating domain motion between PDZ1 and PDZ2 and finite size form factor of 20 Å radius for the FERM domain, PDZ1 and PDZ2 (blue line). $D_0$ at $Q = 0$ Å$^{-1}$ as measured from PFG NMR is shown in red solid square. (See color plate 6).

significantly different from that of the rigid-body model (Fig. 13B). This can be explained as due to the relatively large contribution to Eq. (2) of the effects of rotational diffusion of the overall object, which dominates and obscures the effects of internal motion when no deuteration is performed. For the partially deuterated complex, both the docked domain calculation (Fig. 11) and the four-point model (Fig. 12B–D) show that $D_{\text{eff}}(Q)$ of the rigid-body complex is significantly different from that of the complex with internal domain motion. These analyses demonstrate that deuterium labeling of a domain in a protein or in a protein complex significantly amplifies the effects of internal motion detected by NSE. Our model calculations (Fig. 13A and B) suggest that deuterium labeling a domain can mask a portion of the form factor, and, as a result, highlight the contribution of the terms of internal domain motion in Eq. (2). Thus, we propose that future NSE experiments will benefit by utilizing the strategy of selective deuteration to highlight the domain motion of interest.

Multidomain proteins and protein complexes are complicated systems containing thousands of atoms, and providing precise answers to questions can be done, if at all, by large-scale simulations. We demonstrate in this chapter that the NSE results can be explained by a detailed docked model, as well by a highly simplified four-point model that is independent of the details of the structural model. By doing so, we have systematically reduced the relevant assumptions, making the results of our calculations progressively more certain.

### E. The Importance of NSE and a Plan for the Future

Protein motion plays several fundamental roles in protein function, from transmitting the flow of energy and allosteric signals to shuttling a protein via biased routes on the energy landscape for folding and catalysis (Miyashita et al., 2003). The timescales of protein motion span from picosecond to seconds, and the length scales range from local Angstrom motion to nanometer global motion (McCammon, 1984). Understanding nanoscale protein motions is essential, for thermal fluctuations on fast timescales, such as on picosecond to nanosecond timescales, ultimately inspire and dictate the kinetics of large conformational changes necessary for protein function (Bu et al., 2000, 2001, 2005; Lindorff-Larsen et al., 2005; Ishikawa et al., 2008).

The genomic sequence reveals that number of genes, which encode the synthesis of proteins, is far less than the number of proteins in high eukaryotes. In order to make the large number of different proteins that are required to perform diverse cellular functions, one may mix-and-match modular domains in a single polypeptide to make new proteins. As a result, a large number of cellular proteins are multidomain proteins. Protein motions in multidomain proteins, on nanoscales comparable to their overall dimension, are indispensible for relaying signals allosterically in the cellular networks (Ma and Nussinov, 2009). This concept is emerging as a powerful theme in cell signaling. Although we know much about the structure and function of individual modules that composes adapters and scaffolding proteins, their overall dynamic architecture, and in particular the long-range motion properties of scaffolding proteins that control the function of assembled macromolecular signaling complexes, remains a largely unexplored territory.

NHERF1 is a multidomain scaffolding protein that assembles membrane protein clusters, regulate the dynamic trafficking of receptors and ion channels, and organize protein–protein interactions that influence multiple cell signaling pathways. As we proposed in an earlier study, the allosteric regulation of NHERF1 by ezrin to assemble membrane protein complexes could provide a means to effectively control the strength and duration of signaling at the membrane–cytoskeleton (Li et al., 2009). Besides propagating allosteric signals, the long-range "activated" interdomain motions in NHERF1 may serve other functional roles during the assembly of macromolecular complexes. The activated interdomain motion may allow the PDZ domains to sample certain conformational space and to search the target membrane proteins effectively (Windisch et al., 2006).

Using NSE, we have found that the long-range domain motions in the adapter proteins, which are in the nanospatial–temporal regime, relay signals between the F-actin cytoskeletal network and the cell membranes, and exercise control of the assembly of protein complexes in the cell membrane. Our NSE study demonstrates the activation of long-range coupled domain motion on submicrosecond and on nanometer length scales, which influences the long-range allosteric couplings of the different functional domains for binding to target proteins. Remarkably, the changes in protein domain motion are associated with propagating allosteric signals from a binding site to a remote domain that is a distance of 110 Å away.

Protein motion on nanoscales is, at best, difficult to observe by other experimental techniques. The deuterium labeling approach and the theoretical analyses that we presented therefore should pave the way for using NSE to study protein motions in multidomain proteins. We expect NSE to fill an important nanoscale spatial–temporal gap in our ability to characterize protein motion and function.

## VI. Summary and Perspective

Allosteric transduction of cellular signals by multiple protein–protein interactions has emerged as an important theme to elucidate the mechanisms of hierarchy signaling pathways and networks. Nanoscale protein domain motions on length scales comparable to protein dimensions hold the key to understanding how signals are relayed through multiple protein–protein interactions. We have explained in this review our view of the present state of protein dynamics, and described our view of the future of this essential field. Protein dynamics (and, in particular, long-range allostery) presents us with a unique paradigm for cell-signaling: the idea that proteins can communicate *within themselves* to effect long-range information transfer. This idea has not been explored, in large part because of a paucity of experimental techniques that can address the necessary questions. We have shown how this burgeoning field can be developed by utilizing NSE spectroscopy to demonstrate long-range coupled protein domain motion. It is essential to recognize the absolute need for interdisciplinary approaches to the study of this very complex problem. For example, one must utilize the concept of a mobility tensor, which in turn is derived from nonequilibrium statistical mechanics, a field which, in itself, is challenging and little explored. We believe that it is impossible not to be excited about the challenges ahead, and the rewards for their successful solution. Nanoscale protein dynamics could hold the key to manipulate protein–protein interactions that comprise the cellular signaling network for therapeutic intervention.

### Acknowledgment

This work is supported by National Institutes of Health Grants R01 HL086496.

# References

Akcasu, Z., Gurol, H. (1976). Quasi-elastic scattering by dilute polymer-solutions. *J. Polym. Sci., B: Polym. Phys.* **14**, 1–10.

Bates, I. R., Hébert, B., Luo, Y., Liao, J., Bachir, A. I., Kolin, D. L., et al. (2006). Membrane lateral diffusion and capture of CFTR within transient confinement zones. *Biophys. J.* **91**, 1046–1058.

Benkovic, S. J., Hammes-Schiffer, S. (2003). A perspective on enzyme catalysis. *Science* **301**, 1196–1202.

Berne, B. J., Pecora, R. (1976). Dynamic Light Scattering with Applications to Chemistry, Biology and Physics. Dover Publications, Inc., Mineola, New York.

Bhattacharya, S., Dai, Z., Li, J., Baxter, S., Callaway, D. J. E., Cowburn, D., et al. (2010). A conformational switch in the scaffolding protein NHERF1 controls autoinhibition and complex formation. *J. Biol. Chem.* **285**, 9981–9994.

Bossard, F., Robay, A., Toumaniantz, G., Dahimene, S., Becq, F., Merot, J., et al. (2007). NHE-RF1 protein rescues {Delta}F508-CFTR function. *Am. J. Physiol. Lung Cell. Mol. Physiol.* **292**, L1085–L1094.

Bretscher, A., Chambers, D., Nguyen, R., Reczek, D. (2000). ERM-merlin and EBP50 protein families in plasma membrane organization and function. *Annu. Rev. Cell Dev. Biol.* **16**, 113–143.

Bretscher, A., Edwards, K., Fehon, R. G. (2002). ERM proteins and merlin: integrators at the cell cortex. *Nat. Rev. Mol. Cell Biol.* **3**, 586–599.

Bu, Z., Biehl, R., Monkenbusch, M., Richter, D., Callaway, D. J. (2005). Coupled protein domain motion in Taq polymerase revealed by neutron spin-echo spectroscopy. *Proc. Natl. Acad. Sci. USA* **102**, 17646–17651.

Bu, Z., Cook, J., Callaway, D. J. (2001). Dynamic regimes and correlated structural dynamics in native and denatured alpha-lactalbumin. *J. Mol. Biol.* **312**, 865–873.

Bu, Z., Engelman, D. M. (1999). A method for determining transmembrane helix association and orientation in detergent micelles using small angle X-ray scattering. *Biophys. J.* **77**, 1064–1073.

Bu, Z., Neumann, D. A., Lee, S. H., Brown, C. M., Engelman, D. M., Han, C. C. (2000). A view of dynamics changes in the molten globule-native folding step by quasielastic neutron scattering. *J. Mol. Biol.* **301**, 525–536.

Bu, Z. M., Perlo, A., Johnson, G. E., Olack, G., Engelman, D. M., Wyckoff, H. W. (1998). A small-angle X-ray scattering apparatus for studying biological macromolecules in solution. *J. Appl. Crystallogr.* **31**, 533–543.

Cao, T. T., Deacon, H. W., Reczek, D., Bretscher, A., von Zastrow, M. (1999). A kinase-regulated PDZ-domain interaction controls endocytic sorting of the beta2-adrenergic receptor. *Nature* **401**, 286–290.

Chung, I., Akita, R., Vandlen, R., Toomre, D., Schlessinger, J., Mellman, I. (2010). Spatial control of EGF receptor activation by reversible dimerization on living cells. *Nature* **464**, 783–787.

Cooper, A., Dryden, D. T. F. (1984). Allostery without conformational change, a plausible model. *Eur. Biophys. J.* **11**, 103–109.

Creighton, T. E. (1993). Proteins: Structures and Molecular Properties. Freeman and Company, New York.

Cunningham, R. E. X., Steplock, D., Shenolikar, S., Weinman, E. J. (2005). Defective PTH regulation of sodium-dependent phosphate transport in NHERF-1$^{-/-}$ renal proximal tubule cells and wild-type cells adapted to low-phosphate media. *Am. J. Physiol. Ren. Physiol.* **289**, F933–F938.

Curto, M., McClatchey, A. I. (2004). Ezrin…a metastatic detERMinant? *Cancer Cell* **5**, 113–114.

Denker, S. P., Barber, D. L. (2002). Ion transport proteins anchor and regulate the cytoskeleton. *Curr. Opin. Cell Biol.* **14**, 214–220.

Doi, M., Edwards, S. F. (1986). The Theory of Polymer Dynamics. Oxford University Press, Oxford.

Donowitz, M., Cha, B., Zachos, N. C., Brett, C. L., Sharma, A., Tse, C. M., et al. (2005). NHERF family and NHE3 regulation. *J. Physiol.* **567**, 3–11.

Doyle, D. A., Lee, A., Lewis, J., Kim, E., Sheng, M., MacKinnon, R. (1996). Crystal structures of a complexed and peptide-free membrane protein-binding domain: molecular basis of peptide recognition by PDZ. *Cell* **85**, 1067–1076.

Eisenmesser, E. Z., Millet, O., Labeikovsky, W., Korzhnev, D. M., Wolf-Watz, M., Bosco, D. A., et al. (2005). Intrinsic dynamics of an enzyme underlies catalysis. *Nature* **438**, 117–121.

Elliott, B. E., Meens, J. A., SenGupta, S. K., Louvard, D., Arpin, M. (2005). The membrane cytoskeletal crosslinker ezrin is required for metastasis of breast carcinoma cells. *Breast Cancer Res.* **7**, R365–R373.

Farago, B., Li, J., Cornilescu, G., Callaway, D. J., Bu, Z. (2010). Activation of nanoscale allosteric protein domain motion revealed by neutron spin echo spectroscopy. *Biophys. J.* **99**, 3473–3482.

Favia, M., Guerra, L., Fanelli, T., Cardone, R. A., Monterisi, S., Di Sole, F., et al. (2009). Na$^+$/H$^+$ exchanger regulatory factor 1 overexpression-dependent increase of cytoskeleton organization is fundamental in the rescue of F508del cystic fibrosis transmembrane conductance regulator in human airway CFBE41o-cells. *Mol. Biol. Cell* **21**, 73–86.

Fehon, R. (2006). Cell biology: polarity bites. *Nature* **442**, 519–520.

Fehon, R. G., McClatchey, A. I., Bretscher, A. (2010). Organizing the cell cortex: the role of ERM proteins. *Nat. Rev. Mol. Cell Biol.* **11**, 276–287.

Ferrand, M., Dianoux, A. J., Petry, W., Zaccaï, G. (1993). Thermal motions and function of bacteriorhodopsin in purple membranes: effects of temperature and hydration studied by neutron scattering. *Proc. Natl. Acad. Sci. USA* **90**, 9668–9672.

Fersht, A. R. (1998). Structure and Mechanism in Protein Science: a Guide to Enzyme Catalysis and Protein Folding. W.H. Freeman and Company, New York.

Fievet, B. T., Gautreau, A., Roy, C., Del Maestro, L., Mangeat, P., Louvard, D., et al. (2004). Phosphoinositide binding and phosphorylation act sequentially in the activation mechanism of ezrin. *J. Cell Biol.* **164**, 653–659.

Fievet, B., Louvard, D., Arpin, M. (2007). ERM proteins in epithelial cell organization and functions. *Biochim. Biophys. Acta Mol. Cell Res.* **1773**, 653–660.

Fixman, M. (1983). Variational bounds for polymer transport coefficients. *J. Chem. Phys.* **78**, 1588.

Garbett, D., LaLonde, D. P., Bretscher, A. (2010). The scaffolding protein EBP50 regulates microvillar assembly in a phosphorylation-dependent manner. *J. Cell Biol.* **191**, 397–413.

Garcia De La Torre, J., Huertas, M. L., Carrasco, B. (2000). Calculation of hydrodynamic properties of globular proteins from their atomic-level structure. *Biophys. J.* **78**, 719–730.

Gary, R., Bretscher, A. (1995). Ezrin self-association involves binding of an N-terminal domain to a normally masked C-terminal domain that includes the F-actin binding site. *Mol. Biol. Cell* **6**, 1061–1075.

Gautreau, A., Louvard, D., Arpin, M. (2002). ERM proteins and NF2 tumor suppressor: the Yin and Yang of cortical actin organization and cell growth signaling. *Curr. Opin. Cell Biol.* **14**, 104–109.

Gerstein, M., Lesk, A. M., Chothia, C. (1994). Structural mechanisms for domain movements in proteins. *Biochemistry* **33**, 6739–6749.

Gisler, S. M., Stagljar, I., Traebert, M., Bacic, D., Biber, J., Murer, H. (2001). Interaction of the Type IIa Na/Pi cotransporter with PDZ proteins. *J. Biol. Chem.* **276**, 9206–9213.

Guerra, L., Fanelli, T., Favia, M., Riccardi, S. M., Busco, G., Cardone, R. A., et al. (2005). $Na^+/H^+$ exchanger regulatory factor isoform 1 overexpression modulates cystic fibrosis transmembrane conductance regulator (CFTR) expression and activity in human airway 16HBE14o-cells and rescues {Delta}F508 CFTR functional expression in cystic fibrosis cells. *J. Biol. Chem.* **280**, 40925–40933.

Haggie, P. M., Kim, J. K., Lukacs, G. L., Verkman, A. S. (2006). Tracking of quantum-dot-labeled CFTR shows immobilization by C-terminal PDZ interactions. *Mol. Biol. Cell* **17**, 4937–4945.

Haggie, P. M., Stanton, B. A., Verkman, A. S. (2004). Increased diffusional mobility of CFTR at the plasma membrane after deletion of its C-terminal PDZ binding motif. *J. Biol. Chem.* **279**, 5494–5500.

Hall, R. A., Ostedgaard, L. S., Premont, R. T., Blitzer, J. T., Rahman, N., Welsh, M. J., et al. (1998). A C-terminal motif found in the beta2-adrenergic receptor, P2Y1 receptor and cystic fibrosis transmembrane conductance regulator determines binding to the $Na^+/H^+$ exchanger regulatory factor family of PDZ proteins. *Proc. Natl. Acad. Sci. USA* **95**, 8496–8501.

Hall, R. A., Premont, R. T., Chow, C. W., Blitzer, J. T., Pitcher, J. A., Claing, A., et al. (1998). The beta2-adrenergic receptor interacts with the $Na^+/H^+$-exchanger regulatory factor to control $Na^+/H^+$ exchange. *Nature* **392**, 626–630.

Harris, B. Z., Lim, W. A. (2001). Mechanism and role of PDZ domains in signaling complex assembly. *J. Cell Sci.* **114**, 3219–3231.

Hawkins, R. J., McLeish, T. C. B. (2004). Coarse-grained model of entropic allostery. *Phys. Rev. Lett.* **93**, 098104.

Hernando, N., Deliot, N., Gisler, S. M., Lederer, E., Weinman, E. J., Biber, J., et al. (2002). PDZ–domain interactions and apical expression of type IIa Na/P(i) cotransporters. *Proc. Natl. Acad. Sci. USA* **99**, 11957–11962.

Higgins, J. S., Benoit, H. C. (1994). Polymers and Neutron Scattering. Clarendon Press, Oxford.

Howard, J. (2001). Mechanics of Motor Proteins and the Cytoskeleton. Sinauer Associates, Sunderland, MA.

Ishikawa, H., Kwak, K., Chung, J. K., Kim, S., Fayer, M. D. (2008). Direct observation of fast protein conformational switching. *Proc. Natl. Acad. Sci. USA* **105**, 8619–8624.

James, M. F., Beauchamp, R. L., Manchanda, N., Kazlauskas, A., Ramesh, V. (2004). A NHERF binding site links the betaPDGFR to the cytoskeleton and regulates cell spreading and migration. *J. Cell Sci.* **117**, 2951–2961.

Karim, Z., Gerard, B., Bakouh, N., Alili, R., Leroy, C., Beck, L., et al. (2008). NHERF1 mutations and responsiveness of renal parathyroid hormone. *N. Engl. J. Med.* **359**, 1128–1135.

Karthikeyan, S., Leung, T., Birrane, G., Webster, G., Ladias, J. A. (2001). Crystal structure of the PDZ1 domain of human Na(+)/H(+) exchanger regulatory factor provides insights into the mechanism of carboxyl-terminal leucine recognition by class I PDZ domains. *J. Mol. Biol.* **308**, 963–973.

Karthikeyan, S., Leung, T., Ladias, J. A. (2002). Structural determinants of the $Na^+/H^+$ exchanger regulatory factor interaction with the beta 2 adrenergic and platelet-derived growth factor receptors. *J. Biol. Chem.* **277**, 18973–18978.

Kerjaschki, D., Sharkey, D. J., Farquhar, M. G. (1984). Identification and characterization of podocalyxin—the major sialoprotein of the renal glomerular epithelial cell. *J. Cell Biol.* 1591–1596.

Kern, D., Zuiderweg, E. R. (2003). The role of dynamics in allosteric regulation. *Curr. Opin. Struct. Biol.* **13**, 748–757.

Khanna, C., Wan, X., Bose, S., Cassaday, R., Olomu, O., Mendoza, A., et al. (2004). The membrane–cytoskeleton linker ezrin is necessary for osteosarcoma metastasis. *Nat. Med.* **10**, 182–186.

Ko, S. B. H., Zeng, W., Dorwart, M. R., Luo, X., Kim, K. H., Millen, L., et al. (2004). Gating of CFTR by the STAS domain of SLC26 transporters. *Nat. Cell Biol.* **6**, 343–350.

Kwon, S. H., Pollard, H., Guggino, W. B. (2007). Knockdown of NHERF1 enhances degradation of temperature rescued DeltaF508 CFTR from the cell surface of human airway cells. *Cell. Physiol. Biochem.* **20**, 763–772.

LaLonde, D. P., Bretscher, A. (2009). The scaffold protein PDZK1 undergoes a head-to-tail intramolecular association that negatively regulates its interaction with EBP50. *Biochemistry* **48**, 2261–2271.

LaLonde, D. P., Garbett, D., Bretscher, A. (2010). A regulated complex of the scaffolding proteins PDZK1 and EBP50 with ezrin contribute to microvillar organization. *Mol. Biol. Cell* **21**, 1519–1529.

Lamprecht, G., Seidler, U. (2006). The emerging role of PDZ adapter proteins for regulation of intestinal ion transport. *Am. J. Physiol. Gastrointest. Liver Physiol.* **291**, G766–G777.

Lazar, C. S., Cresson, C. M., Lauffenburger, D. A., Gill, G. N. (2004). The $Na^+/H^+$ exchanger regulatory factor stabilizes epidermal growth factor receptors at the cell surface. *Mol. Biol. Cell* **15**, 5470–5480.

Lederer, E. D., Khundmiri, S. J., Weinman, E. J. (2003). Role of NHERF-1 in regulation of the activity of Na–K ATPase and sodium–phosphate co-transport in epithelial cells. *J. Am. Soc. Nephrol.* **14**, 1711–1719.

Lee, H., Cheng, Y.-C., Fleming, G. R. (2007). Coherence dynamics in photosynthesis: protein protection of excitonic coherence. *Science* **316**, 1462–1465.

Li, J., Callaway, D. J., Bu, Z. (2009). Ezrin induces long-range interdomain allostery in the scaffolding protein NHERF1. *J. Mol. Biol.* **392**, 166–180.

Li, J., Dai, Z., Jana, D., Callaway, D. J., Bu, Z. (2005). Ezrin controls the macromolecular complexes formed between an adapter protein $Na^+/H^+$ exchanger regulatory factor and the cystic fibrosis transmembrane conductance regulator. *J. Biol. Chem.* **280**, 37634–37643.

Li, Q., Nance, M. R., Kulikauskas, R., Nyberg, K., Fehon, R., Karplus, P. A., et al. (2007). Self-masking in an intact ERM-merlin protein: an active role for the central alpha-helical domain. *J. Mol. Biol.* **365**, 1446–1459.

Li, J., Poulikakos, P. I., Dai, Z., Testa, J. R., Callaway, D. J., Bu, Z. (2007). Protein kinase C phosphorylation disrupts $Na^+/H^+$ exchanger regulatory factor 1 autoinhibition and promotes cystic fibrosis transmembrane conductance regulator macromolecular assembly. *J. Biol. Chem.* **282**, 27086–27099.

Lindorff-Larsen, K., Best, R. B., Depristo, M. A., Dobson, C. M., Vendruscolo, M. (2005). Simultaneous determination of protein structure and dynamics. *Nature* **433**, 128–132.

Liu, Y., Belkina, N. V., Shaw, S. (2009). HIV infection of T cells: actin-in and actin-out. *Science* 2pe23.

Liu-Chen, L. Y. (2004). Agonist-induced regulation and trafficking of kappa opioid receptors. *Life Sci.* **75**, 511–536.

Ma, B., Nussinov, R. (2009). Amplification of signaling via cellular allosteric relay and protein disorder. *Proc. Natl. Acad. Sci. USA* **106**, 6887–6888, Epub 2009 Apr 6822.

Mahon, M. J. (2008). Ezrin promotes functional expression and parathyroid hormone-mediated regulation of the sodium–phosphate cotransporter 2a in LLC-PK1 cells. *Am. J. Physiol. Ren. Physiol.* **294**, F667–F675.

Mahon, M. J., Donowitz, M., Yun, C. C., Segre, G. V. (2002). Na(+)/H(+) exchanger regulatory factor 2 directs parathyroid hormone 1 receptor signalling. *Nature* **417**, 858–861.

Mahon, M. J., Segre, G. V. (2004). Stimulation by parathyroid hormone of a NHERF-1-assembled complex consisting of the parathyroid hormone I receptor, phospholipase Cbeta, and actin increases intracellular calcium in opossum kidney cells. *J. Biol. Chem.* **279**, 23550–23558.

Mangia, A., Chiriatti, A., Bellizzi, A., Malfettone, A., Stea, B., Zito, F. A., et al. (2009). Biological role of NHERF1 protein expression in breast cancer. *Histopathology* **55**, 600–608.

Matsui, T., Maeda, M., Doi, Y., Yonemura, S., Amano, M., Kaibuchi, K., et al. (1998). Rho-kinase phosphorylates COOH-terminal threonines of ezrin/radixin/moesin (ERM) proteins and regulates their head-to-tail association. *J. Cell Biol.* **140**, 647–657.

Matsui, T., Yonemura, S., Tsukita, S. (1999). Activation of ERM proteins *in vivo* by Rho involves phosphatidyl-inositol 4-phosphate 5-kinase and not ROCK kinases. *Curr. Biol.* **9**, 1259–1262.

Maudsley, S., Zamah, A. M., Rahman, N., Blitzer, J. T., Luttrell, L. M., Lefkowitz, R. J., et al. (2000). Platelet-derived growth factor receptor association with Na(+)/H(+) exchanger regulatory factor potentiates receptor activity. *Mol. Cell. Biol.* **20**, 8352–8363.

McCammon, J. A. (1984). Protein dynamics. *Rep. Prog. Phys.* **47**, 1–46.

McClatchey, A. I. (2003). Merlin and ERM proteins: unappreciated roles in cancer development? *Nat. Rev. Cancer* **3**, 877–883.

McClatchey, A. I., Fehon, R. G. (2009). Merlin and the ERM proteins–regulators of receptor distribution and signaling at the cell cortex. *Trends Cell Biol.* **19**, 198–206, Epub 2009 Apr 2001.

Mezei, F. (1980). The Principles of Neutron Spin Echo, Neutron Spin Echo: Proceedings of a Laue-Langevin Institut Workshop. Springer, Heidelberg.

Mi, H., Endo, T., Schreiber, U., Ogawa, T., Asada, K. (1994). NAD(P)H dehydrogenase-dependent cyclic electron flow around photosystem I in the *Cyanobacterium synechocystis* PCC 6803: a study of dark-starved cells and spheroplasts. *Plant Cell Physiol.* **35**, 163–173.

Miyashita, O., Onuchic, J. N., Wolynes, P. G. (2003). Nonlinear elasticity, proteinquakes, and the energy landscapes of functional transitions in proteins. *Proc. Natl. Acad. Sci. USA* **100**, 12570–12575.

Morales, F. C., Takahashi, Y., Momin, S., Adams, H., Chen, X., Georgescu, M. M. (2007). NHERF1/EBP50 head-to-tail intramolecular interaction masks association with PDZ domain ligands. *Mol. Cell. Biol.* **27**, 2527–2537.

Moyer, B. D., Duhaime, M., Shaw, C., Denton, J., Reynolds, D., Karlson, K. H., et al. (2000). The PDZ-interacting domain of cystic fibrosis transmembrane conductance regulator is required for functional expression in the apical plasma membrane. *J. Biol. Chem.* **275**, 27069–27074.

Naren, A. P., Cobb, B., Li, C., Roy, K., Nelson, D., Heda, G. D., et al. (2003). A macromolecular complex of beta 2 adrenergic receptor, CFTR, and ezrin/radixin/moesin-binding phosphoprotein 50 is regulated by PKA. *Proc. Natl. Acad. Sci. USA* **100**, 342–346.

Palmer, A. G., 3rd. (2004). NMR characterization of the dynamics of biomacromolecules. *Chem. Rev.* **104**, 3623–3640.

Perez-Moreno, M., Jamora, C., Fuchs, E. (2003). Sticky business: orchestrating cellular signals at adherens junctions. *Cell* **112**, 535–548.

Petoukhov, M. V., Svergun, D. I. (2006). Joint use of small-angle X-ray and neutron scattering to study biological macromolecules in solution. *Eur. Biophys. J.* **35**, 567–576.

Pettersen, E. F., Goddard, T. D., Huang, C. C., Couch, G. S., Greenblatt, D. M., Meng, E. C., et al. (2004). UCSF chimera—a visualization system for exploratory research and analysis. *J. Comput. Chem.* **25**, 1605–1612.

Pollard, T. D., Cooper, J. A. (2009). Actin, a central player in cell shape and movement. *Science* **326**, 1208–1212.

Prat, A. G., Cunningham, C. C., Jackson, G. R., Jr., Borkan, S. C., Wang, Y., Ausiello, D. A., et al. (1999). Actin filament organization is required for proper cAMP-dependent activation of CFTR. *Am. J. Physiol.* **277**, C1160–C1169.

Pujuguet, P., Del Maestro, L., Gautreau, A., Louvard, D., Arpin, M. (2003). Ezrin regulates Ecadherin-dependent adherens junction assembly through Rac1 activation. *Mol. Biol. Cell* **14**, 2181–2191, Epub 2003 Feb 2186.
Raghuram, V., Mak, D. D., Foskett, J. K. (2001). Regulation of cystic fibrosis transmembrane conductance regulator single-channel gating by bivalent PDZ-domain-mediated interaction. *Proc. Natl. Acad. Sci. USA* **98**, 1300–1305.
Reczek, D., Berryman, M., Bretscher, A. (1997). Identification of EBP50: a PDZ-containing phosphoprotein that associates with members of the ezrin–radixin–moesin family. *J. Cell Biol.* **139**, 169–179.
Reczek, D., Bretscher, A. (1998). The carboxyl-terminal region of EBP50 binds to a site in the amino-terminal domain of ezrin that is masked in the dormant molecule. *J. Biol. Chem.* **273**, 18452–18458.
Roch, F., Polesello, C., Roubinet, C., Martin, M., Roy, C., Valenti, P., et al. (2010). Differential roles of PtdIns(4,5)P2 and phosphorylation in moesin activation during *Drosophila* development. *J. Cell Sci.* **123**, 2058–2067.
Roumier, A., Olivo-Marin, J. C., Arpin, M., Michel, F., Martin, M., Mangeat, P., et al. (2001). The membrane-microfilament linker ezrin is involved in the formation of the immunological synapse and in T Cell activation. *Immunity* **15**, 715–728.
Schlessinger, J. (1988). Signal transduction by allosteric receptor oligomerization. *Trends Biochem. Sci.* **13**, 443–447.
Schmieder, S., Nagai, M., Orlando, R. A., Takeda, T., Farquhar, M. G. (2004). Podocalyxin activates RhoA and induces actin reorganization through NHERF1 and ezrin in MDCK cells. *J. Am. Soc. Nephrol.* **15**(9), 2289–2298.
Scott, J. D., Pawson, T. (2009). Cell signaling in space and time: where proteins come together and when they're apart. *Science* **326**, 1220–1224.
Sheng, M., Sala, C. (2001). PDZ domains and the organization of supramolecular complexes. *Annu. Rev. Neurosci.* **24**, 1–29.
Shenolikar, S., Voltz, J. W., Cunningham, R., Weinman, E. J. (2004). Regulation of ion transport by the NHERF family of PDZ proteins. *Physiology (Bethesda)* **19**, 362–369.
Short, D. B., Trotter, K. W., Reczek, D., Kreda, S. M., Bretscher, A., Boucher, R. C., et al. (1998). An apical PDZ protein anchors the cystic fibrosis transmembrane conductance regulator to the cytoskeleton. *J. Biol. Chem.* **273**, 19797–19801.
Singh, A. K., Riederer, B., Krabbenhöft, A., Rausch, B., Bonhagen, J., Lehmann, U., et al. (2009). Differential roles of NHERF1, NHERF2, and PDZK1 in regulating CFTR-mediated intestinal anion secretion in mice. *J. Clin. Investig.* **119**, 540–550.
Sizemore, S., Cicek, M., Sizemore, N., Ng, K. P., Casey, G. (2007). Podocalyxin increases the aggressive phenotype of breast and prostate cancer cells *in vitro* through its interaction with ezrin. *Cancer Res.* **67**, 6183–6191.
Sneddon, W. B., Syme, C. A., Bisello, A., Magyar, C. E., Rochdi, M. D., Parent, J. L., et al. (2003). Activation-independent parathyroid hormone receptor internalization is regulated by NHERF1 (EBP50). *J. Biol. Chem.* **278**, 43787–43796.
Somasiri, A., Nielsen, J. S., Makretsov, N., McCoy, M. L., Prentice, L., Gilks, C. B., et al. (2004). Overexpression of the anti-adhesin podocalyxin is an independent predictor of breast cancer progression. *Cancer Res.* **64**(15), 5068–5073.

Svergun, D. I. (1999). Restoring low resolution structure of biological macromolecules from solution scattering using simulated annealing. *Biophys. J.* **76**, 2879–2886.
Takahashi, Y., Morales, F. C., Kreimann, E. L., Georgescu, M. M. (2006). PTEN tumor suppressor associates with NHERF proteins to attenuate PDGF receptor signaling. *EMBO J.* **25**, 910–920.
Takeda, T., McQuistan, T., Orlando, R. A., Farquhar, M. G. (2001). Loss of glomerular foot processes is associated with uncoupling of podocalyxin from the actin cytoskeleton. *J. Clin. Invest.* **108**, 289–301.
Taouil, K., Hinnrasky, J., Hologne, C., Corlieu, P., Klossek, J. M., Puchelle, E. (2003). Stimulation of beta 2-adrenergic receptor increases cystic fibrosis transmembrane conductance regulator expression in human airway epithelial cells through a cAMP/protein kinase A-independent pathway. *J. Biol. Chem.* **278**, 17320–17327.
Terawaki, S.-i., Maesaki, R., Hakoshima, T. (2006). Structural basis for NHERF recognition by ERM proteins. *Structure* **14**, 777–789.
Thelin, W. R., Hodson, C. A., Milgram, S. L. (2005). Beyond the brush border: NHERF4 blazes new NHERF turf. *J. Physiol.* **567**, 13–19.
Verpy, E., Leibovici, M., Zwaenepoel, I., Liu, X. Z., Gal, A., Salem, N., et al. (2000). A defect in harmonin, a PDZ domain-containing protein expressed in the inner ear sensory hair cells, underlies Usher syndrome type 1C. *Nat. Genet.* **26**, 51–55.
Wang, C., Pan, L., Chen, J., Zhang, M. (2010). Extensions of PDZ domains as important structural and functional elements. *Protein Cell* **1**, 737–751.
Weinman, E. J., Hall, R. A., Friedman, P. A., Liu-Chen, L. Y., Shenolikar, S. (2006). The association of NHERF adaptor proteins with g protein-coupled receptors and receptor tyrosine kinases. *Annu. Rev. Physiol.* **68**, 491–505.
Weinman, E. J., Wang, Y., Wang, F., Greer, C., Steplock, D., Shenolikar, S. (2003). A C-terminal PDZ motif in NHE3 binds NHERF-1 and enhances cAMP inhibition of sodium-hydrogen exchange. *Biochemistry* **42**, 12662–12668.
Windisch, B., Bray, D., Duke, T. (2006). Balls and chains—a mesoscopic approach to tethered protein domains. *Biophys. J.* **91**, 2383–2392.
Wong, W., Gough, N. R. (2009). Focus issue: the protein dynamics of cell signaling. *Sci. STKE* **2**, eg4.
Yao, S., Babon, J. J., Norton, R. S. (2008). Protein effective rotational correlation times from translational self-diffusion coefficients measured by PFG-NMR. *Biophys. Chem.* **136**, 145–151.
Yap, A. S., Brieher, W. M., Gumbiner, B. M. (1997). Molecular and functional analysis of cadherin-based adherens junctions. *Annu. Rev. Cell Dev. Biol.* **13**, 119–146.
Yonemura, S., Matsui, T., Tsukita, S. (2002). Rho-dependent and -independent activation mechanisms of ezrin/radixin/moesin proteins: an essential role for polyphosphoinositides *in vivo*. *J. Cell Sci.* **115**, 2569–2580.
Yu, Y., Khan, J., Khanna, C., Helman, L., Meltzer, P. S., Merlino, G. (2004). Expression profiling identifies the cytoskeletal organizer ezrin and the developmental homeoprotein Six-1 as key metastatic regulators. *Nat. Med.* **10**, 175–181.
Zaccai, G. (2000). How soft is a protein? A protein dynamics force constant measured by neutron scattering. *Science* **288**, 1604–1607.

# STRUCTURAL DIVERSITY OF CLASS I MHC-LIKE MOLECULES AND ITS IMPLICATIONS IN BINDING SPECIFICITIES

By MD. IMTAIYAZ HASSAN AND FAIZAN AHMAD

Centre for Interdisciplinary Research in Basic Sciences, Jamia Millia Islamia, New Delhi, India

| | | |
|---|---|---|
| I. | Introduction | 224 |
| II. | Sequence Analysis | 227 |
| | A. Ligand-Binding Residues | 232 |
| | B. TCR-Binding Residues | 236 |
| | C. Light Chain-Binding Residues | 238 |
| III. | Structure Analysis | 239 |
| | A. Structure of Ligand-Binding Groove | 244 |
| | B. Structural Basis of TCR Binding | 251 |
| | C. Structural Basis of Light Chain Binding | 255 |
| IV. | Glycosyalation | 258 |
| V. | Conclusion | 261 |
| | References | 262 |

## Abstract

The binding groove of class I major histocompatibility complex (MHC) class is essentially important for antigen binding and presentation on T cells. There are several molecules that have analogous conformations to class I MHC. However, they bind specifically to varying types of ligands and cell-surface receptors in order to elicit an immune response. To elucidate how such recognition is achieved in classical MHC-I like molecules, we have extensively analyzed the structure of human leukocyte antigen (HLA-1), neonatal Fc receptor (FcRn), hereditary hemochromatosis protein (HFE), cluster of differentiation 1 (CD1), gamma delta T cell receptor ligand (22), zinc-$\alpha$2-glycoprotein (ZAG), and MHC class I chain-related (MIC-A) proteins. All these molecules have analogous structural anatomy, divided into three distinct domains, where $\alpha 1-\alpha 2$ superdomains form a groove-like structure that potentially bind to certain ligand, while the $\alpha 3$ domain adopts a fold resembling immunoglobulin constant domains, and holds this $\alpha 1-\alpha 2$ platform and the light chain. We have observed many remarkable features of $\alpha 1-\alpha 2$ platform, which provide specificities to these proteins toward a particular class of ligands. The relative orientation of α1, α2, and α3 domains is primarily responsible for the specificity to the light chain. Interestingly, light chain of all these proteins is $β_2$-microglobulin ($β_2$M), except ZAG which has prolactin-induced protein (PIP). However, MIC-A is devoid of any light chain. Residues on $β_2$M recognize a sequence motif on the α3 domain that is essentially restricted to specific heavy chain of MHC class I molecules. Our analysis suggests that the structural features of class I molecules determine the recognition of different ligands and light chains, which are responsible for their corresponding functions through an inherent mechanism.

## I. Introduction

Major histocompatibility complex I (MHC-I) is a set of molecules anchored to cell surfaces and are essentially important for lymphocyte recognition and presentation through the T cell receptor (TCR; Bjorkman and Parham, 1990; Madden, 1995). The class I molecules consist of two polypeptide chains including one heavy chain while the light chain is $β_2$-microglobulin ($β_2$M; Vitiello et al., 1990). Class I MHC molecules usually bind to peptides (8–10 residues long; Madden, 1995) and subsequently interact with the TCR of CD8+T cells leading to the CD8+T cell activation, proliferation, differentiation, and effector function (Salter et al., 1990). The sequence and structure analysis of class I MHC molecules suggests that it is a heterodimer, consisting of a heavy alpha chain of ~45 kDa and a light chain, $β_2$M of ~12 kDa (Bjorkman et al., 1987b; Bjorkman and Parham, 1990). In addition to the transmembrane and cytoplasmic domains, the MHC heavy chain comprises three extracellular domains, α1, α2, and α3. The α1 and α2 domains are highly polymorphic and this polymorphism is mostly localized around the peptide-binding groove. The α1–α2 domain groove is occupied by a peptide or mixture of peptides in all class I MHC structures solved to date, and continuous electron-density representing peptide(s) is always seen in the α1–α2 groove. The α3 domain is highly conserved and closely interacts with $β_2$M. Both α3 domain and $β_2$M are homologous to the CH3 domain of human immunoglobulin (Ig; Bjorkman and Parham, 1990; Madden, 1995).

The various isoforms of HLA are encoded by a small number of genes, that is, HLA-A, -B, and -C in man and H2-K, -D, and -L in mouse, located within the MHC on human chromosome 6p21.3 and murine chromoso-

17 (The MHC Sequencing Consortium, 1999). MHCs are further classified into classical (MHC-Ia) and nonclassical genes (MHC-Ib), such as HLA-E, HLA-F, and HLA-G in man and the H2-Q, -T, and -M loci in mice (Pease et al., 1991). All these show a restricted pattern of tissue expression and unique immunologic functions. These functions include presentation of *N*-formylated peptides by H2-M3 and leader sequence peptides by HLA-E and many others (Lindahl et al., 1997). Despite these apparently special features, the high degree of sequence similarity between these molecules defines them as members of the same gene lineage and suggests that they are derived from a common ancestor through gene duplication and minimal diversification (Riegert et al., 1998).

Despite of antigen presenting MHC-I, a large number of proteins have been identified to have a close sequence and structural similarity with class I molecules and hence these members are considered as a MHC class-I related proteins (Ojcius et al., 2002). All these molecules have structurally similar groove but chemical character of the groove is different and depends upon the nature of their ligands (Table I). For example, the largely hydrophobic grooves of CD1 (Zeng et al., 1997; Porcelli and Modlin, 1999), class Ib MHC molecules Qa-2 (He et al., 2001), and HLA-E (O'Callaghan et al., 1998) allow binding of lipids (including CD1) and hydrophobic peptides (class Ib proteins). HFE is also an MHC-related protein that is mutated in the iron overload disease hereditary hemochromatosis. HFE binds to transferrin receptor (TfR) and reduces its affinity for iron-loaded transferrin, implicating HFE in the iron metabolism (Feder et al., 1996; Lebron et al., 1998). The other member, neonatal Fc receptor (FcRn) for IgG, expressed in monocytes, macrophages, and dendritic cells, originally characterized as a transport receptor involved in the uptake of maternal IgG by an intestinal route in rodents (Simister and Mostov, 1989). Its major function is to transport IgG across polarized epithelial cells and protect IgG from degradation (Burmeister et al., 1994a,b). Many MHC class I-like molecules have distinct ligand-binding capabilities, suggesting that they have evolved for specific tasks that are distinct from those of MHC class. The zinc-$\alpha$2-glycoprotein (ZAG) is involved in the lipid mobilization and is overexpressed in certain human malignant tumors to stimulate lipolysis in adipocytes (Bing et al., 2004; Hassan et al., 2008b). Further, ZAG directly influences the uncoupling protein and in turn enhances the lipid utilization during cancer cachexia (Sanders and Tisdale, 2004). Murine T10 and T22 are highly

TABLE I
Important Features of MHC Class I and Class I-related Proteins

| Protein name | PDB code | Ligand | Receptor | Light chain | Functions | References |
|---|---|---|---|---|---|---|
| HLA A2 | 1HLA | Peptides | αβ-TCR | β$_2$M | Antigen presentation | Bjorkman et al. (1987a,b) |
| HLA b27 | 1HSA | Peptides | αβ-TCR | β$_2$M | Antigen presentation | Madden et al. (1992) |
| HLA H-2Dd | 1BII | Peptides | αβ-TCR, Ly94 | β$_2$M | Activation/inhibition of CD8 +T or NK cells | Achour et al. (1998) |
| HLA-I H-2Kb | 2VAA | Peptides | αβ-TCR | β$_2$M | Antigen presentation | Fremont et al. (1992) |
| HLA-E | 3BZE | Peptides | CD94 | β$_2$M | Activation/inhibition of NK cells | Hoare (2008) |
| HLA-G | 3KYO | Peptides | LIR2 | β$_2$M | Inhibition of NK cells | Walpole (2010) |
| Mouse CD1 | 1CD1 | Glycolipids | Vα14-Jα281 TCR | β$_2$M | Lipidic antigen presentation | Zeng et al. (1997) |
| Rat FcRn | 3FRU | | Fc of IgG | β$_2$M | IgG transport | Burmeister et al. (1994a,b) |
| Human HFE | 1A6Z | | TfR | β$_2$M | Iron metabolism | Lebron et al. (1998) |
| Murine T10/T22 | 1c16 | | γδ-TCR | β$_2$M | Signaling for intracellular infections | Wingren et al. (2000) |
| Human ZAG | 1ZAG 3ES6 | Fatty acids | β$_3$-Adrenogenic receptor | PIP | Lipid mobilization | Sanchez et al. (1999a,b), Hassan et al. (2008a,b) |
| MIC-A | 1B3J | | γδ-TCR NKG2D | – | Signal for stress, activation of NK cells | Li et al. (1999) |
| MIC-B | 1JE6 | | γδ-TCR NKG2D | – | Signal for stress, activation of NK cells | Holmes et al. (2002) |
| RAE-1 | 1JFM | | NKD2D | – | Activation of NK cells | Li et al. (2002) |

related nonclassical MHC class Ib proteins that bind to certain γδ-TCR in the absence of other components (Wingren et al., 2000). Recognition and stimulation by T22 and T10 presumably do not require peptide or any other ligand for cell-surface expression or recognition by γδ-T cells (Schild et al., 1994). Another γδ-TCR-binding MHC class I homolog is MIC-A which functions as a stress-inducible antigen and its recognition is independent of $\beta_2$M and bound peptides (Li et al., 1999).

Despite of the fact that all these proteins have class I MHC-like fold structure and sequence, they perform various unrelated functions (Table I). We searched protein data bank (PDB) for finding proteins that belongs to class I family, and observed a large number of structures of class I molecules and its complexes with peptide, potential ligands, TCR, and CD8 receptor have been determined (Table II). Interestingly, all these proteins have $\beta_2$M as a light chain except MIC-A and ZAG. Recently, we have observed that ZAG accommodates prolactin-induced protein (PIP), a molecule which is identical to $\beta_2$M in their size and fold despite of least sequence identity (3%; Hassan et al., 2008a). All these findings reveal that a minor change in the sequences of α1, α2, and α3 domains of MHC class I protein leads to a significant change in their functions. The main function of these proteins is binding/transport and is directly associated with the sequences of α1 and α2 domains. However, the α3 domain is solely responsible for binding to the light chains, $\beta_2$M (PIP in ZAG), and for holding the α1–α2 platform. In this section, our aim is to focus on the importance of sequence and three-dimensional structure of class I proteins in their ligands and light chain binding, in order to get an insight to structure–function relationship in class I molecules.

## II. Sequence Analysis

In the past few years, more than 500 primate MHC genes or parts thereof have been sequenced. The extraordinary sequence information is used here to draw conclusions about structure, function, diversity, and evolution of class I proteins (Bjorkman and Parham, 1990; Klein et al., 1993). The sequence analysis of class-I mutants have provided a model system for understanding the generation of diversity of the genes encoding various histocompatibility molecules, and the relationship of their structure to function (Nathenson et al., 1986; Bjorkman and Parham, 1990). We have aligned the sequences of all heavy chain from MHC-I-

TABLE II
List of MHC Class I Molecules and Class I-Like Molecules in Protein Data Bank

| S. no. | MHC molecules | PDB code | Ligand | Receptor |
|---|---|---|---|---|
| 1 | HLA-A1 | 1W72, 3BO8 | Peptide | |
| 2 | HLA-AW68 | 2HLA, 1HSB | Peptide | |
| 3 | HLA-B*1402 | 3BVN, 3BXN | Peptide | |
| 4 | HLA-B*1501 | 1XR8, 1XR9 | Peptide | |
| 5 | HLA-A*1101 | 2HN7, 1X7Q, 1Q94, 1QVO | Peptide | |
| 6 | HLA-B21 | 3BEV, 3BEW | Peptide | |
| 7 | HLA-B*27 | 1HSA | Peptide | |
| 8 | HLA-B*2705 | 3LV3, 1OGT, 1UXS, 3B6S, 3DTX, 3BP4, 2A83, 1W0V, 1JGE | Peptide | |
| 9 | HLA-B*2709 | 3HCV, 1OF2, 3B3I, 3BP7, 1JGD, 1K5N, 1UXW, 3CZF, 1W0W | Peptide | |
| 10 | HLA-B*4103 | 3LN4 | Peptide | |
| 11 | HLA-B*4104 | 3LN5 | Peptide | |
| 12 | HLA B*3501 | 2CIK, 1ZSD, 1A1N | Peptide | |
| 13 | HLA B*3501 | 2NX5 | Peptide | TCR |
| 14 | HLA B*3508 | 3KWW, 2NW3 | Peptide | |
| 15 | HLA-B*3508 | 2AK4, 3KXF | Peptide | TCR |
| 16 | HLA, B*4403 | 1SYS | Peptide | |
| 17 | HLA-B*4405 | 1SYV | Peptide | |
| 18 | HLA-B*5703 | 2BVO, 2BVP, 2BVQ | Peptide | |
| 19 | HLA-DP2 | 3LQZ | | |
| 20 | HLA B*44 | 1N2R | Peptide | |
| 21 | HLA-B*4402 | 3L3D, 1M6O, 3L3G, 3L3I, 3L3J, 3L3K, 3KPL, 3KPM | Peptide | |
| 22 | HLA-B*4403 | 3KPN, 3KPO | Peptide | |
| 23 | HLA B*4405 | 3KPP, 3KPQ, 3KPS | Peptide | |
| 24 | HLA B*4405 | 3KPR | Peptide | TCR |
| 25 | HLA-B*5301 | 1A1M, 1A1O, 2VLJ, 2VLK, 2VLR, 2J8U | | |
| 26 | HLA-B*8 | 1M05 | Peptide | |
| 27 | HLA-A2.1 | 2X4N, 2X4O, 2X4P, 2X4Q, 2X4R, 2X4S, 2X4T, 2X4U, 2V2W, 2V2X, 1EEY, 1I1Y | Peptide | |
| 28 | HLA-A2.1 | 3HLA | | |
| 29 | HLA-A2.1 | 2JCC, 2P5E, 2P5W, 2PYE, 2UWE, 2BNQ, 2BNR, 1OGA, 1G6R, 1FO0, 2CKB | Peptide | TCR |
| 30 | HLA-A2.1 | 2C7U | | CD8 |

TABLE II (*Continued*)

| S. no. | MHC molecules | PDB code | Ligand | Receptor |
|---|---|---|---|---|
| 31 | HLA-A2 | 2GUO, 2GIT, 2FO4, 1G7Q, 2AV1, 2AV7, 1S8D, 1T1W, 1T1X, 1T1Y, 1T1Z, 1T20, 1T21, 1T22, 1I7T, 1I7U, 1IM3, 2VLL | Peptide | |
| 32 | HLA-A2 | 1P7Q | Peptide | LIR |
| 33 | HLA-A*0201 | 1DUZ, 1JF1, 1JHT, 1I4F, 1HHJ | Peptide | |
| 34 | HLA-Ib M10.5 | 1ZS8 | | |
| 35 | H2-M3 | 1MHC | Peptide | |
| 36 | HLA-E | 3BZE, 3BZF, 1KPR, 1KTL, 1MHE | Peptide | |
| 37 | HLA-E | 2ESV | Peptide | TCR |
| 38 | HLA-H2 | 3CVH | Peptide | TCR |
| 39 | H2-DB | 2VE6, 1BZ9, 2CII, 1YN6, 1YN7, 1ZHB, 1WBX, 1WBY, 1WBZ, 1N5A, 1FFN, 1FFO, 1FFP, 1QLF, 1HOC | Peptide | |
| 40 | H2-D$^D$ | 1DDH, 1BII | Peptide | |
| 41 | H2-L$^D$ | 1LDP, 1LD9 | Peptide | |
| 42 | HLA-D$^R$ | 1H15, 1FV1, 1IIE | Peptide | |
| 43 | HLA-DR$^{2a}$ | 1ZGL | Peptide | TCR |
| 44 | H-2K$^b$ | 3C8K | Peptide | L49C |
| 45 | H-2K$^b$ | 1BQH, | Peptide | TCR |
| 46 | H-2Kb | 1S7Q, 1S7R, 1S7S, 1S7T, 1KPU, 1KPV, 1N59, 1LEG, 1LEK, 1KJ3, 1FZJ, 1FZK, 1FZM, 1FZO, 1KBG, 1OSZ, 1VAA, 1VAB, 1VAD, 1VAC, 2MHA | Peptide | |
| 47 | H-2KbM3 | 1MWA, 1JTR, | Peptide | TCR |
| 48 | H-2K$^b$ | 3C8J | | L49C |
| 49 | H-2K$^d$ | 1VGK | Peptide | |
| 50 | H-2D$^b$ | 1S7U, 1S7V, 1S7W, 1S7X | Peptide | |
| 51 | HLA-D$^{R1}$ | 1SJE, 1SJH, 1T5X | Peptide | |
| 52 | HLA-G | 3KYN, 3KYO, 2D31 | Peptide | |
| 53 | HLA-G | 2DYP | Peptide | LIR |
| 54 | H-2Dd | 3ECB, 3DMM | Peptide | CD8 |
| 55 | HLA-T10 | 1R3H | | |
| 56 | HLA-T22 | 1C16 | | |
| 57 | HLA-T22 | 1YPZ | Peptide | γδ-TCR |

(*Continued*)

TABLE II (*Continued*)

| S. no. | MHC molecules | PDB code | Ligand | Receptor |
|---|---|---|---|---|
| 58 | ZAG | 3ES6, 1ZAG, 1T7V, 1T7W, 1T7X, 1T7Y, 1T7Z, 1T80 | | |
| 59 | FcRn | 3M17, 3M1B | Peptide inhibitor | |
| 60 | FcRn | 1I1A, 1FRT | | Fc |
| 61 | FcRn | 3FRU | | |
| 62 | CD1 | 1CD1, 3JVG, 3DBX | | |
| 63 | CD1a | 1ONQ | Sulfatide | |
| 64 | CD1a | 1XZ0 | Peptide | |
| 65 | CD1b | 1UQS, 2H26, 1GZP, 1GZQ | Glycolipids | |
| 66 | CD1d | 2GAZ, 3ILQ, 3GML, 3GMN, 3GMM, 3GMP, 3GMQ, 3GMR, 2Q7Y, 2FIK, 2AKR, 1Z5L, 1ZHN, 1ZT4 | Lipid | |
| 67 | CD1d | 3HE6, 3HE7, 2CDF, 2CDG | Galactoceramide | TCR |
| 68 | CD1d | 2PO6, 2EYR, 2EYS, 2EYT, 2CDE | | NK cells |
| 69 | HFE | 1A6Z | | |
| 70 | HFE | 1DE4 | | TfR |
| 71 | MIC-A | 1B3J | | |
| 72 | MIC-A | 1HYR | | NK receptor |
| 73 | MIC-B | 1JE6 | | |
| 74 | MIC-B | 2WY3 | UL16 | |
| 75 | RAE-1 BETA | 1JFM | | |
| 76 | RAE-1 BETA | 1JSK | | NKG2D |
| 77 | ULBP3 | 1KCG | | NKG2D |

like proteins and observed a remarkable sequence identity especially in the sequence of functional relevance evident from crystal structure analysis (Fig. 1; Bjorkman et al., 1987b; Bjorkman and Parham, 1990; Fremont et al., 1992; Burmeister et al., 1994a,b; Madden, 1995; Gao et al., 1997; Zeng et al., 1997; Lebron et al., 1998; O'Callaghan et al., 1998; Li et al., 1999; Sanchez et al., 1999a; Burdin et al., 2000; Zajonc et al., 2003; Koch et al., 2005; Hassan et al., 2008a). The residues binding to light chain as well as the ligands are found to be highly conserved in all members of class I MHC proteins. These finding suggests that all these proteins belong to a common ancestor and adapted different conformations in order to bind suitable ligands to perform corresponding functions, including antigen

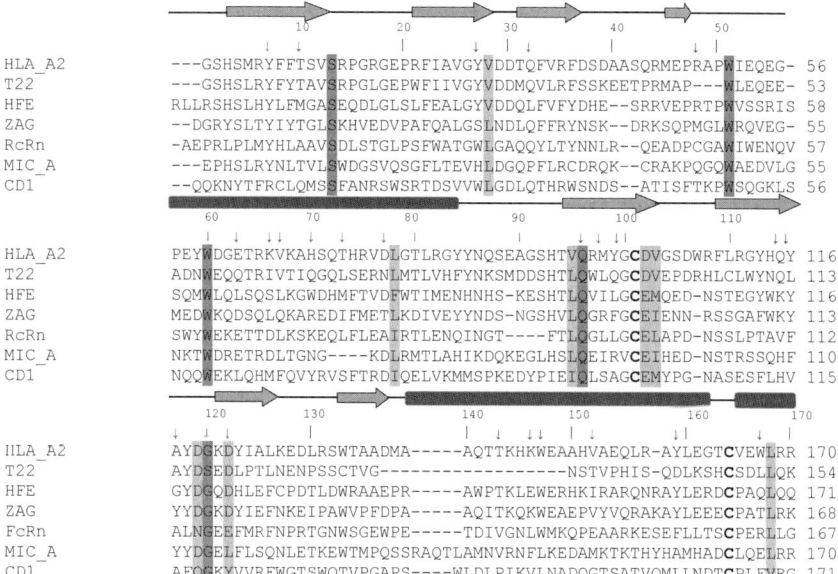

FIG. 1. Multiple sequence alignment of class I MHC proteins showing conserved residues. The sequence was taken from protein data bank with their PDB codes as HLA_A2, 2CLR; T22, 1C16; HFE, 1A6Z; ZAG, 1ZAG; FcRn, 3FRU; MIC_A, 1B3J; and CD1, 1CD1. The identical and homologous residues are highlighted in dark and light gray, respectively. Conserved cysteine residues are shaded in yellow. The secondary structure elements are shown at the top of sequences (taken from PDB code 2CLR) where loop is represented by black line, β-strand by filled arrow (green), and α helices are in filled cylinder (red). Residues of HLA_A2 showing interaction to peptide and β$_2$M are marked by blue and red arrow (↓), respectively, on the top of sequences. (For interpretation of the references to color in this figure legend, the reader is referred to the Web version of this chapter.)

presentation (Nathenson et al., 1986; Bjorkman and Parham, 1990; Klein et al., 1993; Madden, 1995). The localization of nonconserved amino acids in the various MHC molecules further provides in insight to complementary recognition regions in the MHC sequences that play a major role in determining their ligands and light chain associative recognition (Fig. 1). A comprehensive sequence analysis may be useful to answer how do these molecules interact with ligands, TCR, CD8/CD4, and β$_2$M.

## A. Ligand-Binding Residues

The antigen-binding cleft of MHC class I molecules is formed by α1 and α2 domains, in which two α-helices overlie the floor of eight antiparallel β-stranded sheets (residues 1–180), which is positioned on top of the immunoglobulin (Ig) constant-like domain, α3 (residues 181–275) and $\beta_2 M$ (Nathenson et al., 1986; Bjorkman et al., 1987b; Bjorkman and Parham, 1990; Madden, 1995; Fig. 2). It has been observed that the substitution of a few amino acids in the proposed antigen (ligand) recognition regions located in the α1 domain (residues 70–90) and/or α2 domain (residues 150–180) can have marked effects on biological functions of class I molecules (Nathenson et al., 1986). Structurally the proposed antigen-binding site of MHC-I (Bjorkman et al., 1987a,b) is similar to those of FcRn (Burmeister et al., 1994a,b), HFE (Bennett et al., 2000), ZAG (Sanchez et al., 1999a), T22 (Wingren et al., 2000), CD1 (Zeng et al., 1997), and MIC-A (Li et al., 1999). Structure analysis of peptide–MHC (pMHC) complex (Fremont et al., 1992; Collins et al., 1994; Madden, 1995; Achour et al., 1998) demonstrates that the peptide-binding sequences can be categorized into two types of residues: constant and variable (polymorphic or dimorphic). The constant residues are identical in all MHC class I, which provides a common conserved framework for the binding to the peptide in a gene- or allele-independent manner. These constant residues are Tyr59, Tyr84, Tyr123, and Tyr159 (Reche and Reinherz, 2003). In fact, these residues are involved in forming a network of H bonds to the N- and C-termini of the peptide. It has been observed in a complex structure of HLA with the HIV-1-derived peptide (RGPGRAFVTI) that the N-terminus of peptide binds strongly through hydrogen bonds to the side chains of Tyr171 and Tyr7 in the A pocket, while the C-terminal is anchored by a network of hydrogen bonds formed by Tyr84, Thr143, and Lys146 in the F pocket (Achour et al., 1998), similar to other peptides in MHC class I complexes (Madden, 1995). Moreover, many hydrophobic residues are present in the antigen-binding clef which offers hydrophobic interactions to hold the peptide strongly. The hydrophobic residues are Tyr123, Phe116, Ile24, Trp147, Trp97, Ala99, Trp114, Tyr159, and Leu95 (Fremont et al., 1992; Achour et al., 1998).

The FcRn is structurally similar to the class I molecules, involved in transporting IgG from mother to her fetus, and providing the passive humoral immunity (Roopenian and Akilesh, 2007). Further, it protects

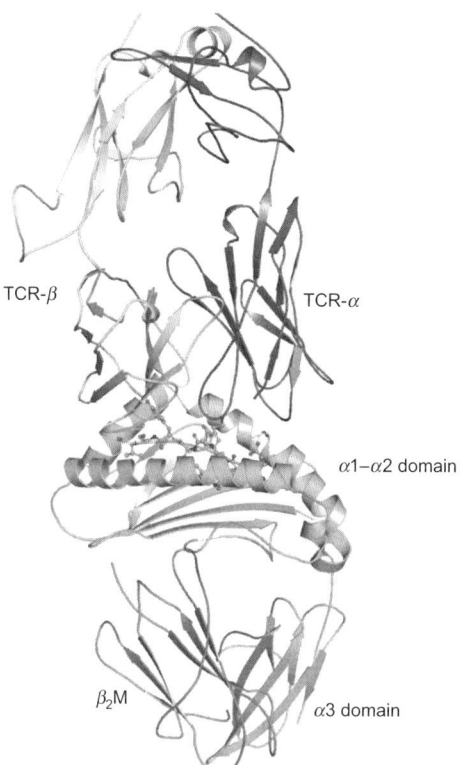

FIG. 2. Architecture of MHC-like molecules showing anatomy of complex formed between MHC class I proteins and αβ-TCR. Class I molecules consist of a heavy chain (light green) and a light β₂M chain (orange). The peptide-binding site is formed exclusively by elements of the heavy chain formed by α1α2 platform, whereas the α3 domain is mainly responsible for β₂M binding. A view from the TCR binding onto the class I peptide-binding groove where peptide is shown in ball and stick model (reddish). The α (sky blue) and β (yellow) chains of TCR are shown in ribbon diagram. The structure was drawn in PyMOL using the atomic coordinates of pMHC–TCR complex with PDB code 2CKB (Garcia et al., 1998). (See color plate 7).

IgG from degradation, thereby providing increase in half-life of this class of antibody in the serum. As evident from its crystal structure, the first contact zone of the heavy chain of the rat FcRn molecule can be found at the end of the α1 domain (residues 84–86 and 90). The second and third contact zones involve residues of α2 domains (113–119 and 131–137) (Burmeister et al., 1994a,b). The second contact zone, which is part of the α2 domain, is well conserved, emphasizing its importance in IgG binding (Burmeister et al., 1994b). Zhou et al. (2003) performed an experiment to investigate the molecular basis for the difference in binding specificity between human and mouse FcRn. They suggested that a few residues of FcRn are of great importance in the IgG binding, such as Leu137, Asp132, loop121–132, loop79–89, and loop79–89.

The HFE does not bind peptides or perform any known immune function because it has a narrower and shallower version of the class I peptide-binding groove (Lebron et al., 1998; Bennett et al., 2000). However, HFE binds to the TfR, a macromolecule associated with iron homeostasis (Townsend and Drakesmith, 2002). Crystal structure analysis suggests that Val78 and Trp81 from HFE pairs Leu619 and Val622 from TfR (Bennett et al., 2000). This finding was further strengthened by site-directed mutants of HFE at positions 78 and 81, for either shows no detectable binding to TfR (Barton et al., 1999). Further, binding may be affected by two residues near this position in hereditary hemochromatosis (Ile83Thr and Gly71Arg). Another important residue that takes part in TfR binding is Leu22, showing a decrease in binding affinity upon mutation (Lebron and Bjorkman, 1999). TfR binds to HFE in a pH-dependent manner, and subsequently facilitates release of iron from Fe–Tf complex. Thus, residues on HFE and/or TfR must be responsible for the sharp pH dependence. It was suggested that His74 and His150 of HFE are essentially important in hydrophobic contacts with TfR residues (Bennett et al., 2000). Lebron and Bjorkman (1999) carried out site-directed mutagenesis studies to determine the critically important residues in HFE. They observed many important residues which are quite important for TfR binding, such as Leu22, Val78, Trp81, His87, His89, His94, and His123. However, there are many residues such as His41, Arg44, Ser43, Trp141, and Pro166 which do not strongly affect the TfR binding on mutation.

CD1 is also an antigen-presenting molecule; especially the lipidic antigen is distantly related to class I MHC molecules (Porcelli and Modlin, 1999; Zajonc et al., 2003). It appears from sequence similarity and domain topology that CD1 is more closely related to MHC class Ia and Ib molecules (Zeng et al., 1997; Koch et al., 2005). The CD1 family can be divided into two groups by sequence homology, that is, group I (CD1a, b, and c) and group II (CD1d, and e), suggesting that each group of CD1 molecules could have a distinct function (Porcelli and Modlin, 1999). The groove of CD1 is slightly narrower and deeper as compared to other class I molecules (Moody et al., 2005). Three highly conserved residues Phe49, Trp40, and Phe18 appear to be responsible for the elevation of helices of β-sheet platform (Zeng et al., 1997). CD1 groove is predominated by hydrophobic residues, Val28, His38, Leu114, Leu162, and Phe169 in the A′ pocket. However, the F′ pocket contains Trp14, Ile96, and Leu148, and it becomes wider at the surface of the binding groove due to the presence of Val98, Leu116, Phe144, and Val147 (Zajonc et al., 2003). Further, three conserved residues (Tyr7, Tyr59, and Tyr171) in MHC class I form hydrogen-bonding network with the NH2-terminus of the bound peptide (Bjorkman et al., 1987b; Bjorkman and Parham, 1990; Fremont et al., 1992). While, in the case of CD1d1, the same position is occupied by entirely hydrophobic residues (Phe10, Trp63, and Phe171), this indicates preferential binding with lipidic antigen instead of peptide in MHC-I (Fig. 1; Zeng et al., 1997). Moreover, the residues that interact with the carboxy terminus of class I peptides (Thr80, Tyr84, Thr143, Lys146, and Trp147) are also substituted by hydrophobic residues (Asp83, Met87, Pro146, Val149, and Leu150) in CD1. All these findings strengthen the possibility of binding of lipid or glycolipid antigens to the CD1 groove which is compatible due to the hydrophobic character.

The closely related nonclassical MHC class I molecules, T10 and T22, have been identified as a ligand for γδ-TCR (Moriwaki et al., 1993). The increased expression of T22 and/or T10 might trigger immunoregulatory γδ-T cells during immune responses (Crowley et al., 2000). T22 and T10 are specifically recognized by γδ-T cells, which do not require any peptide or ligand for cell-surface expression (Schild et al., 1994). A deletion of three amino acid in the α1 domain (residues 46–48) was observed with respect to class Ia molecules (Collins et al., 1994). The α2 domain has the least structural similarity to any known classical or nonclassical MHC-I-like molecule (Fig. 1). T22 has two potential N-linked glycosylation sites at (Asn86 and Asn150) in the α1–α2 domains. Interestingly, Asn86 highly conserved MHC class Ia molecules, while the second site is not conserved (Collins et al., 1994). Residues Arg6, Tyr7, Val25, and Arg35 are highly conserved in the sequence of T22b and that of MHC class Ia (Fig. 1). The structure of T22b is stabilized by two potential salt bridges formed between Arg6 and Asp102, and Arg35 and Glu40/Asp53. Further, position of a few residues is critically important and mutation of these residues (Leu5, Arg6, Tyr7, Val25, or Arg35) may cause a local disruption of the T22 structure (Wingren et al., 2000).

ZAG is the another member of MHC class I, and it appears to be responsible for the fat-depletion component of cachexia (Hassan et al., 2008b). Although, the helices of α1–α2 domains are almost identically positioned to their counterparts in peptide-binding class I MHC molecules forming an open groove, ZAG does not bind to any peptides. However, electron density of different ligands observed in the groove of ZAG, reflects its possible role in fatty acid depletion. The groove is predominated by hydrophobic residues Tyr14, Arg73, Ile76, Phe101, Trp115, Trp148, and Tyr161 (Hassan et al., 2008a). Among these residues, Arg73 is critically important for ZAG's activity while same position is occupied by His70 in HLA-A2 which points into the groove to separate the central and left pockets (Sanchez et al., 1999a; Hassan et al., 2008b).

MIC-A and -B are divergent membrane-bound members of the class I family, expressed in response to stress and activating receptor NKG2D, and provide signaling to the immune system cells (Li et al., 1999; Hue et al., 2004). Sequence analysis suggests that MIC-A has diverged significantly from the MHC class I family (Bahram and Spies, 1996) and shares sequence identities of 28%, 23%, and 35% in the α1, α2, and α3 domains, respectively, when aligned with the human MHC class I protein HLA-B27 (Fig. 1). The surface lining this pocket is not particularly hydrophobic as in the ZAG and CD1, while a number of charged or polar residues Asn69, Thr76, Glu92, Arg94, Arg108, and Trp146 present in the groove are conserved with class I like proteins (Holmes et al., 2002).

### B. TCR-Binding Residues

Their major known function of class I MHC molecules is antigen presentation during viral infections. It is now believed that the class I antigens associate with small fragments derived from viral products and present these complexes (pMHC) to cytotoxic T cells (CTL) in a form recognizable by TCR (Stroynowski, 1990; Rudolph and Wilson, 2002; Fig. 2). Recognition of pMHC by the TCR is necessary for the initiation and propagation of cellular immune responses, as well as the development and maintenance of the T cell repertoire (Margulies, 1997; Hennecke and Wiley, 2001). Structure analysis of HLA–TCR complex reveals that many residues are conserved among the molecules encoded by the HLA class I (Willcox et al., 1999; Marrack et al., 2008). These contact residues Glu58, Tyr59, Asp61, Lys68, Gln72, Lys146, Glu154, Gln155, Arg157, Tyr159,

Gly162, and Arg170 are constant and conserved in all three subgroups of HLA molecules (Reche and Reinherz, 2003). Baxter et al. (2004) showed that HLA-A2 have a "functional hot spot" that comprises Arg65 and Lys66 and is involved in recognition by most peptide-specific HLA-A2-restricted TCRs. They showed that the Lys66 is involved in the dual recognitions of both peptides and TCRs. Mutations in the α2 domain of HLA class I molecules can also interfere with CD8α/α binding suggesting the possibility that α2 domain on class I may be involved in CD8 interaction, perhaps including activated CD8 binding (Sun et al., 1995).

CD1 isotypes are meant for the presentation of specialized sets of lipidic antigens. Residues on the exposed surfaces of the α1H2, α2H1, and α2H2 of α helices are among the most variable in the collection of CD1 sequences and may confer T cell specificity for different isotypes (Burdin et al., 2000). It has been reported that a single mutation of Arg79Glu reduces proliferation of some mouse iNKT hybridomas (Porcelli and Modlin, 1999; Burdin et al., 2000). A similar result was observed in the double mutation (Arg79Glu79 and Asp80Ala) of CD1 (Sidobre et al., 2002). Another mutation of Ser76 to Gly showed a reduced ability of a few iNKT hybridomas to recognize α-GalCer, a ligand of CD1 (Burdin et al., 2000). Koch et al. (2005) suggested that these residues on the α1 helix are exposed in the putative TCR recognition surface and are probably involved in maintaining contact with the TCR.

The structural differences between T22$^b$ and MHC class Ia molecules clearly indicate that γδ- and αβ-TCRs recognize antigens differently (Schild et al., 1994). Moreover, the γδ-TCRs may be more like immunoglobulins in their recognition properties. To know the importance of T22$^b$ residues involved in TCR binding, Moriwaki et al. (1993) performed Ala mutation studies followed by KN6-stimulatory assay. They observed that four mutations (Tyr9Ala, Val25Ala, Ile23Ala, and Gln115Ala) did not affect the KN6-stimulatory activity appreciably. However, another four mutations (Tyr7Ala, Leu95Ala, Leu98Ala, and Leu116Ala) showed a drastic decrease in the KN6-stimulatory activity. Finally, it was suggested that mutations at the floor of the proposed antigen-binding groove of T22$^b$ affect recognition by the γδ-TCR. It is also evident from the crystal structure analysis that patches on the exposed floor of T22$^b$ may be in direct contact site for a complementarity determining region (CDR) loop of the γδ-TCR (Wingren et al., 2000). The ZAG and MHC share high sequence homology (sequence identities: ZAG with HLA-A2, 36%; FcRn, 27%; mouse CD1d, 23%; HFE,

36%, and MIC-A, 29%). However, ZAG is unlikely to associate with the T cell coreceptor CD8 because, of 15 CD8-binding residues of class I heavy chain, only one is conserved between class I and ZAG sequences (class I Asp122, ZAG Asp123).

The classical function of MHC molecule is antigen processing and presentation. However, the MIC-A and -B are ligands for the C-type lectin-like activating immunoreceptor, NKG2D (Bahram et al., 1994; Bahram and Spies, 1996). These proteins are expressed on the surface of cells in response to stress where they are recognized by γδ-T cells, CD8+ αβ-T cells, and NK cells through the NKG2D receptor (Bauer et al., 1999; Wu et al., 1999). Interestingly, NKG2D-MIC recognition events stimulate effectors response and serving a costimulatory function because NK cells and γδ-T cells positively modulates CD8+ αβ-T cell responses (Groh et al., 1998; Bauer et al., 1999). The exact mode of recognition at the sequence and structure level are not described.

## C. Light Chain-Binding Residues

X-ray structural studies suggest that the α3 domain of class I molecules and $\beta_2$M are structurally related to immunoglobulin domains and interact with each other, specifically having a role in enhancing peptide-binding *in vitro* (Bjorkman et al., 1987a,b; Vitiello et al., 1990; Fig. 2). The α3 domain is showing high conservation of sequences with classical molecules such as HFE, FcRn, and CD1 (Fig. 1). Functional constraints imposed by the binding of $\beta_2$M to the α3 domain, appear to be responsible for the amino acid sequence conservation (Govindarajan et al., 2003). Other site-directed mutagenesis studies suggest that the negatively charged loop (residues 222–229) in the highly conserved α3 domain of MHC class I molecules plays an important role in CD8-dependent CTL recognition and activation (Potter et al., 1989; Salter et al., 1990). Interestingly, in the absence of CD8 binding, this charged loop is highly flexible (Madden, 1995), whereas in the CD8αα/HLA-A2/peptide complex, the loop is clamped between the CDR1- and CDR3-like loops from both of the CD8 subunits (Gao et al., 1997). This observation was further supported by the inability of murine CD8-dependent CTL response by the α3 mutated class I molecules, which was attributed to the disruption of CD8 interaction with class I (Potter et al., 1989; Salter et al., 1990).

In the FcRn, the critical residues of the α3 domains (amino acids Lue216, Lys242, His248, and His249) involved in the FcRn/Fc interaction are conserved among the different species thus far analyzed (Burmeister et al., 1994b). HFE is also associated with β$_2$M. However, in the case of hereditary hemochromatosis, the mutation (Gly845Ala) leads to similar structural changes as in Cys260Tyr mutation which eliminates a disulphide bond in the α3 domain of HFE and prevents association with β$_2$M and cell-surface expression in cell-culture models (Lebron et al., 1998). In class I molecules, α3 domain contains a highly acidic loop comprising residues 220–229, which is essentially important for CD8 binding (Salter et al., 1990). The similar structure organization is observed in CD1 (on change) and even in FcRn which does not bind accessory molecules (Bjorkman and Burmeister, 1994; Burmeister et al., 1994a). The α3 domain of human ZAG contains an RGD sequence (Arg231, Gly232, and Asp233) thought to be involved in cell adhesion (Sanchez et al., 1999a). ZAG does not bind with β$_2$M. Instead it binds with new protein, PIP has been recently discovered in the cleft of ZAG (Hassan et al., 2008a). The displacement of the ZAG α3 domain compared to its class I counterpart results in the inability of β$_2$M to optimally contact the α3 and α1–α2 domains of ZAG which contributes to lack of affinity of ZAG for the β$_2$M. The most important interactions between ZAG and PIP occur through their adjacent β4 strands. These two strands are almost perfectly aligned in the antiparallel manner and generate a number of interactions between them. The potential backbone hydrogen bonds include residues Glu229, Arg231, Asp233, and His236 from ZAG with residues Lys41, Lys57, Thr59, Thr60, and Lys68 from PIP. These also include two salt bridges between pairs of residues, Lys41–Asp233 and Lys68–Glu229. Similarly, MIC proteins do not require either peptide or β$_2$M for folding, stability, or cell-surface expression (Li et al., 1999). Structurally, the α3 domains of MIC-A are joined through a flexible linker to α1–α2 platform, allowing considerable interdomain flexibility. This unique feature of MIC-A among MHC class I proteins and homologs may be a reason for its existence without β$_2$M (Tysoe-Calnon et al., 1991).

## III. Structure Analysis

The MHC superfamily comprised class I MHC and MHC-like proteins that are involved in a large variety of biological processes. The crystal structure determination of classical and nonclassical MHCs has been a prime focus of structure biologists as well as immunologists to explain the structural basis of antigen recognition and presentation. Structural basis of antigen recognition by various HLAs has been reviewed by many authors independently (Nathenson et al., 1986; Bjorkman and Parham, 1990; Stroynowski, 1990; Janeway, 1992; Germain and Margulies, 1993; Klein et al., 1993; Shawar et al., 1994; Madden, 1995; Lindahl et al., 1997). Class I MHC molecules acts as a peptide receptor and are capable of binding different peptides. However, each MHC molecule has a sequence and structure specific for a set of peptides. On the other hand are various molecules such as ZAG, HFE, FcRn, CD1, etc., which have class I MHC-like fold and considerable r.m.s. deviations (Table III) but cannot bind peptide; instead, they bind several specific ligands. Hence there is still a need to compare the structural features of all these proteins which have MHC-I-like fold, to gain an insight for the structural basis of protein and peptide interaction.

Crystallographic studies of all these molecules reveal that the two membrane distal domains ($\alpha$ and $\alpha 2$) of MHC form a superdomain comprising two $\alpha$ helices supported on a platform of eight $\beta$-pleated sheets, forming a groove-like structure, which has been postulated to be the antigen/ligand-binding function (Fig. 2; Bjorkman et al., 1987a,b; Madden, 1995). The enormous sequence variability has been reported in this region and are therefore likely to be involved in the selection of specific sets of peptides/ligand that are bound to the specific MHC-I-like molecules (Fig. 1). These findings provide a basis for the known diversity in immune responsiveness by different molecules (Stroynowski, 1990; Madden, 1995). Further, ligands should assume a specific conformation in the groove, and the availability of space for the ligand and the presence of charged, polar, and hydrophobic amino acids at certain positions in the groove would determine whether a given peptide or ligand can fit (Figs. 3A, B and 4A, B; Garrett et al., 1989). Another important features of class I molecule is binding with $\beta_2 M$ which binds noncovalently to the $\alpha 1$, $\alpha 2$, and $\alpha 3$ domains of the heavy chain of class I molecules. There are in fact four distinct regions that form the contact points between the $\alpha$-chain and $\beta_2 M$: one on each of the $\alpha$, $\alpha 2$ and two on the $\alpha 3$ domain (Fig. 2). Both $\alpha 3$ domain and $\beta_2 M$ resembles to the Ig constant domain. $\beta_2 M$ does not bind to MHC until peptide is first bound to the heavy chain in order to facilitate the formation of $\beta_2 M$-binding site on it (Ljunggren et al., 1990). Third important point is the binding of pMHC complex to the TCR, which is a

TABLE III
Structural Comparisons of Class I MHC Molecules, by Calculating Root Mean Square Deviations (RMSD) (Å²) for Their Backbone Carbon Atoms

| Class I molecules | HLA (2CLR) | CD1 (1CD1) | T22 (1C16) | FcRN (3FRU) | HFE (1A6Z) | ZAG (1ZAG) | MICA (1B3J) |
|---|---|---|---|---|---|---|---|
| HLA | 0 | 3.88 | 1.51 | 2.84 | 2.72 | 2.89 | 1.94 |
|     | 0 | 0.64 | 9.81 | 3.11 | 1.50 | 1.78 | 2.32 |
|     | 0 | 1.86 | 1.28 | 1.53 | 1.75 | 2.54 | 1.36 |
|     | 100 | 15 | 55 | 27 | 39 | 36 | 29 |
| CD1 |  | 0 | 4.94 | 1.45 | 1.32 | 2.37 | 2.32 |
|     |  | 0 | 7.04 | 2.95 | 2.86 | 1.72 | 3.15 |
|     |  | 0 | 2.16 | 1.19 | 1.32 | 2.31 | 2.31 |
|     |  | 100 | 20 | 21 | 20 | 20 | 15 |
| T22 |  |  | 0 | 3.64 | 2.31 | 2.59 | 1.81 |
|     |  |  | 0 | 3.44 | 1.18 | 3.49 | 2.07 |
|     |  |  | 0 | 3.64 | 1.34 | 2.54 | 1.62 |
|     |  |  | 100 | 24 | 30 | 30 | 26 |
| FcRn |  |  |  | 0 | 3.21 | 2.79 | 4.07 |
|      |  |  |  | 0 | 2.22 | 2.78 | 4.07 |
|      |  |  |  | 0 | 0.90 | 2.15 | 1.41 |
|      |  |  |  | 100 | 28 | 28 | 22 |
| HFE |  |  |  |  | 0 | 3.16 | 6.49 |
|     |  |  |  |  | 0 | 1.85 | 2.79 |
|     |  |  |  |  | 0 | 2.21 | 1.42 |
|     |  |  |  |  | 100 | 35 | 26 |
| ZAG |  |  |  |  |  | 0 | 3.54 |
|     |  |  |  |  |  | 0 | 3.54 |
|     |  |  |  |  |  | 0 | 2.51 |
|     |  |  |  |  |  | 100 | 27 |

The comparisons are shown in blue[a] row for whole A chain, white row for α1–α2 superdomains, and green row for α3 domain. The sequence identities are given in yellow box. RMSD and sequence identity were calculated by Swiss PDB Viewer and clustalW, respectively.

[a] For interpretation of the references to color in this figure legend, the reader is referred to the Web version of this chapter.

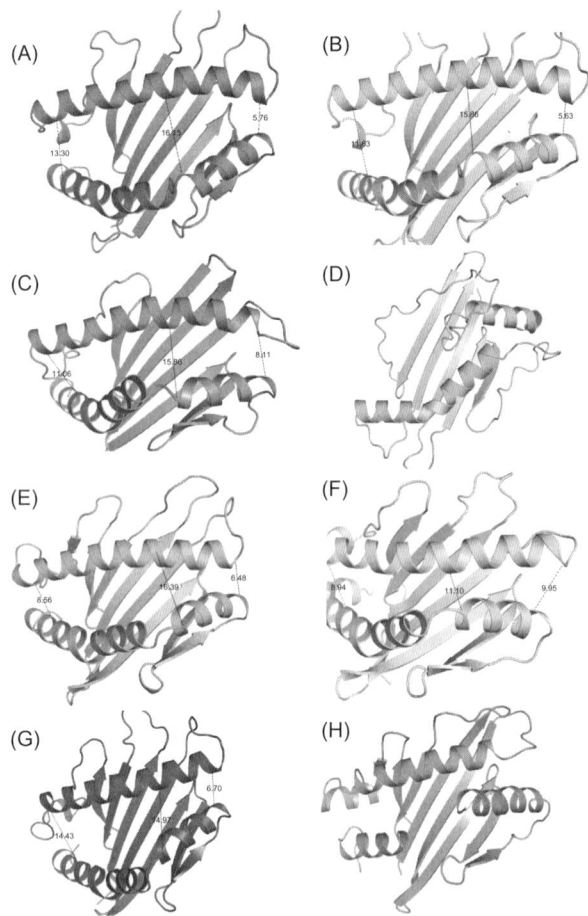

FIG. 3. Cartoon diagram showing top view of the ligand binding α1 and α2 domains of various MHC-like molecules. (A) HLA-A2 is involved in the presentation of peptides (orange); (B) HLA-B27 also binds to peptides (cyan); (C) a hydrophobic binding groove of the CD1 (pink); (D) the nonclassical MHC molecule T22, which is a γδ-T cell ligand (light green); (E) Fc transported molecules FcRn (green); (F) TfR-binding protein HFE (yellow); (G) a view of the α1 and α2 domains of ZAG showing the proposed site for fatty acid binding (sky blue); and (H) the α1 and α2 domains of nonclassical MHC-like molecule MICA showing that its groove is structurally analogous to a class I molecule. The distance between the helices was calculated in PyMol. The structure was drawn using atomic coordinates of PDB code, 2CLR (HLA_A2), 3HLA (HLA_B27), 1C16 (T22), 1A6Z (HFE), 1ZAG (ZAG), 3FRU (FcRn), 1B3J (MIC_A), and 1CD1 (CD1). (See color plate 8).

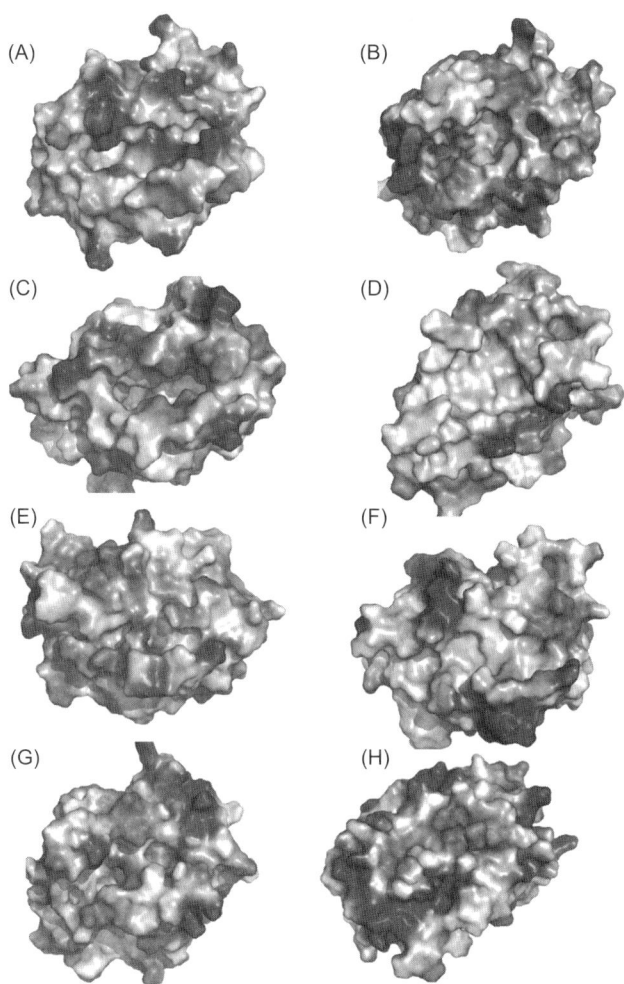

FIG. 4. Comparison of molecular surfaces of the ligand-binding grooves formed by α1 and α2 domains of MHC class I and class I-like proteins. The electrostatic potentials mapped to the surface of α1–α2 superdomains. (A) Size and charge distribution of HLA-A2 groove showing it is sufficient to accommodate peptide; (B) HLA-B27 is also having identical groove but the charge distribution is different; (C) the groove molecular surface of CD1 is neutral and hydrophobic compared to the grooves of the other MHC-like molecules (largest and deepest groove of all molecules) showing its efficiency in lipidic antigen binding; (D) T22 is not having proper groove hence it does not bind to any peptide or ligand; (E) there is essentially no groove in FcRn; (F) HFE binds TfR and

crucial step for antigen processing (Madden, 1995; Margulies, 1997). Structure analysis of pMHC–TCR complex reveals that CDR forms many contacts with highly conserved residues of helices of α1 and α2 domains of MHC-I (Fig. 2). Structurally, CDRs $2_\alpha$ and $2_\beta$ lie directly on top of the α2 and α1 helices, respectively, and, therefore, interact exclusively with the MHC heavy chain (Garcia et al., 1998). The analysis of all three mentioned features of class I MHC-like molecules showed many striking differences in their conformation and topology. Here, many significant structural features of these molecules will be discussed in order to relate functions of these molecules with their structures.

### A.  Structure of Ligand-Binding Groove

The classical MHC molecules are highly polymorphic, especially in the α1 and α2 domains. The hypervariable residues found on the top of helices of the groove attain significant peptide- and TCR-binding properties (Fig. 2). A given antigen-binding groove is capable of binding hundreds or thousands of different peptides that may be identical or homologous at only a few side chain positions (Engelhard, 1994a,b). The size of groove is sufficient to accommodate peptide antigen (Fig. 3A and B). A collection of hydrogen and van der Waals contacts offered by the main chain and side chain atoms of peptide as well as groove help to stabilize the binding of any peptide in the corresponding groove (Madden, 1995). The size limitation of peptides binding to MHC class I is mainly determined by the amino- and carboxy-terminal ends of the antigen-binding cleft (Fremont et al., 1992; Madden et al., 1992). Structurally, the antigen-binding groove can be divided into three distinct regions: a narrow deep cleft responsible for binding to the N-terminal part of the peptide, a more shallow broad mid-cleft depression over which 5–9 amino acid residues of the peptide are

---

regulates iron homeostasis having a shallow groove; (G) the molecular surface of the central portion of the ZAG groove is nearly neutral except for the contribution of Arg73 (positively charged); and (H) MIC is showing a big shallow groove essentially negative in charge that may be suitable to bind NK cell receptor NKG2D. Electrostatics potentials were calculated in PyMol with full formal charge. Positive potential ($\geq 15$ mV), neutral potential (0 mV), and negative potential ($\leq 210$ mV) are shown in blue, gray, and red, respectively. The atomic coordinates are same as used in Fig. 3. (For interpretation of the references to color in this figure legend, the reader is referred to the Web version of this chapter.)

binding, and the characteristic deep hydrophobic F pocket which binds to the carboxy terminal of peptide (Achour et al., 1998). It is evident from the crystal structure and sequence analysis that certain residues in the antigen-binding cleft found to be highly conserved, such as Tyr7, Tyr59, and Tyr171, form a network of hydrogen bonds directly to the N-terminus of peptide (Fig. 1). However, Tyr84, Thr143, and Lys146 interact with C-terminus of peptide. Hence these are considered as essential residues for holding the peptide in the cleft despite of sequence specificity (Engelhard, 1994a,b; Madden, 1995). Moreover, these conserved residues are found to be present in two relatively shallow and open pockets (A and F) that are specific for binding to the amino- and carboxy-terminal of peptides. In addition, the central part of the cleft contains many pockets (B, C, D, and E), which presumably bind to the side chain of complementary peptides (Madden et al., 1991, 1992; Fremont et al., 1992; Matsumura et al., 1992). Further, residues contributing to these pockets are polymorphic, and therefore, these molecules are characterized by differences in the precise nature (charge) and position in the pockets (Fig. 4A and B; Matsumura et al., 1992). Interestingly, a single-residue polymorphism distinguishes HLA-B*3508 (Arg156) from HLA-B*3501 (Leu156) and controls differential peptide selection, even though HLA-B*3501 can bind to the LPEP epitope *in vitro* (Tynan et al., 2005a). The $\alpha 1$ and $\alpha 2$ helices of free and bound pMHC superimpose closely (0.72 Å r.m.s. deviation between $C^{\alpha}$ atoms; Garcia et al., 1998).

CD1 may have evolved to present antigens that are predominantly found in microbial pathogens (Porcelli and Modlin, 1999). It is evident from the crystal structure that CD1 folds very much like MHC class I molecules including arrangements of domain and secondary structure elements (Fig. 3C). However, a notable difference in the groove was observed as the first $\alpha 1$ helix (H1) adopts an irregular, extended conformation in CD1 (Zeng et al., 1997). Interestingly, the $\alpha 1$ domain of CD1 is structurally more similar to MHC class I molecules (r.m.s. deviation = 1.9 Å) and FcRn (r.m.s. deviation = 1.8 Å), whereas the $\alpha 2$ domain is more similar to the structure of FcRn (r.m.s. deviation = 0.9 Å) than to class I molecules (r.m.s. deviation = 1.5 Å; see Table III). For the combined $\alpha 1 - \alpha 2$ domain, the r.m.s. deviation is 2.6 Å between FcRn or class Ia molecules (Zeng et al., 1997). The putative ligand binding site of CD1 is quite distinct from other MHC-like molecules (Fig. 4C). The binding groove of CD1 is narrower as compared to class I molecule (14 Å) and is broader than FcRn (~12 Å;

Fig. 3C). The groove of CD1 is 4–6 Å deeper, presumably due to the presence of bulkier residues (Phe49, Trp40, and Phe18). The A′ pocket in CD1a, CD1b, and mouse CD1d is the more conserved region of the lipid-antigen-binding groove (Gadola et al., 2002; Zajonc et al., 2003; Koch et al., 2005). The pocket A′ of MHC class I contains a cluster of tyrosine residues (Tyr7, Tyr59, and Tyr171) to form hydrogen bonds with the amino-terminus of the bound peptide. However, in mouse CD1d1, the corresponding pocket wall contains hydrophobic residues (Phe10, Trp63, and Phe171), emphasizing the role of CD1 in the binding of lipidic antigen (Fig. 4A; Bjorkman et al., 1987b; Zeng et al., 1997). Similarly, the F9 end of class I molecule at the end of the groove usually have polar residues (Thr80, Tyr84, Thr143, Lys146, and Trp147) which interact with the carboxy-terminus of the peptide. However, the same are substituted by mostly hydrophobic residues (Asp83, Met87, Pro146, Val149, Leu150) in CD1. In addition to the hydrophobic residues, CD1 has least number of residues in the groove, which have capacity to form hydrogen bonds to the bound ligands. These residues are Gln14, Ser28, Arg74 (at A′ pocket), Tyr73, Ser76, Asp80, and Thr156 (at the entrance of pocket; Fig. 4C; Zeng et al., 1997). The structure analysis of CD1a in the complex with a sulfatide self antigen (Zajonc et al., 2003) reveals that A′ pocket of CD1a forms a semicircular curve between the $\alpha 1$–$\alpha 2$-helices, limited by Val28, His38, Leu114, Leu162, and Phe169 and a clearly defined terminus. In contrast, the F′ pocket is long and straight, delineated by Trp14, Ile96, and Leu148. It becomes wider as it approaches the surface of the binding groove due to the presence of Val98, Leu116, Phe144, and Val147. The sulfogalactosyl unit is located between Glu154 ($\alpha 2$-helix) and Arg76 ($\alpha 1$-helix). The galactose moiety is stabilized by hydrogen bonding to Arg73 and Ser77, whereas the 3′ sulfate group hydrogen bonds to Arg76 and Glu154 and to a water molecule that is in complex between Arg73 and Glu154. The fatty acid chain of the sulfatide protrudes at the T cell recognition surface and makes van der Waals contacts with Ile81, Phe144, and Val147 of the F′ pocket. The $\alpha 1$–$\alpha 2$ domain of human CD1a has the smallest ligand-binding capacity of the CD1 molecules, with a volume of 1300 $\text{Å}^3$, which is the most distinctive and has only one pocket (A′) that is similar in length and orientation to those of the other CD1 molecules (Koch et al., 2005).

Another MHC-like molecule does not seem to present any antigen such as $\gamma\delta$-TCR ligands T10 and T22 (Wingren et al., 2000). The $\alpha 1$ of MHC class I molecules comprises two segments, $H_1$ (residues 49–54) and $H_2$

(residues 56–85). Interestingly, T22 is devoid of $H_1$ segment and the $H_2$ is shorter (residues 60–86). The α2 domain of T22 has the least structural resemblance to any known classical or nonclassical MHC-like molecule (Fig. 3D). The α2 of MHC class I molecules contains three segments, named $H_1$ (residues 138–150), $H_{2a}$ (residues 153–162), and $H_{2b}$ (residues 164–180). Similarly, T22 does not have the $H_1$ segment and very short $H_{2a}$, whereas $H_{2b}$ is approximately same with a r.m.s. deviation (residues 163–180) of ~0.35 between T22 and class Ia molecules (Wingren et al., 2000; see Table III). Further, the minimum distance between the α1 and α2 helical structure is only about 6–10 Å ($C^α$ to $C^α$), compared with 12–20 Å (class Ia and II), 10–13 Å (FcRn), and 7–10 Å (MIC-A);(Fig. 3). The groove formed by α1–α2 domains are not as precise to accommodate any peptide or ligand (Fig. 4D). These evidences are sufficient to describe the inability of T10 and T22 to bind a peptide.

FcRn transfers IgG from the mother to the fetus across the placenta and the proximal small intestine (Ghetie and Ward, 2000; Roopenian and Akilesh, 2007). In order to understand the mechanism of IgG transport by FcRn, the crystal structure was determined in the native (Burmeister et al., 1994a) and the complex (FcRn–IgG) forms (Burmeister et al., 1994b; Vaughn and Bjorkman, 1998). Like class I MHC molecules, the α1 and α2 domains groove-like structure, where the width of cleft is very low, hence FcRn forms narrowest groove and is devoid of any peptide-binding capability (Fig. 3E). The narrowing of groove is due to the Pro162, conserved in all FcRn, which causes a dramatic movement of the α2 $H_{2a}$ and $H_1$ helices toward α1, leading to an effective collapsing of the groove at one end (Burmeister et al., 1994a). Although, the groove of FcRn primarily consists of Tyr9, Tyr60, Phe156, His168, and Arg164 as similar to the class I molecules, it cannot accommodate any peptide because of Arg164 which effectively obstructs the pocket A (Fig. 4D). It is evident from the FcRn–Fc cocrystal structure that FcRn binds to the Fc portion of IgG at a site that is distinct from the binding sites of the classical component of the complement such as FcγRs or the C1q (Burmeister et al., 1994b). The FcRn binds to the $CH_2$–$CH_3$ hinge region of IgG, a versatile region of Fc responsible to bind with various microbial antigens (Derrick and Wigley, 1992; Roopenian and Akilesh, 2007). Further, FcRn binds to the Fc in a pH-dependent manner (Vaughn and Bjorkman, 1998). Interestingly, at physiological pH 7.4, FcRn does not bind IgG, and at acidic pH of the endosome (pH 6–6.5), FcRn strongly binds to the Fc region of IgG.

Moreover, FcRn does not undergo any significant conformational change when it binds IgG (Vaughn and Bjorkman, 1998). It has been observed from structure analysis (Martin et al., 2001) and site-directed mutational studies (Vaughn et al., 1997) that Glu115 and Asp130 in human FcRn and Glu117 and Glu132 in rat FcRn are responsible for the pH-dependent binding to residues His310 and His435 on Fc (Shields et al., 2001). Apart from hydrogen bonds, the hydrophobic interaction offered by Ile253 of IgG to the Trp133 of FcRn is equally important. Another important interaction between Ile1 of $\beta_2M$ with the hydrophobic residue at position 309 of Fc can also be useful in the Fc–RcRn complex formation. Recently, the crystal structure of FcRn with the peptides which inhibit the binding of the natural ligand IgG has been determined (Mezo et al., 2010). Structure analysis reveals that the peptide acts as competitive inhibitors of the IgG–FcRn interactions and it binds to the same site on FcRn as does Fc.

A considerable sequence and structure similarities between HFE and class I MHC molecules have been noted. However, HFE neither binds peptides and TCR nor serves any known function in the immune system (Beutler, 2006). The crystal structure analysis of HFE clearly explains the reason for HFE unable to bind any peptide. As compared to the groove of class I MHC, its groove is narrowed by a translation of the $\alpha 1$ helix bringing it to $\sim 4\,\text{Å}$ closer to the $\alpha 2$ helix, which is almost identically positioned (Fig. 3F). Each domain of HFE exactly superimposed upon the corresponding domains in HLA-A2 (Bjorkman et al., 1987b), FcRn (Burmeister et al., 1994a), and CD1 (Zeng et al., 1997) with r.m.s. deviations $\sim 1.5\,\text{Å}$ for most $C^{\alpha}$ atoms (see Table III). Amino acid residues in the A pocket of class I molecules are Tyr7, Tyr59, Tyr159, and Tyr171 that are meant to interact with the N terminus of the peptide, including Trp167. Interestingly, only two of the corresponding tyrosines are conserved (Tyr10 and Tyr160) in HFE while the Tyr10 is buried (Lebron et al., 1998). Further, the side chain of Gln168 of HFE points into the groove to occlude pocket A in the same way as Arg164 in FcRn (Burmeister et al., 1994a). The structures of various HLA molecules with the peptide have been superimposed in order to predict the clashing residues of HFE which interfere with the peptide binding. It was observed that a total of 10 side chains (eight from $\alpha 1$ helix and two from the $\alpha 2$ helix) are sterically clashed with the peptide in the binding groove (Lebron et al., 1998). This occurs due to the translation of the $\alpha 1$ helix toward the $\alpha 2$ helix and the presence of larger side chains in HFE as compared to class I (HFE:

Leu69, Trp72, Met75, and Arg153 vs. class I Val67, His70, Thr73, and Val152). In addition, Trp114 on the HFE groove floor is also interfering with peptide binding. Surface area measurements reveal that total surface area of the groove in HFE (415 Å$^2$) is intermediate between those in FcRn (235 Å$^2$) and class I molecules (760 Å$^2$; Lebron et al., 1998; Fig. 4F). However, CD1 has a narrower but deeper groove than class I grooves, with the most extensive surface area of 1440 Å$^2$ (Zeng et al., 1997). All these findings clearly suggest the inability of HFE toward peptide binding (Lebron et al., 1998). However, it is reported that HFE binds to the TfR at the pH of the cell surface (pH 7.4; Feder et al., 1998), a macromolecule involved in the iron homeostasis (Aisen et al., 1999). The HFE–TfR structure was determined and observed that one HFE molecule is bound to each chain of the TfR homodimer to form a twofold symmetric complex (Bennett et al., 2000). The primary intermolecular contacts are formed between HFE α1–α2 domains and the helical domain of TfR, whereas the α3 and β$_2$M domains of HFE do not interact with TfR. The interaction between HFE and TfR covers a large interface; a total of 2000 Å$^2$ of solvent accessible surface area is buried upon binding between these two molecules. It has been demonstrated that HFE binds TfR with high affinity in a pH-dependent manner (Lebron et al., 1999), presumably because of the critical histidine residue. It was postulated that His74 and His150 of HFE are involved in hydrophobic contacts with TfR residues and got disrupted on changing the pH (Lebron et al., 1998; Lebron and Bjorkman, 1999). Other possibility may be arisen due to the histidine residues from TfR which has two histidine (His684 and His475) in each polypeptide chain, which are found to be present on the TfR dimer interface (Bennett et al., 2000).

The α1–α2 superdomain of the heavy chain of ZAG forms a groove-like structure which serves as the binding site, an analogous to the antigen-binding site in class I MHC and CD1 (Fig. 3G; Madden, 1995). The crystal structures of ZAG in the native form (Sanchez et al., 1999a) as well as in complex with PIP (Hassan et al., 2008a) have been determined independently, and in both the studies, it was proven that the groove is open and occupied by some ligand (unidentified electron density) presumably crucial for the biological function of ZAG. The molecular surface of the central part of the groove appears to be neutral in charge, due to the presence of 3 Trp, 4 Tyr, 2 Phe, and 1 Ile residues. However, one Arg73 protrudes its side chain from one side of the groove. Interestingly, replacing Arg73 to alanine has abolished ligand binding (McDermott et al., 2006). The groove of ZAG is similar to the ligand-binding site of CD1 which is also hydrophobic in nature, suggesting that ZAG may bind hydrophobic ligand although its groove is not as much deeper as CD1 groove (Fig. 4G). Structure studies showed that ZAG, CD1, HFE, and FcRn contain prolines in their $\alpha 2$ domain at a position corresponding to Val165 in classical class I MHC molecules (Bjorkman and Parham, 1990). This proline in FcRn and CD1 causes a kink at a position near their proline residues. However, helices in ZAG and HFE are similar to the $\alpha 2$ domain helices of class I molecules without any kink. In addition to crystal structure determination (Sanchez et al., 1999a), various studies have been carried by Bjrokman and coworker (Kennedy et al., 2001; Delker et al., 2004; McDermott et al., 2006) in order to determine the critical residues of ZAG responsible for binding to a particular fatty acid. Therefore, it may be possible to manipulate the biological effects of ZAG with knowledge of an appropriate ZAG-binding ligand. Despite of low sequence similarity with class I molecules, residues closest to the ZAG ligand are completely conserved in between ZAG and class I molecules (Sanchez et al., 1999a).

There are many cell-surface receptors which serve as general stress sensors. Usually they do not present any peptide antigens and are independent of transporter associated with antigen processing (Stephens, 2001). These nonclassical MHC molecules comprising ligands for the NK cell lead to the activation of NKG2D receptor that apparently lacks affinity for small molecule antigens or peptide (Wolan et al., 2001). These molecules are human MIC-A, MIC-B, ULBP, murine Rae1, and H60 (Vivier et al., 2002). Crystal structures of MIC-A (Li et al., 1999) and Rae-1$\beta$ (Li et al., 2002) indicated that the loss of peptide (or any other ligand) binding was due to the reduced distance between the $\alpha 1$ and $\alpha 2$ helices (7–10 Å) which do not form any appropriate groove (Fig. 3H). On superimposition of this groove over class I proteins, many structural differences are observed, especially in the loop between the first two strands of the $\beta$-sheet (residues 13–20) and the hairpin connecting the third and fourth $\beta$-strands (residues 37–40; Li et al., 1999). Further, the helix 1 ($\alpha 1$ domain: residues 45–54) is much longer than any other class I homolog, whereas the $\alpha 2$ domain of MIC-A is quite similar to that of FcRn. The groove is also not sufficient to hold a ligand (Fig. 4H). In Rae-1$\beta$, these helices come close enough to permit formation of a noncanonical disulfide bond with a leucine-rich interface filling the former ligand-binding cavity.

## B. Structural Basis of TCR Binding

MHC–peptide–TCR recognition is the fundamental process required to trigger the cellular immune response. The ability of a pMHC complex to activate a T cell generally correlates with the strength and duration of TCR binding (Davis et al., 1998). An important question is: Does the unique structural properties of TCR–pMHC interface facilitate specificity? To answer this question, residues of MHC bound to the TCR have been extensively studied (Rudolph et al., 2006; Marrack et al., 2008). From crystal structure analysis of TCR/pMHC complexes (Garcia et al., 1998), it was suggested that this interaction is likely to be similar to the well-characterized interactions between antibodies and antigens (Hennecke and Wiley, 2001). Typically up to three upward-facing, solvent-exposed residues from the antigenic peptide contribute directly to the TCR interaction (Rudolph and Wilson, 2002). The TCR binds pMHC in a conserved diagonal orientation that positions the CDR1 and CDR2 loops mainly over the MHC and the CDR3 loops over peptide. TCR recognition is highly sensitive to small numbers of ligands among thousands of different pMHC complexes, and yet it is often highly specific for a particular pMHC combination (Germain and Stefanova, 1999). It is evident from the structures of TCR–pMHC that the architecture of a given peptide-binding groove can impose a recurrent structural signature on the peptides and hence it accommodates and thereby restricts the repertoire of CDR3s with which it can cooperate (Housset and Malissen, 2003). The flexibility in CDR3 and variability in $V_\alpha$–$V_\beta$ pairing angles facilitates the adaptation of the TCR to conformational constrain on pMHC surface (Willcox et al., 1999). Thermodynamic analyses of TCR–pMHC interactions further suggest that the TCR and/or pMHC possess flexible binding surfaces that are stabilized on complexation (Boniface et al., 1999).

Large conformational changes in three of the TCR CDR loops are induced upon binding, providing a mechanism of structural plasticity to accommodate a variety of different peptide antigens (Garcia et al., 1998). It is reported that a structural plasticity or flexibility is essential for the recognition of pMHC (Kuzushima et al., 1995) in response to ligand (Janeway, 1992), and properties for the survival and function of T cells. In order to determine structural alterations in either the TCR or pMHC upon complexation, Garcia et al. (1998) reported the crystal structure of the mouse 2C TCR in complex with mouse MHC class I H-2Kb bound to the self-peptide dEV8 (EQYKFYSV) and H-2Kb–dEV8 unliganded (Garcia et al., 1996). The structure analysis has clearly explained the degenerate specificity of TCR–pMHC interaction in terms of a structural plasticity in the TCR–pMHC interface. About 1876 Å$^2$ of surface is buried in the 2C-Kb interface, of which 900 Å$^2$ is contributed by the TCR and 976 Å$^2$ by the pMHC (Garcia et al., 1996, 1998). Overall, the Vα CDR1 and CDR2 contact residues appear to be more highly conserved, not only within Kb-restricted TCRs but also across other TCRs compared with the corresponding β-chain contact residues. In particular, Ser27 of CDR1α, that forms hydrogen bonds to the conserved Kb residue Glu58, and Ser51 of CDR2α, that contacts Glu166, are the most frequently occurring residues to occur in Vα gene sequences at these positions (Garcia et al., 1996, 1998). These interactions between conserved residues are consistent with recent data indicating a critical role for Vα residues at 27th and 51st in the restriction of particular murine TCRs (Sim et al., 1996). Other potentially conserved contacts appear between α (Tyr31) and β (His29, Glu56) chain of TCR residues to the MHC helices. The TCR interaction with the peptide is mediated directly and indirectly by hydrogen bonds to the functional groups of the upward-facing side chains (P1, P4, P6, and P7) from CDRs $1_\alpha$, $1_\beta$, $3_\alpha$, and $3_\beta$, which are in simultaneous contact with the α1 and α2 helices of the MHC. It was observed that when the 2C TCR in the complex is compared with the unliganded 2C, the individual α and β chains superimpose with r.m.s. deviation of 1.2 and 0.7 Å, respectively, for back bone carbon atoms (Garcia et al., 1996, 1998).

Longer peptides can either bind by extension at the C terminus (Speir et al., 2001) or due to the fixing of their termini, bulge out of the binding groove, providing additional surface area for TCR recognition (Tynan et al., 2005a). It is evident from the crystal structure analysis of pMHC and TCR complex that the CDR1α, CDR2α, and CDR3α loops of the Vα domain of TCR provided the main contacts with HLA molecule onto the α2-helix-spanning residues (150–158: Tynan et al., 2005b). More precisely, Asn50α and Phe52α of TCR fill the void between the TCR and the MHC because the Phe52α has been packed against Arg157, Glu154, and Ala158 of HLA, whereas Asn50α forms many interactions with Glu154. Moreover, Phe95α and Tyr96α form a contact site to interact with the Gln155 of HLA. This Gln155 have multifunctional role in switching of its interactions between peptide and TCR (Ding et al., 1998), suggesting that this residue may act as a critical residue for guiding MHC class I-restricted T cell recognition (Kjer-Nielsen et al., 2003). Further, Leu98$_\beta$ from the CDR3$_\beta$ loop forms many van der Waals interactions with Gln155 and Arg151, whereas Glu101$_\beta$ forms a salt bridge to the Arg151 of the α2-helix of MHC. Taken together, these findings provide a structural basis for the highly constrained αβ-TCR restriction toward the pMHC complex.

However, in the nonclassical MHC molecule such as CD1, the ligand-binding groove is deeper, narrower, and more hydrophobic than in class I MHCs (Zeng et al., 1997). The groove formed in such a way that the lipid tails of glycolipids and lipopeptides are bound in the groove and their polar moieties presented to T cells (Zajonc et al., 2003; Koch et al., 2005; Moody et al., 2005). The structure analysis (Koch et al., 2005) has confirmed that three hydrogen bonds offered by key residues (Asp80, Asp151, and Thr154) to the α-galactoceramide are seemingly crucial for maintaining α-galactoceramide in the correct position and orientation for recognition by the TCR (Marrack et al., 2008). Moreover, the hydrophilic residues of ligands are necessary for the specific TCR-mediated recognition, whereas the hydrophobic moiety is for the presentation by CD1 (Porcelli and Modlin, 1999). The highly acidic loop, α3 residues 220–229 from class I implicated in CD8 binding (Salter et al., 1990), retains some structure similarity in CD1 as well as in FcRn (does not bind antigen) (Burmeister et al., 1994a; Zeng et al., 1997). However, a single residue deletion occurs in the putative binding loop of CD1 compared with other MHC class I molecules (Konig et al., 1992).

Structure analysis revealed that T22$^b$ adopts modified MHC-like fold, whereas it lacks an appropriate classical peptide-binding groove (Wingren et al., 2000). Hence, it was suggested that T22$^b$ and T10 do not require any peptide for recognition by stimulation of γδ-T cells (Schild et al., 1994). The structural differences between αβ- and γδ-TCR and between T22$^b$ and MHC class I molecules support the idea that γδ- and αβ-TCRs recognize antigens differently (Schild et al., 1994; Chien et al., 1996; Allison et al., 2001). Moreover, the γδ-TCRs may be more like immunoglobulins in their recognition properties (Li et al., 1998), not restricted in their choice of ligands or epitopes (Chien et al., 1996). CD8 primarily binds to residues of α3 domain, especially residues from 223 to 229, similar to the class I molecules. The r.m.s. deviation calculated for this region was found to be 0.8–1.2. In addition to that, three residues (Gln115, Asp122, and Glu128) in the α2 domain is also responsible for binding of CD8 (Gao et al., 1997) and are conserved in T22$^b$ and class I molecules.

The α1 and α2 domain of HFE are involved in TfR binding in the same fashion as class I MHC molecules bind to TCRs (Bennett et al., 2000). However, many significant variations are observed in the complex of HFE–TfR as compared to the class I MHC interactions with TCRs. Despite the fact that TCRs and TfR bind across the α1–α2 domain, class I-restricted TCRs bind with a diagonal interaction with α1–α2 domains in order to avoid the kink of the helices (Lebron et al., 1998). However, the α2 helix of HFE, having a kink, is a major contact point for TfR. Further, TCRs recognize α1–α2 helices along with the peptide bound in the groove, whereas the same groove in HFE is narrowed and devoid of any peptide or other ligand. Interestingly, the distance between α1 and α2 helices of HFE is critical for its interaction with TfR. Hence it was suggested that if the HFE groove got broadened, may be due to the binding of peptide, it presumably would prevent the interaction between HFE and TfR (Bennett et al., 2000).

Despite of a considerable sequence and structure similarity between ZAG and class I molecules, structural features of the ZAG α3 domain make it unlikely to associate with the T cell coreceptor CD8. Out of 15 residues responsible for CD8 binding in the HLA-A2/CD8 cocrystal structure (Gao et al., 1997), only one is conserved between class I and ZAG sequences (class I Asp122, ZAG Asp123). Further, any interaction between ZAG and TCR can be possible, because the class I MHC residues that contact these receptors are not particularly conserved with respect to the ZAG (Ding et al., 1998; Garcia et al., 1998). However, it may be suggested that ZAG has a completely different and unidentified $\beta_3$-adrenergic receptor which plays a significant role in lipid metabolism especially in cancer cachexia (Ojcius et al., 2002).

MIC-A and MIC-B are compatible with γδ-TCRs (Bahram et al., 1994; Chien et al., 1996), instead of αβ-TCR (Davis et al., 1998), and are proposed to mediate antimicrobial responses and regulate innate and acquired immunity. The mode of interaction between γδ-TCRs and their ligands are different from that of interaction between αβ-TCRs and their ligands (Germain and Margulies, 1993). Unlike conventional MHC class I proteins, MIC-A requires neither peptide nor $\beta_2$M for stability or cell-surface expression (Groh et al., 1998). However, MIC-A is known to form an unrelated complex of the NK cell receptor NKG2D (Li et al., 2001). Interestingly, MIC-A refolds upon receptor binding from its partially unwound α2-helix structure, whereas, this kind of refolding is not observed with T22 which also binds to γδ-TCR does (Rudolph et al., 2006).

## C. Structural Basis of Light Chain Binding

$\beta_2M$ is noncovalently bound to class I MHC proteins and is required for their surface expression and function (Vitiello et al., 1990). However, MHC-I-like proteins may or may not interact with $\beta_2M$. The membrane proximal end of the class I molecules is formed by the $\alpha3$ and $\beta_2M$ domains, which adopts a standard immunoglobulin-like fold, characterized by a β-sandwich formed by close packing together of a three-stranded and a four-stranded β-sheet (Madden, 1995). It is required to predict the $\beta_2M$ binding (or nonbinding) to the various members of class I MHC superfamily in order to decipher their function and to understand the molecular recognition mechanisms. Structure analysis of class I molecules reveals that the $\alpha3$ and $\beta_2M$ residues that are involved in the contact interface are within their respective four-stranded sheets. There are more extensive contacts between $\beta_2M$ and the $\alpha1-\alpha2$ platform than between $\alpha3$ and the platform (Bjorkman et al., 1987b), because $\beta_2M$ supports the $\alpha1-\alpha2$ platform. Finally, $\alpha3$ domain interacts extensively only to the $\beta_2M$, whereas $\beta_2M$ interacts with all three domains of the heavy chain (Bjorkman and Parham, 1990). It is evident from the structure analysis that 19 residues of the heavy chain form 44 contacts with 18 different residues on human $\beta_2M$ (Tysoe-Calnon et al., 1991). In between $\alpha1$ domain and $\beta_2M$, several side chains pointed into a central pocket, where Asp53 of $\beta_2M$ forms two ionic and two hydrogen bonds with Gln32, Arg35, and Arg48 on the $\alpha1$ domain. Further, Ser55 and Tyr63 of $\beta_2M$ offer hydrogen bonds to the hydroxy group of Tyr27 and Tyr63. Similarly, between $\beta_2M$ and the $\alpha2$ domain, many well-defined hydrogen bonds are observed, connecting His31, Trp60, and Phe62 of the $\beta_2M$ and Gln96 on the $\alpha2$ domain. Further, there are two contact zones observed in the case of $\alpha3$ and $\beta_2M$ interactions. The upper patch (residues near to $\alpha1-\alpha2$ domains) was formed by a large number of closely spaced interactive contacts between these domains (Tysoe-Calnon et al., 1991). There are six interactions formed by the main chain atoms of $\alpha3$ domains (Gln232, Pro235, Ala36, and Gly37) to the side chains of Gln8, Tyr10, Arg12, and Asn24 of the $\beta_2M$. However, the lower patch involves the two C-terminal residues Asp98 and Met99 of $\beta_2M$ and His192, Arg202, Arg234, and Trp244 of the $\alpha3$ domain (Tysoe-Calnon et al., 1991).

Although CD1 molecule has identical structure assembly, the relative position of the $\beta_2M$ and the CD1 $\alpha1-\alpha2$ domains have a slight rotation and translation as compared to the arrangement in class I molecules (Matsumura et al., 1992; Zeng et al., 1997). After aligning all homologous molecules on their $\beta_2$M domains, the transformation necessary to align the $\alpha 1\alpha 2$ domains with CD1 is a 0.4–1° rotation and 3.9–11.4 Å translation for class I molecules, and 1.0° rotation and 5.3 Å translation for FcRn (Burmeister et al., 1994a). The total buried surface area between $\beta_2$M and the heavy chain is slightly less than for class I molecules, but it is more hydrophobic, which may account for the CD1–$\beta_2$M complex being more stable to thermal denaturation. Further, a notable difference in quaternary structure has been observed in $\alpha 3$–$\beta_2$M association. Interestingly, the pseudodyadic rotation necessary to superimpose $\alpha 3$ onto $\beta_2$M in CD1 is more similar to class II than to class I molecules. In the crystal structure of mouse CD1, the outermost $\alpha 3$ S1strand now comes closer to $\beta_2$M and results in a slightly different pattern of interactions.

The quaternary arrangement of the $\alpha 1$–$\alpha 2$ platform, $\alpha 3$ domain, and $\beta_2$M in the case of T10 and T22 are within the range of orientations observed for MHC class Ia molecules (19). The total buried surface area and the number of hydrogen bonds between $\beta_2$M and the heavy chain is slightly less (845 Å$^2$ and 5 for T22) on comparison to class I molecules (850–1036 Å$^2$ and 7–21) that may explain the observation that a closely related T10–$\beta_2$M complex is less stable to thermal denaturation than class Ia molecules.

Apart from slight differences between class I molecules and FcRn, there are many significant differences reported. Most of the interactions formed between heavy chain and light chain of FcRn and MHC class molecules are conserved except the interaction formed between $\beta$-strand (residues 52–55) with $\alpha 1$–$\alpha 2$ domain of FcRn. Furthermore, the heavy chain (residues 16–21) dips downward to contact $\beta_2$M and forms many extra contacts (Burmeister et al., 1994a).

In HFE, as in class I and class I-related proteins, $\alpha 3$ and $\beta_2$M interact with a symmetry that deviates from the pseudodyad symmetry relating antibody constant domains. Specifically, the HFE domains are related by a 161° rotation and a 13 Å translation (Lebron et al., 1998). This relative orientation is in the range seen for the $\alpha 3$–$\beta_2$M relationship in class I molecules (146–160° rotation, 13–14 Å translation; Madden, 1995), CD1 (170° rotation, 12 Å translation; Zeng et al., 1997), and FcRn (157° rotation, 13 Å translation; Burmeister et al., 1994a). Site-directed mutagenesis studies confirmed that on mutating Cys282Tyr mutant protein cannot form disulfide bond with $\beta_2M$, such mutation was observed in hereditary haemochromatosis (Feder et al., 1996; Waheed et al., 1997). Hence in absence of the $\beta_2M$, HFE does not properly fold (Parkkila et al., 1997). However, His63Asp mutant is capable of interaction with $\beta_2M$, but it is unable to regulate cellular iron uptake like the wild-type HFE protein (Waheed et al., 1997).

The quaternary arrangement of the ZAG $\alpha 3$ domain with respect to the $\alpha 1-\alpha 2$ platform differs from that found in $\beta_2M$-binding class I and related proteins. The spatial relationship between its three domains is slightly different from that found in class I MHC proteins, which may explain the lack of affinity of ZAG for $\beta_2M$. Structural features that prevent ZAG from binding $\beta_2M$ include the residues such as Ile13, Thr15, Leu30, Arg40, Gln98, Tyr118, Lys122, Val234, His236, Trp245, which clash with $\beta_2M$ when it is positioned on the ZAG structure either by interacting with $\alpha 3$ or with $\alpha 1-\alpha 2$ domains (Sanchez et al., 1999a). However, ZAG accommodates a molecules of PIP which has similar scaffold to the $\beta_2M$ in the same cleft but in opposite orientation (Hassan et al., 2008a). The interactions between ZAG and PIP are highly specific. There are 12 hydrogen bonds and three well-defined salt bridges formed between ZAG and PIP. The two salt bridges in the ZAG–PIP complex act as anchors at the two ends of vertically aligned β-strands of the $\alpha 3$ domain of ZAG and β-strands of PIP (Hassan et al., 2008a). Further, there are dissimilar charge distributions at the interface of ZAG and HLA in the two respective complexes, which further highlight the specificity of ZAG–PIP and HLA–$\beta_2M$.

The $\alpha 3$ domain of MIC-A adopted a C-type immunoglobulin conformation and is highly similar to the structure to $\alpha 3$ domains of MHC class I proteins, as well as $\beta_2M$ itself (Li et al., 1999). Interestingly, the relative positions of the two domains are significantly different in MIC-A and other MHC class I proteins (Li et al., 1999). The $\alpha 3$ domains of MIC-A contains a linker between the platform, consists of four residues (Leu178-Thr181) leading to an extended conformation that is different from the additional turn of helix found in place of the linkers in other MHC class I homologs (Madden, 1995). Interestingly, there are no contacts observed in between the two domains, whereas the linker is good enough to allow for considerable interdomain flexibility (Li et al., 1999). The important point to be discussed here is the lack of affinity of MIC-A from the $\beta_2M$. The crystal structure analysis reveals that the $\beta_2M$-binding interfaces are oriented in incompatible directions in the extended structure of MIC-A (Li et al.,

1999, 2001). Moreover, N-linked oligosaccharide site (Asn8) is located near the center of the β$_2$M-binding surface on the underside of the platform, which presumably would prevent β$_2$M binding in MIC-A. It is noteworthy that the interface residues of class I proteins (Gln96, Ala117, and Gly120, numbered as in HLA-B27) and of the β$_2$M (His31, Phe56, Trp60, and Phe-62) are highly conserved. Interestingly, the residues corresponding to these positions in MIC-A are also conserved except for Tyr111 (corresponds to Ala117 in HLA-B27; Li et al., 1999). Structure analysis suggests that the side chain of Tyr111 sterically clashes with Trp60 of β$_2$M when MIC-A is superimposed onto a class I structure. This further suggests the possibility of inability of β$_2$M binding to MIC-A. In addition to that, Asp122, a highly conserved residue in class I molecules, forms a hydrogen bond with Trp60 of β$_2$M. Interestingly, substitution of this residue by either Leu116 or Arg116, further support the preventing factor of MIC-A against β$_2$M binding (Collins et al., 1995).

## IV. GLYCOSYALATION

Glycosylation is a major feature of proteins expressed on the surfaces of cells, including molecules involved in the cellular immune system such as MHC class I molecule, and class I-like molecules (Rudd et al., 1999b). Glycosylation of proteins or MHC-I molecules provides a unique structure to modulate its selectivity and to control their interactions with their complementary proteins or ligands (Rudd et al., 1999a). It has been reported that the conformational stability of glycoproteins increases upon addition of glycan chain (Shental-Bechor and Levy, 2008). Although, glycosylation does not affect the average backbone conformation of a folded protein, it has a significant impact on protein stability by decreasing the flexibility of the protein backbone (Shental-Bechor and Levy, 2008). Further, in the immunological interactions, glycosylation is one of the important factors that could affect the protein–protein interactions significantly by conferring rigidity to the protein, which could lead to an increase in affinity and selectivity of proteins for their corresponding receptors (Rudd et al., 1999b; Wormald and Dwek, 1999). Further, it has been observed that glycans are equally important at all three proposed stages such as formation of the cell junction, recruitment of pMHC by the TCRs, and stabilizing the conformation of the complexes involved in the T cell recognition of antigen-presenting cells (Grakoui et al., 1999).

Sequence analysis suggests that there are three highly conserved glycosylation sites in class I molecules, that is, Asn86, Asn176 ($\alpha$2-domain), and Asn256 ($\alpha$3-domain) (Stroynowski, 1990). Interestingly, the class Ia molecules have glycosylation at all three sites, whereas in the class Ib, the position 176 is devoid of glycosylation (Shawar et al., 1994). However, the function of several human class I molecules lacking this site are apparently unaffected (Watts et al., 1989). The Asn86 is present in the loop connecting the $\alpha$1 domain to $\alpha$2 domain (Bjorkman et al., 1987b). Biochemical analysis suggests that glycosylation is really important for the function of class I proteins. MHC class I heavy chains that bind to calnexin through the terminal glucose moiety, which is required for protein folding and assembly pathway, because calnexin and calreticulin are lectins and acts as a chaperone. Subsequently, after association with $\beta_2$M, the fully assembled MHC is released from calnexin and subsequently binds to calreticulin through the same glucose residue. This event is necessary for the loading of antigenic peptide onto MHC class I molecules from the transporter associated with antigen processing (Sadasivan et al., 1996).

Murine CD1 sequence contains five potential N-glycosylation sites: Asn25, Asn38, Asn60, Asn128, and Asn183. However, in the crystal structure, only three sites showed glycan chain electron density. Structurally, glycan chain at $\alpha$1 helix contributes to stabilization of the $H_2$ helix by interaction with conserved Arg74 (Zeng et al., 1997). It was proposed that carbohydrate moiety at these sites provides protection against endosomal proteases. Moreover, glycosylation also plays an important role in the organization and spacing of CD1 on the cell surface by prevention of nonspecific aggregation (Rudd et al., 1999b).

Structure and biochemical analysis suggest that T22$^b$ has two potential N-linked glycosylation sites (Asn86 and Asn150) in the $\alpha$1$-\alpha$2 superdomains, as evident from the surface potential analysis that a negatively charged patch is present on a completely neutral surface (Fig. 4D). Notably, the Asn86 is highly conserved in MHC class I molecules (Madden, 1995), however, Asn150 is not conserved and present in the middle of protruding loop of $\alpha$2 domain (Wingren et al., 2000). Crystal structure suggests that carbohydrate plays a significant role in folding of the heterodimer in MHC class I (Asn86; Parham, 1996). Although, glycan moiety does not provide any specificity in the T22$^b$–$\gamma\delta$-TCR interaction (Crowley et al., 1997), but the bulky glycan chains may be important in guiding the $\gamma\delta$- T cell to its binding site.

Sequence analysis suggests that N-linked glycosylation site at Asn128 is found in rodent but not human or bovine forms of FcRn (Simister and Mostov, 1989; Story et al., 1994). Moreover, in the case of mouse FcRn, differential glycosylation affects the receptor/ligand stoichiometry from typical 1:1 FcRn/Fc complexes (high-mannose forms of FcRn) to 2:1 complexes (complex carbohydrates). Crystal structure analysis revealed that Asn128 is critically important for FcRn/Fc complex formation as evident from surface area calculation that 10–15% of the buried surface area in the FcRn/Fc interface is contributed by three glycans attached to Asn128 form close contact to four Fc residues (Martin et al., 2001). Moreover, it was suggested that maximal Fc binding affinity requires bulkier glycan attachment to FcRn at Asn128 (Sanchez et al., 1999b).

The protein sequence of human HFE contains three putative N-glycosylation sites at residues Asn110, Asn130, and Asn234 (Beutler, 2006), and Asn110 is highly conserved with class I proteins. It has been suggested that the glycosylation is important for the normal intracellular trafficking and functional activity of HFE (Bhatt et al., 2010). It has been shown that glycosylation of HFE provides different conformation as compared to unglycosylated from and subsequently affect their association with the TfR (Salter-Cid et al., 1999). Structurally, Val78 and Trp81 of HFE are mainly responsible for TfR binding, corresponding to a helical and loop region in the α1 domain that possesses both the Asn110 and Asn130 residues. It was presumed that N-glycosylation sites in a region are of functional importance for TfR1 binding. Therefore, it may be possible that the 110/130 double mutations may directly disrupt regulation of TfR1 (Bhatt et al., 2010). It was further confirmed that Asn-linked glycosylation is necessary for the association of HFE with TfR by HeLa cells transfection with HFE (Gross et al., 1998). Studies based on site-directed mutagenesis and biochemical characterization revealed that all three potential glycosylation sites contain glycan chain as evident from their electrophoretic mobility (Bhatt et al., 2010). The triple mutant (functionally identical to Cys282Tyr HFE) showed significantly decreased interactions to the $\beta_2$M, which is believed to be important for HFE folding and stabilization. Moreover, the proper folding of HFE is also essential for transport and exit of the protein from the ER (Bhatt et al., 2010).

Extensive carbohydrate density is found at Asn239 (nine ordered carbohydrate residues) and to a much lesser extent at Asn89 and Asn108 in all four ZAG molecules (Sanchez et al., 1999a). Crystal structures of glycoproteins rarely show more than three ordered carbohydrate residues at each glycosylation site (Vaughn and Bjorkman, 1998). The Asn in the fourth potential N-linked glycosylation site (Asn92) does not show density corresponding to carbohydrate. The bond between Asn92 and Gly93 can be broken by hydroxylamine, confirming that Asn92 is not glycosylated. There are eight N-linked glycosylation sites in MIC-A at Asn8, Asn56, Asn102, Asn187, Asn197, Asn211, Asn238, and Asn266 as evident from sequence and structure analysis. However, no glycan moiety has been observed at Asn56 and Asn197 during crystal structure analysis (Li et al., 1999). It was suggested that in MIC-A, glycosylation was not essential for its structure but it enhances the complex formation with NKG2D (Li et al., 2001).

## V. Conclusion

In the past 2 decades, enormous number of crystal structures of class I MHC-like molecules have been determined, which led to a great progress in our understanding of the structure of the class I proteins and their binding affinity for ligands and TCR. Structural studies showed that all class I MHC molecules have identical fold and forming a groove-like structure by $\alpha 1 - \alpha 2$ domain, which take part in ligand binding. Variation in the groove residues leads to the different binding specificities for their corresponding ligands. Further, pMHC complex binds to the TCR by the residues of $\alpha 1 - \alpha 2$ domain along with few residues from peptides. A notable difference has been observed in the molecules bind to $\alpha\beta$- or $\gamma\delta$-TCR. Except ZAG and MIC-A, all molecules are stabilized by $\beta_2 M$, which is essential for loading the peptides in the case of class I MHC molecules. A new molecule, PIP, has been identified in the cleft of ZAG, in the same place of $\beta_2 M$, but in totally different orientation. Glycosylation of class I molecules is essentially important for their corresponding functions as well as their specific binding and stabilization. Sequence analysis of class I molecules suggests the role critically important residues in the folding, stability, and function. Our comprehensive analysis may be useful for better understanding of mechanism of ligand binding by class I MHC molecules.

## Acknowledgments

Authors are grateful to the funding agencies, Department of Science and Technology, Council of Scientific and Industrial Research, and Indian National Science Academy for financial assistance.

## References

Achour, A., Persson, K., Harris, R. A., Sundback, J., Sentman, C. L., et al. (1998). The crystal structure of H-2Dd MHC class I complexed with the HIV-1-derived peptide P18-I10 at 2.4 A resolution: implications for T cell and NK cell recognition. *Immunity* **9**, 199–208.

Aisen, P., Wessling-Resnick, M., Leibold, E. A. (1999). Iron metabolism. *Curr. Opin. Chem. Biol.* **3**, 200–206.

Allison, T. J., Winter, C. C., Fournie, J. J., Bonneville, M., Garboczi, D. N. (2001). Structure of a human gammadelta T-cell antigen receptor. *Nature* **411**, 820–824.

Bahram, S., Bresnahan, M., Geraghty, D. E., Spies, T. (1994). A second lineage of mammalian major histocompatibility complex class I genes. *Proc. Natl. Acad. Sci. USA* **91**, 6259–6263.

Bahram, S., Spies, T. (1996). The MIC gene family. *Res. Immunol.* **147**, 328–333.

Barton, J. C., Sawada-Hirai, R., Rothenberg, B. E., Acton, R. T. (1999). Two novel missense mutations of the HFE gene (I105T and G93R) and identification of the S65C mutation in Alabama hemochromatosis probands. *Blood Cells Mol. Dis.* **25**, 147–155.

Bauer, S., Groh, V., Wu, J., Steinle, A., Phillips, J. H., et al. (1999). Activation of NK cells and T cells by NKG2D, a receptor for stress-inducible MICA. *Science* **285**, 727–729.

Baxter, T. K., Gagnon, S. J., Davis-Harrison, R. L., Beck, J. C., Binz, A. K., et al. (2004). Strategic mutations in the class I major histocompatibility complex HLA-A2 independently affect both peptide binding and T cell receptor recognition. *J. Biol. Chem.* **279**, 29175–29184.

Bennett, M. J., Lebron, J. A., Bjorkman, P. J. (2000). Crystal structure of the hereditary haemochromatosis protein HFE complexed with transferrin receptor. *Nature* **403**, 46–53.

Beutler, E. (2006). Hemochromatosis: genetics and pathophysiology. *Annu. Rev. Med.* **57**, 331–347.

Bhatt, L., Murphy, C., O'Driscoll, L. S., Carmo-Fonseca, M., McCaffrey, M. W., et al. (2010). N-glycosylation is important for the correct intracellular localization of HFE and its ability to decrease cell surface transferrin binding. *FEBS J.* **277**, 3219–3234.

Bing, C., Bao, Y., Jenkins, J., Sanders, P., Manieri, M., et al. (2004). Zinc-alpha2-glycoprotein, a lipid mobilizing factor, is expressed in adipocytes and is up-regulated in mice with cancer cachexia. *Proc. Natl. Acad. Sci. USA* **101**, 2500–2505.

Bjorkman, P. J., Burmeister, W. P. (1994). Structures of two classes of MHC molecules elucidated: crucial differences and similarities. *Curr. Opin. Struct. Biol.* **4**, 852–856.

Bjorkman, P. J., Parham, P. (1990). Structure, function, and diversity of class I major histocompatibility complex molecules. *Annu. Rev. Biochem.* **59**, 253–288.

Bjorkman, P. J., Saper, M. A., Samraoui, B., Bennett, W. S., Strominger, J. L., et al. (1987a). The foreign antigen binding site and T cell recognition regions of class I histocompatibility antigens. *Nature* **329**, 512–518.

Bjorkman, P. J., Saper, M. A., Samraoui, B., Bennett, W. S., Strominger, J. L., et al. (1987b). Structure of the human class I histocompatibility antigen, HLA-A2. *Nature* **329**, 506–512.

Boniface, J. J., Reich, Z., Lyons, D. S., Davis, M. M. (1999). Thermodynamics of T cell receptor binding to peptide-MHC: evidence for a general mechanism of molecular scanning. *Proc. Natl. Acad. Sci. USA* **96**, 11446–11451.

Burdin, N., Brossay, L., Degano, M., Iijima, H., Gui, M., et al. (2000). Structural requirements for antigen presentation by mouse CD1. *Proc. Natl. Acad. Sci. USA* **97**, 10156–10161.

Burmeister, W. P., Gastinel, L. N., Simister, N. E., Blum, M. L., Bjorkman, P. J. (1994a). Crystal structure at 2.2 A resolution of the MHC-related neonatal Fc receptor. *Nature* **372**, 336–343.

Burmeister, W. P., Huber, A. H., Bjorkman, P. J. (1994b). Crystal structure of the complex of rat neonatal Fc receptor with Fc. *Nature* **372**, 379–383.

Chien, Y. H., Jores, R., Crowley, M. P. (1996). Recognition by gamma/delta T cells. *Annu. Rev. Immunol.* **14**, 511–532.

Collins, E. J., Garboczi, D. N., Karpusas, M. N., Wiley, D. C. (1995). The three-dimensional structure of a class I major histocompatibility complex molecule missing the alpha 3 domain of the heavy chain. *Proc. Natl. Acad. Sci. USA* **92**, 1218–1221.

Collins, E. J., Garboczi, D. N., Wiley, D. C. (1994). Three-dimensional structure of a peptide extending from one end of a class I MHC binding site. *Nature* **371**, 626–629.

Crowley, M. P., Fahrer, A. M., Baumgarth, N., Hampl, J., Gutgemann, I., et al. (2000). A population of murine gammadelta T cells that recognize an inducible MHC class Ib molecule. *Science* **287**, 314–316.

Crowley, M. P., Reich, Z., Mavaddat, N., Altman, J. D., Chien, Y. (1997). The recognition of the nonclassical major histocompatibility complex (MHC) class I molecule, T10, by the gammadelta T cell, G8. *J. Exp. Med.* **185**, 1223–1230.

Davis, M. M., Boniface, J. J., Reich, Z., Lyons, D., Hampl, J., et al. (1998). Ligand recognition by alpha beta T cell receptors. *Annu. Rev. Immunol.* **16**, 523–544.

Delker, S. L., West, A. P., Jr., McDermott, L., Kennedy, M. W., Bjorkman, P. J. (2004). Crystallographic studies of ligand binding by Zn-alpha2-glycoprotein. *J. Struct. Biol.* **148**, 205–213.

Derrick, J. P., Wigley, D. B. (1992). Crystal structure of a streptococcal protein G domain bound to an Fab fragment. *Nature* **359**, 752–754.

Ding, Y. H., Smith, K. J., Garboczi, D. N., Utz, U., Biddison, W. E., et al. (1998). Two human T cell receptors bind in a similar diagonal mode to the HLA-A2/Tax peptide complex using different TCR amino acids. *Immunity* **8**, 403–411.

Engelhard, V. H. (1994a). Structure of peptides associated with class I and class II MHC molecules. *Annu. Rev. Immunol.* **12**, 181–207.

Engelhard, V. H. (1994b). Structure of peptides associated with MHC class I molecules. *Curr. Opin. Immunol.* **6**, 13–23.

Feder, J. N., Gnirke, A., Thomas, W., Tsuchihashi, Z., Ruddy, D. A., et al. (1996). A novel MHC class I-like gene is mutated in patients with hereditary haemochromatosis. *Nat. Genet.* **13**, 399–408.

Feder, J. N., Penny, D. M., Irrinki, A., Lee, V. K., Lebron, J. A., et al. (1998). The hemochromatosis gene product complexes with the transferrin receptor and lowers its affinity for ligand binding. *Proc. Natl. Acad. Sci. USA* **95**, 1472–1477.

Fremont, D. H., Matsumura, M., Stura, E. A., Peterson, P. A., Wilson, I. A. (1992). Crystal structures of two viral peptides in complex with murine MHC class I H-2Kb. *Science* **257**, 919–927.

Gadola, S. D., Zaccai, N. R., Harlos, K., Shepherd, D., Castro-Palomino, J. C., et al. (2002). Structure of human CD1b with bound ligands at 2.3 A, a maze for alkyl chains. *Nat. Immunol.* **3**, 721–726.

Gao, G. F., Tormo, J., Gerth, U. C., Wyer, J. R., McMichael, A. J., et al. (1997). Crystal structure of the complex between human CD8alpha(alpha) and HLA-A2. *Nature* **387**, 630–634.

Garcia, K. C., Degano, M., Pease, L. R., Huang, M., Peterson, P. A., et al. (1998). Structural basis of plasticity in T cell receptor recognition of a self peptide-MHC antigen. *Science* **279**, 1166–1172.

Garcia, K. C., Degano, M., Stanfield, R. L., Brunmark, A., Jackson, M. R., et al. (1996). An alphabeta T cell receptor structure at 2.5 A and its orientation in the TCR-MHC complex. *Science* **274**, 209–219.

Garrett, T. P., Saper, M. A., Bjorkman, P. J., Strominger, J. L., Wiley, D. C. (1989). Specificity pockets for the side chains of peptide antigens in HLA-Aw68. *Nature* **342**, 692–696.

Germain, R. N., Margulies, D. H. (1993). The biochemistry and cell biology of antigen processing and presentation. *Annu. Rev. Immunol.* **11**, 403–450.

Germain, R. N., Stefanova, I. (1999). The dynamics of T cell receptor signaling: complex orchestration and the key roles of tempo and cooperation. *Annu. Rev. Immunol.* **17**, 467–522.

Ghetie, V., Ward, E. S. (2000). Multiple roles for the major histocompatibility complex class I-related receptor FcRn. *Annu. Rev. Immunol.* **18**, 739–766.

Govindarajan, K. R., Kangueane, P., Tan, T. W., Ranganathan, S. (2003). MPID: MHC-Peptide Interaction Database for sequence-structure-function information on peptides binding to MHC molecules. *Bioinformatics* **19**, 309–310.

Grakoui, A., Bromley, S. K., Sumen, C., Davis, M. M., Shaw, A. S., et al. (1999). The immunological synapse: a molecular machine controlling T cell activation. *Science* **285**, 221–227.

Groh, V., Steinle, A., Bauer, S., Spies, T. (1998). Recognition of stress-induced MHC molecules by intestinal epithelial gammadelta T cells. *Science* **279**, 1737–1740.

Gross, C. N., Irrinki, A., Feder, J. N., Enns, C. A. (1998). Co-trafficking of HFE, a nonclassical major histocompatibility complex class I protein, with the transferrin receptor implies a role in intracellular iron regulation. *J. Biol. Chem.* **273**, 22068–22074.

Hassan, M. I., Bilgrami, S., Kumar, V., Singh, N., Yadav, S., et al. (2008a). Crystal structure of the novel complex formed between zinc alpha2-glycoprotein (ZAG) and prolactin-inducible protein (PIP) from human seminal plasma. *J. Mol. Biol.* **384**, 663–672.

Hassan, M. I., Waheed, A., Yadav, S., Singh, T. P., Ahmad, F. (2008b). Zinc alpha 2-glycoprotein: a multidisciplinary protein. *Mol. Cancer Res.* **6**, 892–906.

He, X., Tabaczewski, P., Ho, J., Stroynowski, I., Garcia, K. C. (2001). Promiscuous antigen presentation by the nonclassical MHC Ib Qa-2 is enabled by a shallow, hydrophobic groove and self-stabilized peptide conformation. *Structure* **9**, 1213–1224.

Hennecke, J., Wiley, D. C. (2001). T cell receptor-MHC interactions up close. *Cell* **104**, 1–4.

Hoare, H. L., Sullivan, L. C., Clements, C. S., Ely, L. K., Beddoe, T., Henderson, K. N., et al. (2008). Subtle changes in peptide conformation profoundly affect recognition of the non-classical MHC class I molecule HLA-E by the CD94-NKG2 natural killer cell receptors. *J. Mol. Biol.* **377**, 1297–1303.

Holmes, M. A., Li, P., Petersdorf, E. W., Strong, R. K. (2002). Structural studies of allelic diversity of the MHC class I homolog MIC-B, a stress-inducible ligand for the activating immunoreceptor NKG2D. *J. Immunol.* **169**, 1395–1400.

Housset, D., Malissen, B. (2003). What do TCR-pMHC crystal structures teach us about MHC restriction and alloreactivity? *Trends Immunol.* **24**, 429–437.

Hue, S., Mention, J. J., Monteiro, R. C., Zhang, S., Cellier, C., et al. (2004). A direct role for NKG2D/MICA interaction in villous atrophy during celiac disease. *Immunity* **21**, 367–377.

Janeway, C. A., Jr. (1992). The T cell receptor as a multicomponent signalling machine: CD4/CD8 coreceptors and CD45 in T cell activation. *Annu. Rev. Immunol.* **10**, 645–674.

Kennedy, M. W., Heikema, A. P., Cooper, A., Bjorkman, P. J., Sanchez, L. M. (2001). Hydrophobic ligand binding by Zn-alpha 2-glycoprotein, a soluble fat-depleting factor related to major histocompatibility complex proteins. *J. Biol. Chem.* **276**, 35008–35013.

Kjer-Nielsen, L., Clements, C. S., Purcell, A. W., Brooks, A. G., Whisstock, J. C., et al. (2003). A structural basis for the selection of dominant alphabeta T cell receptors in antiviral immunity. *Immunity* **18**, 53–64.

Klein, J., Satta, Y., O'HUigin, C., Takahata, N. (1993). The molecular descent of the major histocompatibility complex. *Annu. Rev. Immunol.* **11**, 269–295.

Koch, M., Stronge, V. S., Shepherd, D., Gadola, S. D., Mathew, B., et al. (2005). The crystal structure of human CD1d with and without alpha-galactosylceramide. *Nat. Immunol.* **6**, 819–826.

Konig, R., Huang, L. Y., Germain, R. N. (1992). MHC class II interaction with CD4 mediated by a region analogous to the MHC class I binding site for CD8. *Nature* **356**, 796–798.

Kuzushima, K., Sun, R., van Bleek, G. M., Vegh, Z., Nathenson, S. G. (1995). The role of self peptides in the allogeneic cross-reactivity of CTLs. *J. Immunol.* **155**, 594–601.

Lebron, J. A., Bennett, M. J., Vaughn, D. E., Chirino, A. J., Snow, P. M., et al. (1998). Crystal structure of the hemochromatosis protein HFE and characterization of its interaction with transferrin receptor. *Cell* **93**, 111–123.

Lebron, J. A., Bjorkman, P. J. (1999). The transferrin receptor binding site on HFE, the class I MHC-related protein mutated in hereditary hemochromatosis. *J. Mol. Biol.* **289**, 1109–1118.

Lebron, J. A., West, A. P., Jr., Bjorkman, P. J. (1999). The hemochromatosis protein HFE competes with transferrin for binding to the transferrin receptor. *J. Mol. Biol.* **294**, 239–245.

Li, H., Lebedeva, M. I., Llera, A. S., Fields, B. A., Brenner, M. B., et al. (1998). Structure of the Vdelta domain of a human gammadelta T-cell antigen receptor. *Nature* **391**, 502–506.

Li, P., McDermott, G., Strong, R. K. (2002). Crystal structures of RAE-1beta and its complex with the activating immunoreceptor NKG2D. *Immunity* **16**, 77–86.

Li, P., Morris, D. L., Willcox, B. E., Steinle, A., Spies, T., et al. (2001). Complex structure of the activating immunoreceptor NKG2D and its MHC class I-like ligand MICA. *Nat. Immunol.* **2**, 443–451.

Li, P., Willie, S. T., Bauer, S., Morris, D. L., Spies, T., et al. (1999). Crystal structure of the MHC class I homolog MIC-A, a gammadelta T cell ligand. *Immunity* **10**, 577–584.

Lindahl, K. F., Byers, D. E., Dabhi, V. M., Hovik, R., Jones, E. P., et al. (1997). H2-M3, a full-service class Ib histocompatibility antigen. *Annu. Rev. Immunol.* **15**, 851–879.

Ljunggren, H. G., Stam, N. J., Ohlen, C., Neefjes, J. J., Hoglund, P., et al. (1990). Empty MHC class I molecules come out in the cold. *Nature* **346**, 476–480.

Madden, D. R. (1995). The three-dimensional structure of peptide-MHC complexes. *Annu. Rev. Immunol.* **13**, 587–622.

Madden, D. R., Gorga, J. C., Strominger, J. L., Wiley, D. C. (1991). The structure of HLA-B27 reveals nonamer self-peptides bound in an extended conformation. *Nature* **353**, 321–325.

Madden, D. R., Gorga, J. C., Strominger, J. L., Wiley, D. C. (1992). The three-dimensional structure of HLA-B27 at 2.1 A resolution suggests a general mechanism for tight peptide binding to MHC. *Cell* **70**, 1035–1048.

Margulies, D. H. (1997). Interactions of TCRs with MHC-peptide complexes: a quantitative basis for mechanistic models. *Curr. Opin. Immunol.* **9**, 390–395.

Marrack, P., Scott-Browne, J. P., Dai, S., Gapin, L., Kappler, J. W. (2008). Evolutionarily conserved amino acids that control TCR-MHC interaction. *Annu. Rev. Immunol.* **26**, 171–203.

Martin, W. L., West, A. P., Jr., Gan, L., Bjorkman, P. J. (2001). Crystal structure at 2.8 A of an FcRn/heterodimeric Fc complex: mechanism of pH-dependent binding. *Mol. Cell* **7**, 867–877.

Matsumura, M., Fremont, D. H., Peterson, P. A., Wilson, I. A. (1992). Emerging principles for the recognition of peptide antigens by MHC class I molecules. *Science* **257**, 927–934.

McDermott, L. C., Freel, J. A., West, A. P., Bjorkman, P. J., Kennedy, M. W. (2006). Zn-alpha2-glycoprotein, an MHC class I-related glycoprotein regulator of adipose tissues: modification or abrogation of ligand binding by site-directed mutagenesis. *Biochemistry* **45**, 2035–2041.

Mezo, A. R., Sridhar, V., Badger, J., Sakorafas, P., Nienaber, V. (2010). X-ray crystal structures of monomeric and dimeric peptide inhibitors in complex with the human neonatal Fc receptor, FcRn. *J. Biol. Chem.* **285**, 27694–27701.

Moody, D. B., Zajonc, D. M., Wilson, I. A. (2005). Anatomy of CD1-lipid antigen complexes. *Nat. Rev. Immunol.* **5**, 387–399.

Moriwaki, S., Korn, B. S., Ichikawa, Y., van Kaer, L., Tonegawa, S. (1993). Amino acid substitutions in the floor of the putative antigen-binding site of H-2 T22 affect recognition by a gamma delta T-cell receptor. *Proc. Natl. Acad. Sci. USA* **90**, 11396–11400.

Nathenson, S. G., Geliebter, J., Pfaffenbach, G. M., Zeff, R. A. (1986). Murine major histocompatibility complex class-I mutants: molecular analysis and structure-function implications. *Annu. Rev. Immunol.* **4**, 471–502.

O'Callaghan, C. A., Tormo, J., Willcox, B. E., Braud, V. M., Jakobsen, B. K., et al. (1998). Structural features impose tight peptide binding specificity in the nonclassical MHC molecule HLA-E. *Mol. Cell* **1**, 531–541.

Ojcius, D. M., Delarbre, C., Kourilsky, P., Gachelin, G. (2002). MHC and MHC-related proteins as pleiotropic signal molecules. *FASEB J.* **16**, 202–206.

Parham, P. (1996). Functions for MHC class I carbohydrates inside and outside the cell. *Trends Biochem. Sci.* **21**, 427–433.

Parkkila, S., Waheed, A., Britton, R. S., Bacon, B. R., Zhou, X. Y., et al. (1997). Association of the transferrin receptor in human placenta with HFE, the protein defective in hereditary hemochromatosis. *Proc. Natl. Acad. Sci. USA* **94**, 13198–13202.

Pease, L. R., Horton, R. M., Pullen, J. K., Cai, Z. L. (1991). Structure and diversity of class I antigen presenting molecules in the mouse. *Crit. Rev. Immunol.* **11**, 1–32.

Porcelli, S. A., Modlin, R. L. (1999). The CD1 system: antigen-presenting molecules for T cell recognition of lipids and glycolipids. *Annu. Rev. Immunol.* **17**, 297–329.

Potter, T. A., Rajan, T. V., Dick, R. F., II, Bluestone, J. A. (1989). Substitution at residue 227 of H-2 class I molecules abrogates recognition by CD8-dependent, but not CD8-independent, cytotoxic T lymphocytes. *Nature* **337**, 73–75.

Reche, P. A., Reinherz, E. L. (2003). Sequence variability analysis of human class I and class II MHC molecules: functional and structural correlates of amino acid polymorphisms. *J. Mol. Biol.* **331**, 623–641.

Riegert, P., Wanner, V., Bahram, S. (1998). Genomics, isoforms, expression, and phylogeny of the MHC class I-related MR1 gene. *J. Immunol.* **161**, 4066–4077.

Roopenian, D. C., Akilesh, S. (2007). FcRn: the neonatal Fc receptor comes of age. *Nat. Rev. Immunol.* **7**, 715–725.

Rudd, P. M., Wormald, M. R., Harvey, D. J., Devasahayam, M., McAlister, M. S., et al. (1999a). Oligosaccharide analysis and molecular modeling of soluble forms of glycoproteins belonging to the Ly-6, scavenger receptor, and immunoglobulin superfamilies expressed in Chinese hamster ovary cells. *Glycobiology* **9**, 443–458.

Rudd, P. M., Wormald, M. R., Stanfield, R. L., Huang, M., Mattsson, N., et al. (1999b). Roles for glycosylation of cell surface receptors involved in cellular immune recognition. *J. Mol. Biol.* **293**, 351–366.

Rudolph, M. G., Stanfield, R. L., Wilson, I. A. (2006). How TCRs bind MHCs, peptides, and coreceptors. *Annu. Rev. Immunol.* **24**, 419–466.

Rudolph, M. G., Wilson, I. A. (2002). The specificity of TCR/pMHC interaction. *Curr. Opin. Immunol.* **14**, 52–65.

Sadasivan, B., Lehner, P. J., Ortmann, B., Spies, T., Cresswell, P. (1996). Roles for calreticulin and a novel glycoprotein, tapasin, in the interaction of MHC class I molecules with TAP. *Immunity* **5**, 103–114.

Salter, R. D., Benjamin, R. J., Wesley, P. K., Buxton, S. E., Garrett, T. P., et al. (1990). A binding site for the T-cell co-receptor CD8 on the alpha 3 domain of HLA-A2. *Nature* **345**, 41–46.

Salter-Cid, L., Brunmark, A., Li, Y., Leturcq, D., Peterson, P. A., et al. (1999). Transferrin receptor is negatively modulated by the hemochromatosis protein HFE: implications for cellular iron homeostasis. *Proc. Natl. Acad. Sci. USA* **96**, 5434–5439.

Sanchez, L. M., Chirino, A. J., Bjorkman, P. (1999a). Crystal structure of human ZAG, a fat-depleting factor related to MHC molecules. *Science* **283**, 1914–1919.

Sanchez, L. M., Penny, D. M., Bjorkman, P. J. (1999b). Stoichiometry of the interaction between the major histocompatibility complex-related Fc receptor and its Fc ligand. *Biochemistry* **38**, 9471–9476.

Sanders, P. M., Tisdale, M. J. (2004). Effect of zinc-alpha2-glycoprotein (ZAG) on expression of uncoupling proteins in skeletal muscle and adipose tissue. *Cancer Lett.* **212**, 71–81.

Schild, H., Mavaddat, N., Litzenberger, C., Ehrich, E. W., Davis, M. M., et al. (1994). The nature of major histocompatibility complex recognition by gamma delta T cells. *Cell* **76**, 29–37.

Shawar, S. M., Vyas, J. M., Rodgers, J. R., Rich, R. R. (1994). Antigen presentation by major histocompatibility complex class I-B molecules. *Annu. Rev. Immunol.* **12**, 839–880.

Shental-Bechor, D., Levy, Y. (2008). Effect of glycosylation on protein folding: a close look at thermodynamic stabilization. *Proc. Natl. Acad. Sci. USA* **105**, 8256–8261.

Shields, R. L., Namenuk, A. K., Hong, K., Meng, Y. G., Rae, J., et al. (2001). High resolution mapping of the binding site on human IgG1 for Fc gamma RI, Fc gamma RII, Fc gamma RIII, and FcRn and design of IgG1 variants with improved binding to the Fc gamma R. *J. Biol. Chem.* **276**, 6591–6604.

Sidobre, S., Naidenko, O. V., Sim, B. C., Gascoigne, N. R., Garcia, K. C., et al. (2002). The V alpha 14 NKT cell TCR exhibits high-affinity binding to a glycolipid/CD1d complex. *J. Immunol.* **169**, 1340–1348.

Sim, B. C., Zerva, L., Greene, M. I., Gascoigne, N. R. (1996). Control of MHC restriction by TCR Valpha CDR1 and CDR2. *Science* **273**, 963–966.

Simister, N. E., Mostov, K. E. (1989). An Fc receptor structurally related to MHC class I antigens. *Nature* **337**, 184–187.

Speir, J. A., Stevens, J., Joly, E., Butcher, G. W., Wilson, I. A. (2001). Two different, highly exposed, bulged structures for an unusually long peptide bound to rat MHC class I RT1-Aa. *Immunity* **14**, 81–92.

Stephens, H. A. (2001). MICA and MICB genes: can the enigma of their polymorphism be resolved? *Trends Immunol.* **22**, 378–385.

Story, C. M., Mikulska, J. E., Simister, N. E. (1994). A major histocompatibility complex class I-like Fc receptor cloned from human placenta: possible role in transfer of immunoglobulin G from mother to fetus. *J. Exp. Med.* **180**, 2377–2381.

Stroynowski, I. (1990). Molecules related to class-I major histocompatibility complex antigens. *Annu. Rev. Immunol.* **8**, 501–530.

Sun, J., Leahy, D. J., Kavathas, P. B. (1995). Interaction between CD8 and major histocompatibility complex (MHC) class I mediated by multiple contact surfaces that include the alpha 2 and alpha 3 domains of MHC class I. *J. Exp. Med.* **182**, 1275–1280.

The MHC Sequencing Consortium (1999). Complete sequence and gene map of a human major histocompatibility complex. *Nature* **401**, 921–923.

Townsend, A., Drakesmith, H. (2002). Role of HFE in iron metabolism, hereditary haemochromatosis, anaemia of chronic disease, and secondary iron overload. *Lancet* **359**, 786–790.

Tynan, F. E., Borg, N. A., Miles, J. J., Beddoe, T., El-Hassen, D., et al. (2005a). High resolution structures of highly bulged viral epitopes bound to major histocompatibility complex class I. Implications for T-cell receptor engagement and T-cell immunodominance. *J. Biol. Chem.* **280**, 23900–23909.

Tynan, F. E., Burrows, S. R., Buckle, A. M., Clements, C. S., Borg, N. A., et al. (2005b). T cell receptor recognition of a 'super-bulged' major histocompatibility complex class I-bound peptide. *Nat. Immunol.* **6**, 1114–1122.

Tysoe-Calnon, V. A., Grundy, J. E., Perkins, S. J. (1991). Molecular comparisons of the beta 2-microglobulin-binding site in class I major-histocompatibility-complex alpha-chains and proteins of related sequences. *Biochem. J.* **277**(Pt 2), 359–369.

Vaughn, D. E., Bjorkman, P. J. (1998). Structural basis of pH-dependent antibody binding by the neonatal Fc receptor. *Structure* **6**, 63–73.

Vaughn, D. E., Milburn, C. M., Penny, D. M., Martin, W. L., Johnson, J. L., et al. (1997). Identification of critical IgG binding epitopes on the neonatal Fc receptor. *J. Mol. Biol.* **274**, 597–607.

Vitiello, A., Potter, T. A., Sherman, L. A. (1990). The role of beta 2-microglobulin in peptide binding by class I molecules. *Science* **250**, 1423–1426.

Vivier, E., Tomasello, E., Paul, P. (2002). Lymphocyte activation via NKG2D: towards a new paradigm in immune recognition? *Curr. Opin. Immunol.* **14**, 306–311.

Waheed, A., Parkkila, S., Zhou, X. Y., Tomatsu, S., Tsuchihashi, Z., et al. (1997). Hereditary hemochromatosis: effects of C282Y and H63D mutations on association with beta2-microglobulin, intracellular processing, and cell surface expression of the HFE protein in COS-7 cells. *Proc. Natl. Acad. Sci. USA* **94**, 12384–12389.

Walpole, N. G., Kjer-Nielsen, L., Kostenko, L., McCluskey, J., Brooks, A. G., Rossjohn, J., Clements, C. S. (2010) The structure and stability of the monomorphic HLA-G are influenced by the nature of the bound peptide. *J. Mol. Biol.* **397**, 467–480.

Watts, S., Wheeler, C., Morse, R., Goodenow, R. S. (1989). Amino acid comparison of the class I antigens of mouse major histocompatibility complex. *Immunogenetics* **30**, 390–392.

Willcox, B. E., Gao, G. F., Wyer, J. R., Ladbury, J. E., Bell, J. I., et al. (1999). TCR binding to peptide-MHC stabilizes a flexible recognition interface. *Immunity* **10**, 357–365.

Wingren, C., Crowley, M. P., Degano, M., Chien, Y., Wilson, I. A. (2000). Crystal structure of a gammadelta T cell receptor ligand T22: a truncated MHC-like fold. *Science* **287**, 310–314.

Wolan, D. W., Teyton, L., Rudolph, M. G., Villmow, B., Bauer, S., et al. (2001). Crystal structure of the murine NK cell-activating receptor NKG2D at 1.95 A. *Nat. Immunol.* **2**, 248–254.

Wormald, M. R., Dwek, R. A. (1999). Glycoproteins: glycan presentation and protein-fold stability. *Structure* **7**, R155–R160.

Wu, J., Song, Y., Bakker, A. B., Bauer, S., Spies, T., et al. (1999). An activating immunoreceptor complex formed by NKG2D and DAP10. *Science* **285**, 730–732.

Zajonc, D. M., Elsliger, M. A., Teyton, L., Wilson, I. A. (2003). Crystal structure of CD1a in complex with a sulfatide self antigen at a resolution of 2.15 A. *Nat. Immunol.* **4**, 808–815.

Zeng, Z., Castano, A. R., Segelke, B. W., Stura, E. A., Peterson, P. A., et al. (1997). Crystal structure of mouse CD1: an MHC-like fold with a large hydrophobic binding groove. *Science* **277**, 339–345.

Zhou, J., Johnson, J. E., Ghetie, V., Ober, R. J., Ward, E. S. (2003). Generation of mutated variants of the human form of the MHC class I-related receptor, FcRn, with increased affinity for mouse immunoglobulin G. *J. Mol. Biol.* **332**, 901–913.

# AUTHOR INDEX

## A

Abagyan, R., 138
Abanero, J., 4
Abou-Saleh, R. H., 96
Abreu, J. I., 9
Acharya, S. S., 97
Achour, A., 226, 232, 245
Achyuthan, K. E., 86
Ackerman, A. L., 134
Acosta, G., 9
Acton, R. T., 234
Adams, H., 174, 185
Adany, R., 97
Addess, K. J., 3
Agard, D. A., 52
Aguero-Chapin, G., 16–18
Ahmad, F., 225–226, 236
Air, G. M., 4
Aisen, P., 249
Aita, V. M., 56
Ajjan, R. A., 96, 99, 104
Akcasu, Z., 193
Akhremitchev, B. B., 83
Akilesh, S., 232, 247
Akita, R., 170
Akutsu, H., 3
Albage, A., 104
Alexander, K. J., 49
Algra, A., 98
Alili, R., 166, 169, 185
Allali, Y., 106
Allen, N., 104
Allison, T. J., 253
Al Mondhiry, H. A., 101
Al-Rawi, H., 68
Altman, J. D., 259
Altschul, S. F., 21
Amano, M., 171
Amara, J. F., 68

Amore, C., 97
Amrani, D. L., 108
Anderson, G. L., 105
Anderson, J. L., 105
Anderson, K. E., 46, 84
Anderson, M. P., 68
Andrieux, A., 108
Ankri, A., 106
Antovic, A., 104
Antunes, A., 17–18
Aoki, N., 93
Apgar, J. R., 135
Apweiler, R., 8
Ariëns, R. A. S., 75, 84–86, 97, 99–100, 103, 105–108
Arnaout, M. A., 107
Arpin, M., 165, 167, 171–172
Arseniev, A. S., 132–133, 135–136, 139–140, 144, 148, 153
Arveiler, D., 97, 99
Asada, K., 196
Ash, W. L., 133–134
Asselta, R., 97
Astrin, K. H., 54, 59–61
Atochin, D., 107
Aunis, D., 131
Ausiello, D. A., 172
Ausubel, F. M., 4
Averett, L. E., 83

## B

Babon, J. J., 209
Bachir, A. I., 168
Bacic, D., 169
Bacon, B. R., 257
Badger, J., 248
Badimon, J. J., 105
Bagoly, Z., 85, 99

Bahram, S., 225, 236, 238, 254
Bai, F., 16–20
Bailey, A. J., 52
Bajaj, M. S., 80
Bajaj, S. P., 80
Bajzar, L., 94
Baker, D., 3, 52, 138
Baker, M. L., 3
Bakke, A. C., 101
Bakker, A. B., 238
Bakouh, N., 166, 169, 185
Balaban, A. T., 9, 11, 17
Baldwin, J., 4
Bale, M. D., 86
Ballabio, D., 12
Ball, E. V., 54–55
Ball, L. K., 5
Bamford, J. M., 99
Bange, J., 152
Bannan, S., 86, 99
Bao, Y., 225
Baras, Y., 103
Barber, D. L., 178
Bark, N., 104
Barry, E. L., 86
Barth, P., 138
Barton, J. C., 234
Basak, S. C., 4–5, 9–11, 13–14, 18, 28–31
Bates, I. R., 168
Battersby, A. R., 44, 48
Bauer, S., 226–227, 230, 232, 238–239, 250, 254, 257–258, 261
Baumann, R. E., 97
Baumgarth, N., 235
Baxter, S., 167, 178, 180, 183–186, 206–207
Baxter, T. K., 237
Bayer, R., 5
Beauchamp, R. L., 170, 178
Beaulieu, L. M., 91
Beck, J. C., 237
Beck, L., 166, 169, 185
Becktel, W. J., 52
Becq, F., 168
Beddoe, T., 226, 245, 252
Behague, I., 97, 99
Bejarano, L. A., 4
Bekeart, L. S., 108

Belkin, A. M., 108
Belkina, N. V., 172
Bellamy, A. R., 4
Bellizzi, A., 166, 170
Bell, J. I., 236, 251
Bell, W. R., 83
Benjamin, R. J., 224, 238–239, 253
Benkovic, S. J., 186
Bennasroune, A., 131
Bennett, M. J., 225–226, 230, 232, 234, 239, 248–249, 254, 256
Bennett, W. S., 224, 226, 230, 232, 235, 238, 240, 246, 248, 255, 259
Benoit, H. C., 191–192, 195
Bensidhoum, M., 56, 61
Ben-Tal, N., 137–138, 153
Berdon, W. E., 67
Bereczky, Z., 85, 99
Berezovsky, I. N., 22
Berger, B. W., 131, 135, 138, 153–154
Berger, H. Jr., 104
Bergonia, H. A., 49
Berkhout, R. J., 25
Berliner, S. A., 108
Berman, H. M., 3
Berne, B. J., 192
Berryman, M., 165
Bertina, R. M., 80, 101
Best, C., 3
Best, R. B., 189, 211
Beuth, B., 3
Beutler, E., 248, 260
Beving, H., 104
Bhasin, N., 107
Bhattacharya, S., 167, 178, 180, 183–186, 206–207
Bhatt, L., 260
Biber, J., 168–169
Biddison, W. E., 252, 254
Biehl, R., 193, 200, 202, 211
Bilgrami, S., 226–227, 230, 236, 239, 249, 257
Bing, C., 225
Binnie, C. G., 87
Binz, A. K., 237
Birrane, G., 183
Bisello, A., 170, 172
Bishop, D. F., 46–47, 49, 51, 54, 58, 62

Bishop, P. D., 84
Biwersi, J., 68
Bjorkman, P. J., 224–227, 230–236, 238–240, 246–250, 253–257, 259–261
Bjornsson, T. D., 104
Blackburn, G. M., 3
Blitzer, J. T., 166–167, 170
Bloch, I., 135
Blombäck, B., 76, 82–83, 89, 104, 106
Blomback, M., 76, 104, 106
Blom, N., 8
Bluestone, J. A., 238
Bluhm, W. F., 3
Blum, M. L., 225–226, 230, 232–233, 239, 247–248, 253, 256
Board, P. G., 84
Bocharova, O. V., 135, 140
Bocharov, E. V., 133, 135–136, 140, 148, 151
Bockenstedt, P., 86
Bodo, G., 2
Boffa, M., 94
Bogorad, L., 52
Bojak, A., 5
Bolas-Fernandez, F., 17
Bond, P. J., 134, 144
Bonhagen, J., 168
Boniface, J. J., 251, 254
Bonneville, M., 253
Borchert, T. V., 52
Borges, F., 18
Borg, N. A., 245, 252
Borkan, S. C., 172
Bosco, D. A., 186
Bose, S., 172
Boshkov, L. K., 101
Bossard, F., 168
Bossenmaier, I., 67
Bostrom, A., 5
Bottenus, R. E., 84
Boucherle, A., 5
Boucher, R. C., 168, 172
Boulechfar, S., 54
Bourne, P., 32
Boutselakis, C. H., 3
Bouzida, D., 142
Bowie, J. U., 131, 135–136
Brannan, C., 104

Brass, E. P., 83
Brasseur, R., 136
Braud, V. M., 225, 230
Braun, R., 133
Bray, D., 212
Brennan, M., 131
Brenner, M. B., 253
Bresnahan, M., 238, 254
Bretscher, A., 165–168, 171–172, 174, 177, 179
Brett, C. L., 165
Brieher, W. M., 165
Briët, E., 80
Briggs, K., 7
Britton, R. S., 257
Broekmans, A. W., 80
Bromley, S. K., 258
Brooks, A. G., 226, 253
Brooks, C. L. III., 136–137
Brosig, B., 134–135
Brossay, L., 230, 237
Brown, A. E., 94, 96
Brown, C. M., 211
Brown, C. R., 68
Brown, D., 67
Brown, J. H., 81–82, 86
Brown, K. L., 102
Brown, S. G., 80
Broze, G. J. Jr., 80–81
Brozek, J., 103
Brummel, K., 104
Brummel-Ziedins, K. E., 104
Brunak, S., 8
Brunmark, A., 252, 260
Buckle, A. M., 252
Buerke, M., 108
Bui, H. H., 32
Bu, L., 137
Buldyrev, S. V., 10
Bunting, M., 108
Burdin, N., 230, 237
Burjanadze, T. V., 52
Burks, C., 8
Burmeister, W. P., 225–226, 230, 232–234, 239, 247–248, 253, 256
Burnett, R. T., 105
Burrows, S. R., 252

Burton, R. A., 83
Busco, G., 168
Bush, L. A., 100
Butcher, G. W., 252
Butina, D., 9, 15–16
Buus, S., 32
Buxton, S. E., 224, 238–239, 253
Bu, Z. M., 163, 167, 171, 174, 177–181, 185, 187, 193, 196, 200, 202–206, 211–212
Byers, D. E., 225, 240

## C

Caballero, J., 9
Caen, J. P., 106
Cai, Y. D., 18
Cai, Z. L., 225
Callaway, D. J. E., 163, 167, 171, 174, 177–181, 183–187, 193, 196, 200, 202–207, 211–212
Calle, E. E., 105
Cao, T. T., 166
Cao, Y., 101
Cardinal, R., 67
Cardone, R. A., 168, 172
Carlisle, C. R., 90, 95–96
Carlsson, K., 89
Carlsson, M., 80
Carmo-Fonseca, M., 260
Carrasco, B., 201
Carrell, R. W., 91
Carr, M. E., 76
Cartailler, J.-P.h., 153
Carter, A. M., 86, 91, 97–99
Carvalho, F. A., 108
Casey, G., 165, 169
Caso, R., 80
Caspary, E. A., 104
Cassaday, R., 172
Castano, A. R., 225–226, 230, 232, 234–235, 245–246, 248–249, 253, 256, 259
Castaño, D., 43, 51–53, 57, 60, 64, 66, 68–69
Castro-Palomino, J. C., 246
Catto, A. J., 86, 97, 99–100
Cellier, C., 236
Cha, B., 165

Chaires, J. B., 5
Chamberlain, A. K., 136
Chambers, C. E., 101
Chambers, D., 171
Chapel, A., 108
Charlton, A. J., 105
Charo, I. F., 108
Chayen, N. E., 3
Cheng, Y.-C., 173
Chen, J., 30, 185
Chen, R., 89
Chen, X., 174, 185
Chen, Y., 24
Cheresh, D. A., 108
Cherezov, V., 132
Chernysh, I., 91
Cheung, C. L., 23
Cheung, E. Y., 101
Chicault, M., 5
Chien, Y. H., 226–227, 232, 235, 237, 246, 253–254, 259
Chikenji, G., 9
Chipot, C., 142
Chiriatti, A., 166, 170
Chirino, A. J., 225–226, 230, 232, 234, 236, 239, 248–250, 254, 256–257, 260
Chodera, J. D., 2–3
Choi, C. H., 6
Choi, J. S., 6
Chothia, C., 4, 187, 190
Chouery, E., 56, 60
Chou, K. C., 8, 17–18
Chou, P. Y., 2, 8
Chow, C. W., 170
Christiansen, V. J., 93
Chugunov, A. O., 132, 139
Chung, D. W., 88
Chung, I., 170
Chung, J. K., 211
Chung, K. H., 93
Chun, J. G., 93
Chupin, V. V., 135, 140
Church, F. C., 91
Cicek, M., 165, 169
Ciepluch, K., 107
Ciesla-Dul, M., 107
Cilia la corte, A. L., 75

Claing, A., 170
Clarke, D. M., 68
Clavero, S., 54
Clements, C. S., 226, 252–253
Cobb, B., 168
Cobley, N., 3
Coggan, M., 84
Cohen, C., 81–82
Colby, B. R., 54
Colgrove, J., 5
Collen, D., 91–93
Collet, J. P., 84, 94–95, 106
Collins, E. J., 232, 235, 258
Collins, P. J., 3
Comp, P. C., 80
Conard, J., 80
Connell, S. D., 96, 108
Conn, P. M., 69
Consonni, V., 12
Conti, E., 95–96
Cook, J., 196, 211
Cook, M., 97
Cooper, A. V., 100, 186, 250
Cooper, D. N., 54
Cooper, J. A., 177
Cordeiro, M. N., 18
Corlieu, P., 168
Cornilescu, G., 187, 202
Corti, R., 105
Cote, H. C., 88
Cottrell, B. A., 90
Cottrell, J. S., 8, 23
Couch, G. S., 203–204, 206
Covic, N., 103
Cowburn, D., 167, 178, 180, 183–186, 206–207
Cox, D., 131
Crabtree, G. R., 81
Creasy, D. M., 8, 23
Credo, R. B., 97, 102
Creighton, T. E., 187
Cremel, G., 131
Cresson, C. M., 170
Cresswell, P., 259
Crick, F. H. C., 137
Crowley, M. P., 226–227, 232, 235, 237, 246, 253–254, 259

Cruz-Monteagudo, M., 18
Cullis, A. F., 2
Cunha, L. F., 47, 49, 51, 54, 58
Cunningham, C. C., 172
Cunningham, R. E. X., 165–166, 169
Curtis, C. G., 97, 102
Curto, M., 172
Cymer, F., 131, 133, 150, 153

## D

Dabhi, V. M., 225, 240
Dahimene, S., 168
Dahlback, B., 80
Dailey, H. A., 45
Dai, S., 236, 251, 253
Dai, Z., 167, 171, 174, 177–180, 183, 186, 206–207
Damus, P. S., 79–80
Dang, C. V., 83
Danielyan, K., 107
Dasari, M., 131
Da Silva, V., 54
Daugherty, C. C., 109
Davidson, J. M., 109
Davie, E. W., 80, 84, 88
Davies, D. R., 2
Davis-Harrison, R. L., 237
Davis, J. H., 52
Davis, M. M., 227, 235, 237, 251, 253–254, 258
Dea-Ayuela, M. A., 17
Deacon, H. W., 166
Deber, C. M., 132, 137
Deck, R. R., 5
De Cristofaro, R., 101
De Filippis, V., 101
Degano, M., 226–227, 230, 232–233, 235, 237, 244–246, 251–254, 259
Degen, J. L., 109
DeGrado, W. F., 131–132, 135, 153
de Groot, P. G., 98
Delarbre, C., 225, 254
de la Riva, G. A., 17–18
Deliot, N., 168–169
Delker, S. L., 250

Del Maestro, L., 165, 171
de Maat, M. P., 108–109
Deml, L., 5
Deng, Y. M., 30
Denis, M. M., 108
Denker, S. P., 178
Denning, G. M., 68
Denton, J., 167
Dent, R. M., 76
Depristo, M. A., 189, 211
Der, L. C., 96
de Ronde, H., 80
Derrick, J. P., 247
Deshpande, N., 3
Desnick, R. J., 46, 54, 56, 59–61, 67
Devasahayam, M., 258
de Verneuil, H., 54, 56, 60–61, 67
De Visser, M. C., 101
de Willige, S. U., 105
Dewitt, C., 5
Deybach, J. C., 54
Dianoux, A. J., 196
Di Castelnuovo, A., 97
Di Cera, E., 100
Dickerson, R. E., 2
Dick, R. F. II., 238
Dill, K. A., 2–3
Dimichele, D. M., 97
Dinauer, P. A., 83
Ding, B. S., 107
Dingermann, T., 6
Ding, Y. H., 252, 254
Dintzis, H. M., 2
Diorio, J. P., 89, 102
Dippel, D. W., 101
Dirrig-Grosch, S., 131
Dirven, R. J., 80
Discher, D. E., 94, 96
Di Sole, F., 168, 172
Dixon, A. M., 131
Dobson, C. M., 3, 189, 211
Dodds, S., 104
Doherty, G. J., 4
Doh, G., 32
Doi, M., 192, 194, 196, 199–200
Doi, Y., 171
Dolja, V. V., 22

Dolling, S., 96, 99
Domanik, R. A., 102
Dominguez, E. R., 18
Dominguez, R., 81, 86
Donati, M. B., 97
Donowitz, M., 165, 169–170
Doolittle, L. R., 81–82
Doolittle, R. F., 81–82, 88–90
D'Orazio, A., 97
Dorwart, M. R., 166
Downey, J., 84
Doxastakis, M., 145
Doyle, D. A., 183
Dragomir, A., 167, 179, 202
Drakesmith, H., 234
Drew, A. F., 109
Druker, B. J., 131
Dryden, D. T. F., 186
Duardo-Sanchez, A., 9, 18
Dubowitz, M., 97
Ducarme, P., 136
Ducatez, M. F., 23
Dudley, J., 19
Duga, S., 97
Duhaime, M., 167
Duke, T., 212
Duncan, R. F., 65
Dunn, E. J., 103
Dutta, S., 3
Duy, C., 53–54
Dwek, R. A., 258

# E

Eaton, D. L., 94
Eddy, S. R., 8
Edholm, O., 150
Edwards, J. P., 30
Edwards, K., 167, 171
Edwards, R. V., 83
Edwards, S. F., 192, 194, 196, 199–200
Efremov, R. G., 129, 132–133, 135–136, 140, 144, 148, 153
Egan, M. E., 68
Ehrich, E. W., 227, 235, 237, 253
Eilers, M., 132, 137

Eisenbraun, J., 7
Eisenmesser, E. Z., 186
Eitzman, D. T., 93
Ekman, G. J., 104
El-Hassen, D., 245, 252
Elliott, B. E., 172
Elsliger, M. A., 230, 234–235, 246, 253
Elsner, M., 138
Ely, L. K., 226
Emami, H., 32
Endo, T., 196
Engelhard, V. H., 244–245
Engelman, D. M., 15, 131, 133–134, 138, 140, 151, 179, 211
Engelse, M. A., 109
Enghild, J. J., 94
Enns, C. A., 260
Enriquez de Salamanca, R., 56, 61
Eramian, D., 8
Eriksson, P., 101
Ermolyuk, Y. S., 133, 135–136, 140, 148, 153
Esmon, C. T., 80
Estes, M. K., 3
Estrada, E., 17
Eswar, N., 8
Evans, A., 97, 99
Evans, G., 5
Evatt, B., 80
Everse, S. J., 81, 90

## F

Fahrer, A. M., 235
Falcón, J., 64, 66, 69
Falcon-perez, J. M., 43
Falls, L. A., 101
Falvo, M. R., 95
Fanelli, T., 168, 172
Fanning, T. G., 4
Farago, B., 187, 202
Farquhar, M. G., 166, 169, 178
Farrar, D., 3
Farrell, D. H., 101, 108
Fasman, G. D., 2, 8
Fatah, K., 76, 89, 104, 106

Favia, M., 168, 172
Fayad, Z. A., 105
Fayer, M. D., 211
Fayos, R., 69
Fay, W. P., 93
Feder, J. N., 225, 249, 257, 260
Fehon, R. G., 167, 171, 179–181, 189, 202
Feig, M., 136
Feldman, H. J., 2
Ferguson, E. W., 90
Fernandez, L., 9
Fernandez, M., 9
Ferrand, M., 196
Ferrell, R. E., 102
Ferry, J. D., 86
Fersht, A. R., 177
Fickova, M., 131
Fields, B. A., 253
Fievet, B. T., 171
Finger, C., 132
Finlayson, J. S., 100–101
Fischetti, R. F., 3
Fitter, J., 53–54
Fixman, M., 195–196
Flanagan, J. M., 134
Fleishman, S. J., 137–138, 153
Fleming, G. R., 173
Fleming, K. G., 134
Fleming, P. J., 137
Folk, J. E., 84, 90
Fontanellas, A., 56, 61
Forman, W. B., 83
Fornace, A. J. Jr., 81
Forrest, L. R., 133
Fortian, A., 43, 52–53, 57, 60, 64, 66, 68–69
Foskett, J. K., 168
Foulks, J. M., 108
Fournie, J. J., 253
Fowler, P. W., 144
Fox, R. I., 69
Francis, C. W., 83, 90, 94, 101, 108
Franco, R. F., 99
Frankenberg, N., 46
Frank, J., 56
Frank, M. B., 98
Fredenburgh, J. C., 101

Freel, J. A., 250
Fremont, D. H., 226, 230, 232, 235, 244–245, 256
Frere, F., 46
Fretto, L. J., 90
Frieden, C., 3
Friedman, A., 5
Friedman, P. A., 165–166
Frokjaer, S., 6
Fuchs, E., 165
Fujikawa, K., 84
Fujitsuka, Y., 9
Fuster, V., 105
Futers, T. S., 86, 100
Fu, Y., 101–102

## G

Gachelin, G., 225, 254
Gadola, S. D., 230, 234, 237, 246, 253
Gaffney, P. J., 90
Gagnon, S. J., 237
Gal, A., 185
Galli, S. J., 80
Gammeltoft, S., 8
Ganguly, K., 107
Gan, L., 248, 260
Gao, G. F., 230, 236, 238, 251, 253–254
Gao, P., 24
Gapin, L., 236, 251, 253
Garavito, R. M., 104
Garbett, D., 174, 177, 179
Garboczi, D. N., 232, 235, 252–254, 258
Garcia De La Torre, J., 201
Garcia, K. C., 225, 233, 237, 244–245, 251–252, 254
Gardin, A., 131
Garnier, N., 138
Garrett, T. P., 224, 238–240, 253
Gary, R., 171
Gascoigne, N. R., 237, 252
Gastinel, L. N., 225–226, 230, 232–233, 239, 247–248, 253, 256
Gates, M. A., 10, 14
Gautreau, A., 165, 171
Gawinowicz, M. A., 90

Ged, C., 56, 60–61, 67
Geigert, J., 6
Geiss, K. T., 18
Geissler, H., 48
Geliebter, J., 227, 231–232, 240
Genest, M., 138
Georgescu, M. M., 165, 174, 185
Geraghty, D. E., 238, 254
Gerard, B., 166, 169, 185
Gerber, D., 131, 135
Gerder, M., 4
Gerdes, V. E., 103
Gerloff, N. A., 23
Germain, R. N., 240, 251, 253–254
Geronimi, F., 67
Gersh, K. C., 96, 100–101, 108
Gerstein, M., 187, 190
Gerth, U. C., 230, 238, 253–254
Gervais, C., 136
Ghetie, V., 234, 247
Ghosh, A., 1, 4, 14, 17–19, 21, 23, 26–31
Ghosh, K., 2–3
Gibbs, A. J., 8–9, 19
Gilks, C. B., 169
Gill, G. N., 170
Gilliland, G. L., 87
Gilman, P. B., 83
Ginsburg, D., 93
Girard, T. J., 80
Gisler, S. M., 168–169
Gnirke, A., 225, 257
Godal, H. C., 83
Goddard, T. D., 203–204, 206
Godlewski, J., 104, 106
Goerz, G., 56
Goldbaum, D. M., 81–82
Goldberg, A. L., 69
Goldberger, A. L., 10
Gold, H. S., 5
Goldman, A., 15
Goldsby, R. A., 5
Goncharuk, M. V., 135–136, 140, 148, 153
Gonzalez, C., 4
González-Díaz, H., 9, 16–18, 20
Gonzalez-Diaz, Y., 9, 16–18
Gonzalez, E., 43, 64, 66, 69
Gonzalez-Vlla, T., 18

Goodenow, R. S., 259
Gorbalenya, A. E., 24–25
Gore, D. B., 3
Gorga, J. C., 226, 244–245
Gorkun, O. V., 83–84, 89
Gorman, J. J., 90
Gotch, F. M., 30
Gotoh, O., 21
Gotoh, T., 84
Gottesman, M. M., 131
Gough, N. R., 172
Govindarajan, K. R., 238
Graf, M., 5
Grakoui, A., 258
Grandchamp, B., 54
Grant, P. J., 85–86, 91, 97–99, 103–107
Gratkowski, H., 132
Gray, A., 102
Greenbaum, J. A., 32
Greenberg, C. S., 85–86
Greenblatt, D. M., 203–204, 206
Greene, M. I., 252
Green, W. N., 131
Greer, C., 172
Greilich, P. E., 76
Grieninger, G., 101–102
Griffin, J. H., 80
Groh, V., 238, 254
Gross, C. N., 260
Grundy, J. E., 239, 255
Guarini, L., 67
Guerra, L., 168, 172
Guggino, W. B., 165
Gui, M., 230, 237
Gulyaev, D. I., 139
Gumbiner, B. M., 165
Gunn, P. R., 4
Gupta, R., 8
Gurezka, R., 134–135
Gurol, H., 193
Gurunathan, S., 5
Gute, B. D., 4, 9, 14, 18, 28–31
Gutgemann, I., 235
Guthold, M., 90, 95–96
Guttman, D. E., 6
Gutwin, K. N., 135

Guyonnet-Duperat, V., 68
Gwozdzinski, K., 104

# H

Hackeng, T. M., 81
Hada, M., 86
Haggie, P. M., 168
Hainfeld, J. F., 89, 102
Haire, L. F., 3
Hai, T., 69
Hajjar, R. J., 2
Hakoshima, T., 185
Halfmann, P., 24
Hall, C. E., 81
Hall, R. A., 165–167, 170
Hamaguchi, M., 83
Hamalainen, E., 99
Hammes-Schiffer, S., 186
Hamori, E., 10, 14
Hampl, J., 235, 251, 254
Hamsten, A., 76, 106
Han, C. C., 211
Hanson, M. A., 132
Hantgan, R. R., 83, 95
Harada, F. A., 67
Harding, D. R., 4
Harle, M., 10–11, 14
Harlos, K., 246
Harrand, R., 96, 99
Harris, B. Z., 165
Harris, K. W., 80
Harris, R. A., 226, 232, 245
Hart, R. G., 2
Harvey, D. J., 258
Hashimoto, S., 80
Hassan, M. I., 225–227, 230, 236, 239, 249, 257
Hatta, M., 24
Havlin, S., 10
Hawiger, J., 107
Hawker, C. J., 48
Hawkins, D. M., 18
Hawkins, R. J., 186
Head-Gordon, T., 138

Hébert, B., 168
Hecht, K., 137
Heda, G. D., 168
Heier, M., 105
Heikema, A. P., 250
Helman, L., 172
Helms, V., 138
Henderson, K. N., 226
Henin, J., 142
Hennecke, J., 236, 251
Henrick, K., 3
Henschen, A. H., 84, 89, 97
Henschen-Edman, A. H., 81–82
Heringa, J., 137
Hermjakob, H., 8
Hernandez, I., 87, 100–102
Hernando, N., 168–169
Heroux, A., 49–51
He, S., 104
Hessel, B., 89
He, X., 225
Hicks, M., 79–80
Higgins, J. S., 191–192, 195
Highsmith, R. F., 79
Hildebrand, P. W., 131
Hilgenfeld, R., 85
Hill, C. P., 49–51
Hill, R. L., 84
Hinnrasky, J., 168
Hinssen, H., 2
Hjemdahl, P., 104
Hoang, A. D., 80
Hoare, H. L., 226
Hodson, C. A., 165
Hoffman, M., 103–104
Hoglund, P., 240
Hogue, C. W., 2
Ho, J., 225
Holmes, M. A., 226, 236
Hologne, C., 168
Homandberg, G. A., 108
Hon, C. C., 23
Hong-Brown, L. Q., 68
Hong, K., 248
Hoof, I., 32
Horellou, M. H., 80

Hormann, A., 105
Horne, B. D., 105
Horton, R. M., 225
Housset, D., 251
Houwing-Duistermaat, J. J., 101
Hovik, R., 225, 240
Howard, J., 187–189, 204
Howells, K., 54
Ho, Y. K., 12, 18
Hristova, K., 131, 133
Hrycyna, C. A., 131
Hsieh, L., 99
Huang, C. C., 203–204, 206
Huang, L. Y., 253
Huang, M., 233, 244–245, 251–252, 254, 258–259
Huber, A. H., 225–226, 230, 232–234, 239, 247
Hudry-Clergeon, G., 108
Hudson, R. P., 51
Huertas, M. L., 201
Hue, S., 236
Hughes, D. A., 102
Humphries, S. E., 97
Huntington, J. A., 91
Husain, S. S., 93

# I

Iacoviello, L., 97
Ibarra-Molero, B., 52
Ichikawa, Y., 235, 237
Ichinose, A., 84
Iijima, H., 230, 237
Im, W., 136–137, 142, 144
Ingham, K. C., 90
Ippen, H., 69
Irie, A., 107
Irrinki, A., 249, 260
Ishikawa, H., 211
Ito, J., 69
Ito, N., 3
Ivert, T., 104
Iwanaga, S., 108
Iwata, H., 102

## J

Jackman, R. W., 80
Jackson, G. R. Jr., 172
Jackson, K. W., 93
Jackson, M. R., 252
Jacobsen, A., 84
Jacoby, E., 139
Jahn, D., 46
Jähnig, F., 150
Jakobsen, B. K., 225, 230
Jakubowski, H., 103
James, M. F., 170, 178
Jamora, C., 165
Jana, D., 174, 177, 179–180
Janczewski, T. A., 4
Janeway, C. A. Jr., 240, 251
Jankowski, M., 103
Janmey, P. A., 95–96
Janosi, L., 145
Janovick, J. A., 69
Janus, T. J., 84
Jaswal, S. S., 52
Jawerth, L. M., 95
Jebbink, M. F., 25
Jedlinski, I., 107
Jeffrey, H. J., 10–11
Jenkins, J., 225
Jeon, T.-J., 135
Jiang, H., 108
Jiang, M., 9
Jiang, T., 135
Jiang, W., 3
Jin, N., 12, 18
Johannesson, M., 167, 179, 202
Johansson, A., 62
Johnson, G. E., 179
Johnson, J. E., 234
Johnson, J. L., 248
Johnson, L. N., 152
Johnson, R. M., 132, 137
Joly, E., 252
Jones, D. T., 138
Jones, E. P., 225, 240
Jores, R., 253–254
Jouihan, S. A., 101

Jugert, F. K., 56
Jun, G., 81–82
Jungerius, C., 109

## K

Kaibuchi, K., 171
Kaijzel, E. L., 109
Kamata, T., 107
Kamboh, M. I., 102
Kaminski, M., 86
Kang, H. I., 69, 95–96
Kangueane, P., 238
Kant, J. A., 81
Kaplan, A. P., 80
Kappler, J. W., 236, 251, 253
Karim, Z., 166, 169, 185
Karlson, K. H., 167
Karpe, F., 76, 106
Karp, G., 2
Karplus, P. A., 179–181, 202
Karpusas, M. N., 258
Karthikeyan, S., 183
Katona, E., 85, 99
Kavathas, P. B., 237
Kawaoka, Y., 24
Kay, L. E., 51
Kazlauskas, A., 170, 178
Kazmierczak, S. C., 101
Keane, P. M., 83
Keating, A. E., 135
Kelleher, D. P., 102
Kemkes-Matthes, B., 102
Kendrew, J. C., 2
Kennedy, M. W., 250
Kerjaschki, D., 169
Kern, D., 186
Khan, J., 172
Khanna, C., 172
Khundmiri, S. J., 172
Kim, E., 183
Kim, J. K., 168
Kim, K. H., 166
Kim, P. I., 6
Kim, S., 135–136, 211

Kimura, S., 93
Kindt, T. J., 5
King, G., 131
Kinjo, A. R., 3
Kisiel, W., 80
Kissel, T., 7
Kitano, T., 102
Kjer-Nielsen, L., 226, 253
Klein, J., 227, 231, 240
Kleiss, A. J., 80
Kletzel, M., 67
Klinman, D. M., 5
Kloczewiak, M., 107
Klossek, J. M., 168
Kobilka, T. S., 132
Koch, M., 230, 234, 237, 246, 253
Koch-Weser, J., 6
Koeleman, B. P., 80
Kohler, H. P., 86, 97, 99
Kokubo, H., 136
Kolibaba, K. S., 131
Kolin, D. L., 168
Kollman, J. M., 81–82
Kollman, P. A., 142
Komanasin, N., 100
Konig, R., 253
Konshina, A. G., 133
Koolwijk, P., 108–109
Koonin, E. V., 22
Kordich, L. C., 96, 103
Korn, B. S., 235, 237
Korzhnev, D. M., 186
Kosakovsky Pond, S. L., 23
Kosakowska-Cholody, T., 131
Ko, S. B. H., 166
Kostenko, L., 226
Koster, T., 80
Kourilsky, P., 225, 254
Kovacs, J. A., 138
Krabbenhöft, A., 168
Kraker, J., 18
Kramer, L. D., 5
Kreda, S. M., 168, 172
Kreimann, E. L., 165
Kremer, J. R., 23
Kretzmer, K. K., 80
Krewski, D., 105

Kristensen, D. M., 22
Kudryk, B., 81
Kulikauskas, R., 179–181, 202
Kulke, M., 2
Kumar, S., 19, 142
Kumar, V., 226–227, 230, 236, 239, 249, 257
Kuppuswamy, M. N., 80
Kusunoki, M., 3
Kuti, M., 47, 49, 51, 58
Kuzushima, K., 251
Kwak, K., 211
Kwon, S. H., 165

# L

Laage, R., 134–135
Labeikovsky, W., 186
Labotka, R. J., 104
Ladbury, J. E., 236, 251
Ladias, J. A., 183
Lai, M. M., 25
Laín, A., 43, 52–53, 57, 60, 64, 66, 68–69
Lai, T. S., 85
Lalanne, M., 56, 60, 68
LaLonde, D. P., 174, 177, 179
Lam, H. M., 56
Lamprecht, G., 165
Lamrissi-Garcia, I., 67–68
Lam, T. T., 23
Lancellotti, S., 101
Landau, D. P., 136–137
Lane, D. A., 80, 86, 99
Langer, B. G., 83
Langosch, D., 134–135, 137
Lanyi, J. K., 153
Larionov, S., 14
Lauffenburger, D. A., 170
Laurens, N., 108–109
Lauricella, A. M., 96, 103
Lau, S. Y., 24
Laver, W. G., 4
Lawrence, S. O., 83
Lazar, C. S., 170
Lazaridis, T., 136, 141–142, 146
Lazorova, L., 167, 179, 202
Leadbeater, R., 57

Leahy, D. J., 237
Leake, M. C., 2
Lear, J. D., 132
Lebedeva, M. I., 253
Lebron, J. A., 225–226, 230, 232, 234, 239, 248–249, 254, 256
Lederer, E. D., 168–169, 172
Ledger, R. E., 95
Lee, A., 183
Leebeek, F. W., 101
Lee, C. S., 93
Lee, D. G., 6
Lee, D. H., 68–69
Lee, E. H., 95
Lee, H., 173
Lee, J., 142, 144
Lee, K. N., 93
Leeper, F. J., 44, 48
Lee, S. H., 95, 211
Lee, V. K., 249
Lefkowitz, R. J., 170
Lehmann-Kopydlowska, A., 107
Lehmann, U., 168
Lehner, P. J., 259
Leibold, E. A., 249
Leibovici, M., 185
Lengauer, T., 8
Lenkowski, A., 103
Leong, P. M., 10
Leroy, C., 166, 169, 185
Lers, N., 9, 11
Lesk, A. M., 187, 190
Lesty, C., 94, 106
Le Trong, I., 84
Leturcq, D., 260
Leung, T., 183
Levy, Y., 258
Lewis, J., 183
Lewis, S. D., 84
Liang, H. Y., 95
Liang, J., 132–133
Liang, T. J., 101
Liao, B., 14, 16, 25, 28–29
Liao, J., 168
Li, C., 15, 17, 19, 22, 26, 168
Li, E., 131, 133
Li, H., 253

Li, J., 5, 167, 171, 174, 177–181, 183–187, 202–207, 212
Lijnen, H. R., 93
Likert, K. M., 80
Lim, B. B., 95
Lim, B. C., 86, 96, 99
Lim, H. W., 67
Lim, W. A., 165
Lindahl, K. F., 225, 240
Lindan, O., 83
Lindemann, S., 108
Lindorff-Larsen, K., 189, 211
Linenberger, M. L., 98
Lin, Y. P., 3
Li, O. T., 23
Li, P., 226–227, 230, 232, 236, 239, 250, 254, 257–258, 261
Li, Q., 179–181, 202
Li, R., 14, 16, 25, 28–29
Litvinovich, S., 81, 86
Litvinov, R. I., 94, 96, 131, 138, 154
Litzenberger, C., 227, 235, 237, 253
Liu-Chen, L. Y., 165–166, 170
Liu, H., 109
Liu, M. A., 5
Liu, W., 95, 132
Liu, X. Z., 185
Liu, Y. Z., 12, 14, 16, 25, 28–29, 172
Li, Y., 260
Li, Z., 3
Ljunggren, H. G., 240
Llera, A. S., 253
Locatelli, F., 7
Lockhart, E., 103
Lohse, M. J., 131
Loll, P. J., 104
London, R. E., 104
Longstreth, W. T., 98
Loo, T. W., 68–69
Lopez, J. A., 4
Lopez-Mendez, B., 52
Lorand, L., 84, 89, 97
Lord, S. T., 84, 87, 100–101
Loskutov, A., 14
Lourenco, D., 99
Louvard, D., 165, 167, 171–172
Lovely, R. S., 101

Lowik, C. W., 109
Lu, C., 12, 18
Luc, G., 97, 99
Lucking, A. J., 105
Luecke, H., 153
Lu, H., 106
Lukacs, G. L., 168
Lukas, T. J., 107
Luo, X., 166
Luo, Y., 102, 168
Luttrell, L. M., 170
Luu Duc, C., 5
Ly, B., 83
Lynch, G. W., 84
Lyons, D. S., 251, 254

# M

Ma, B., 178, 187, 212
MacCallum, J. L., 133–134
MacCallum, R. M., 4
MacDonald, K. A., 83
MacKenzie, K. R., 132, 134–135, 138, 140–141, 151
MacKinnon, R., 183
MacKintosh, F. C., 95–96
Madden, D. R., 224, 226, 230–232, 238, 240, 244–245, 249, 255–257, 259
Madhusudhan, M. S., 8
Madison, E. L., 93
Madrazo, J., 81, 86–87
Maeda, M., 171
Maesaki, R., 185
Maffei, F. H., 99
Magalhaes, A. L., 17
Magyar, C. E., 170, 172
Mahajan, R., 95
Mahon, M. J., 169–170
Majerus, P. W., 104
Mak, D. D., 168
Makowski, L., 3, 95
Makretsov, N., 169
Maldonado, A. J., 4
Malfettone, A., 166, 170
Malissen, B., 251
Manchanda, N., 170, 178

Mancini, A. J., 67
Mandava, S., 3
Mangeat, P., 171–172
Mangia, A., 166, 170
Manieri, M., 225
Mannila, M. N., 101
Mann, K. G., 104
Mansfield, M. W., 98, 106
Marcum, J. A., 80
Marder, V. J., 90, 94, 101, 108
Margulies, D. H., 236, 240, 244, 254
Markley, J. L., 3
Marrack, P., 236, 251, 253
Marrink, S. J., 145–146
Marshall, D. P., 133–134
Marshall, J., 68
Marsh, D., 154
Martin, A. C., 4
Martinez, D., 5
Martínez, J. L., 83, 152
Martin, M., 171–172
Martin, S. E., 94
Martin, W. L., 248, 260
Marti-Renom, M. A., 8
Marx, P. F., 103
Ma, S. K., 23
Masova, L., 91
Mathew, B., 230, 234, 237, 246, 253
Mathews, M. A., 49
Matsuda, M., 80, 89
Matsui, T., 171
Matsuka, Y. V., 90
Matsumura, M., 226, 230, 232, 235, 244–245, 256
Matthews, E. E., 131
Mattsson, N., 258–259
Maudsley, S., 170
Maurer, M. C., 99
Mauri, A., 12
Mavaddat, N., 227, 235, 237, 253, 259
May, H. T., 105
Mayzel, M. L., 133, 135, 140, 153
Mazurier, F., 67
McAlister, M. S., 258
McBride, O. W., 81
McCaffrey, M. W., 260
McCammon, J. A., 211

McClatchey, A. I., 167, 171–172
McCluskey, J., 226
McConnell, J. P., 67
McCoy, M. L., 169
McDermott, G., 226, 250
McDermott, L. C., 250
McDonagh, J., 84, 86
McDonnell, A. V., 135
McIntyre, G. A., 8–9, 19
McKee, P. A., 84, 90, 93
McKenna, P. W., 79
McKenney, J. B., 80
McLeish, T. C. B., 186
McMahon, H. T., 4
McManus, R., 102
McMichael, A. J., 230, 238, 253–254
McMullen, B. A., 84
McNeill, A. M., 68
McQuistan, T., 169, 178
Meade, T. W., 97
Medved, L. V., 81, 83–84, 86–87, 89–90, 101, 108
Meens, J. A., 172
Megarbane, A., 56, 60
Megarbane, H., 56, 60
Meh, D. A., 84, 89, 91
Meijers, J. C., 103
Mellman, I., 170
Meltzer, P. S., 172
Memoli, V. A., 80
Mendoza, A., 172
Meneses-Marcel, A., 17
Meng, E. C., 203–204, 206
Meng, Y. G., 248
Menne, K. M., 8
Mention, J. J., 236
Merino-Ott, J. C., 3
Merlino, G., 172
Merot, J., 168
Metassan, S., 105
Metzner, H. J., 85
Meyer, M. C., 6
Mezei, F., 192
Mezei, M., 47, 49, 51, 58
Mezo, A. R., 248
Michejda, C. J., 131
Michel, F., 172

Migliorini, M. M., 90
Mi, H., 196
Mikkola, H., 86, 97, 99
Mikulska, J. E., 260
Milburn, C. M., 248
Miles, C. A., 52
Miles, J. J., 245, 252
Miletich, J. P., 80
Milgram, S. L., 165
Millar, N. S., 131
Millen, L., 166
Miller, K. A., 105
Millet, O., 43, 51–53, 57, 60, 64, 66, 68–69, 186
Mills, D., 18
Mills, J. D., 106
Mills, N. L., 105
Miltenberger-Miltenyi, G., 108
Minder, E. I., 54
Mineeva, E. A., 148
Mineev, K. S., 133, 135, 140, 148
Minh, D. D., 3
Minning, S., 52
Mirishahi, M., 106
Mitsutake, A., 136
Miyashita, N., 137
Miyashita, O., 211
Mizuguchi, J., 108
Mochalkin, I., 82, 88
Modlin, R. L., 225, 234, 237, 245
Modyanov, N. N., 139
Moellering, R. C. Jr., 5
Moen, J. L., 84
Molina, R., 16–18, 20
Momin, S., 174, 185
Monkenbusch, M., 193, 200, 202, 211
Monroe, D. M., 104
Montalescot, G., 84, 94
Monteiro, R. C., 236
Monterisi, S., 168, 172
Moody, D. B., 235, 253
Moore, D. T., 131, 135, 153
Moore, M. R., 69
Mora-Jensen, H., 69
Morales, F. C., 165, 174, 185
Moran, N., 131
Moreau-Gaudry, F., 67

Morel, C., 67
Morelli, V., 99
Moreno, P. R., 105
Morgelin, M., 103
Morgenthaler, S., 10
Moriwaki, S., 235, 237
Morris, D. L., 226–227, 230, 232, 239, 250, 254, 257–258, 261
Morrow, M. P., 5
Morse, R., 259
Morser, J., 94
Mort, M., 54–55
Moruno Tirado, A., 56, 61
Moscona, A., 3, 24
Moser, T. L., 86
Mosesson, M. W., 84, 87, 89, 91, 96, 100–102, 108
Mosher, D. F., 86
Mostov, K. E., 225, 260
Mothe, B. R., 7
Mount, D., 10
Moyer, B. D., 167
Muhlestein, J. B., 105
Muirhead, H., 2
Muller, C. P., 23
Munteanu, C. R., 17
Munter, L. M., 131
Murayama, H., 80
Murciano, J. C., 107
Murer, H., 169
Murgich, J., 4
Murphy, C., 260
Murthy, S. N., 89
Mushegian, A. R., 22
Musial, J., 104
Muszbek, L., 84–85, 97, 99

# N

Nachman, R. L., 108
Nagai, M., 166, 169, 178
Nagasaki, K., 22
Nagashima, M., 94
Nagaswami, C., 83, 86, 90, 95–97, 99–100, 103, 105, 108
Nagel, G. M., 88

Naidenko, O. V., 237
Nakamura, H., 3
Namenuk, A. K., 248
Nance, M. R., 179–181, 202
Nandy, A., 1, 4–5, 9–15, 17–19, 21, 23, 26–32
Nandy, P., 4, 13–14, 17–19, 21, 23, 26–31
Napier, J. R., 4
Naren, A. P., 168
Natarajan, R., 18
Nathenson, S. G., 227, 231–232, 240, 251
Nawroth, P. P., 80
Ndonwi, M., 81
Neagoe, C., 2
Needleman, S. B., 8
Neefjes, J. J., 240
Neier, R., 46
Nei, M., 19
Nelson, D., 168
Nesheim, M., 93–94
Ness, J., 131
Neumann, D. A., 211
Ng, K. P., 165, 169
Nguyen, R., 171
Nicaud, V., 97, 99
Nielsen, J. S., 169
Nienaber, V., 248
Nierenberg, J., 104
Nieuwdorp, M., 103
Nieuwenhuizen, W., 83, 89
Ni, H., 101
Niknejad, A., 12, 18
Nilsson, H. E., 167, 179, 202
Nishioka, J., 80
Nishiyama, K., 108
Nixon, D. F., 30
Nolde, D. E., 132–133, 136, 144
Nordfang, O., 80
Nordmann, Y., 54
North, A. C., 2
Norton, R. S., 209
Novic, M., 11, 17
Novoseletsky, V. N., 132
Novotny, W. F., 80
Nugent, D., 99
Nugent, T., 138
Nussinov, R., 178, 187, 212
Nyberg, K., 179–181, 202

## O

Ober, R. J., 234
O'Callaghan, C. A., 225, 230
Oddoux, C., 101
O'Driscoll, L. S., 260
Odrljin, T. M., 83
Ogawa, T., 196
Ogunlesi, A. O., 30
Ohba, Y., 104
Ohlen, C., 240
Ohtaki, S., 108
O'HUigin, C., 227, 231, 240
Ojcius, D. M., 225, 254
Okamoto, Y., 136
Olack, G., 179
Oldfield, C. J., 22
Olivo-Marin, J. C., 172
Olomu, O., 172
Onuchic, J. N., 211
Opitz, C. A., 2
Orlando, R. A., 166, 169, 178
Ortega, G., 52–53, 57, 60, 64, 66, 68–69
Ortmann, B., 259
Osborne, B. A., 5
Ossei-Gerning, N., 86, 99
Ostedgaard, L. S., 166–167
Oster, C. G., 7
Osterud, B., 77
Otvos, L. Jr., 6
Otzen, D. E., 6
Owens, J., 6
Owoade, A. A., 23
Ozkan, S. B., 2–3

## P

Packhaeuser, C. B., 7
Palascak, J. E., 83
Palmer, A. G., 190
Pandi, L., 81–82
Paniagua, E., 9, 18
Pan, L., 185
Pappin, D. J., 8, 23
Parameswaran, K. N., 89
Parent, J. L., 170, 172
Parham, P., 224, 227, 230–232, 235, 240, 250, 255, 259
Park, D., 94
Parker, C. J., 86
Parkkila, S., 257
Park, Y., 6, 138
Parrish, R. G., 2
Parry, D. J., 107
Patel, A. B., 132
Paul, P., 250
Pavlov, K. V., 135–136, 140, 148, 153
Pawson, T., 178
Pease, L. R., 225, 233, 244–245, 251–252, 254
Pechik, I., 83, 87
Pecora, R., 192
Pedersen, L. C., 84
Peerschke, E. I., 101
Peltonen, L., 99
Pendurthi, U. R., 80
Peng, C. K., 10
Penny, D. M., 226, 248–249, 260
Penttila, A., 99
Pereira, S. V., 108
Perez-Castillo, Y., 17
Perez-Galan, P., 69
Perez-Montoto, L. G., 9, 18
Perez-Moreno, M., 165
Perkins, D. N., 8, 23
Perkins, S. J., 239, 255
Perlo, A., 179
Perneby, C., 104
Perola, M., 99
Perry, C. C., 3
Persson, K., 226, 232, 245
Perutz, M. F., 2
Peters, A., 105
Peters, B., 32
Petersdorf, E. W., 226, 236
Petersen, P. M., 48
Peterson, P. A., 225–226, 230, 232–235, 244–246, 248–249, 251–254, 256, 259–260
Petoukhov, M. V., 182, 205
Petryka, Z. J., 67
Petry, W., 196
Pettersen, E. F., 203–204, 206
Pfaffenbach, G. M., 227, 231–232, 240
Phelps, R., 62

Philippou, H., 75, 86, 99–100, 103, 105
Phillips, A. D., 54–55
Phillips, D. C., 2
Phillips, D. R., 108
Phillips, J. D., 49–51
Phillips, J. H., 238
Phillips, R. E., 30
Phoenix, F., 96, 99
Picot, D., 104
Pieculewicz, M., 107
Pieper, U., 8
Pieters, M., 103
Piomelli, S., 67
Pippen, A. M., 86
Pitard, V., 68
Pitcher, J. A., 170
Pizzo, S. V., 84
Plant, P. W., 101
Plavsic, D., 9, 11
Plaza del Pino, I. M., 52
Plow, E. F., 94
Podda, G., 17
Podolec, P., 107
Poh-Fitzpatrick, M. B., 67
Pohorille, A., 142
Poirier, O., 97, 99
Polesello, C., 171
Pollard, H., 165
Pollard, T. D., 177
Polyansky, A. A., 129
Ponnamperuma, K., 51
Pons, M., 52–53, 57, 60, 64, 66, 68–69
Poon, L. L., 23
Pope, C. A. III, 105
Porcelli, S. A., 225, 234, 237, 245
Pospisil, C. H., 101
Potter, T. A., 224, 238, 255
Poulikakos, P. I., 171
Powell, K. F., 4
Powers, D. L., 3
Powers, E. T., 3
Pozzi, N., 101
Prado-Prado, F., 18
Prakash, A., 145
Prasad, B. V., 3
Prat, A. G., 172
Pratt, K. P., 88

Premont, R. T., 166–167, 170
Prentice, L., 169
Prestegard, J. H., 134, 138, 140, 151
Priestle, J. P., 139
Procyk, R., 83, 89
Prodohl, A., 132
Pruissen, D. M., 98
Psachoulia, E., 133–134, 144
Psaty, B. M., 98
Puchelle, E., 168
Pujol, F. H., 4
Pujuguet, P., 165
Pullen, J. K., 225
Purcell, A. W., 253
Purohit, P. K., 94, 96
Pustovalova, J. E., 136, 144
Pustovalova, Y. E., 135–136, 140, 148
Pybus, O. G., 23
Pyrc, K., 25
Pyrkov, T. V., 139

## Q

Quintana, I. L., 103

## R

Rae, J., 248
Raghuram, V., 168
Rahman, N., 166–167, 170
Raife, T. J., 101
Rajan, T. V., 238
Ramachandran, G. N., 15
Ramakrishnan, C., 15
Ramesh, V., 170, 178
Ramirez, M. C., 62
Ramos, M. J., 48
Randic, M., 9, 11–13, 15–17
Ranganathan, S., 238
Rao, L. V., 80
Rapaport, S. I., 77, 80
Rasi, V., 99
Rasmussen, S. G., 132
Rath, A., 132
Rau, J. C., 91

Rausch, B., 168
Raychaudhury, C., 13
Rayment, I., 4
Rayner, J. M., 24
Read, R. J., 91
Reche, P. A., 232, 237
Reczek, D., 165–166, 168, 171–172
Redlitz, A., 94
Redman, C. M., 81, 102
Reich, Z., 251, 254, 259
Reid, A. H., 4
Rein, C. M., 101
Reiner, A. P., 98
Reinherz, E. L., 232, 237
Reitsma, P. H., 99
Remijn, J. A., 98
Rendell, M., 104
Renlund, D. G., 105
Retief, J. D., 19
Reynolds, D., 167
Riccardi, S. M., 168
Rice, W. G., 131
Richard, E., 67–68
Rich, R. R., 240, 259
Richter, D., 193, 200, 202, 211
Richter, H.-T., 153
Richter, L., 131
Riederer, B., 168
Riegert, P., 225
Rijken, D. C., 94
Riley, M., 81–82
Rinkel, G. J., 100
Risenmay, B. W., 108
Roach, P., 3
Robay, A., 168
Robert-Richard, E., 67–68
Roberts, A. G., 56, 61
Roberts, H. R., 104
Rocca, B., 101
Rochdi, M. D., 170, 172
Roch, F., 171
Rodgers, J. R., 240, 259
Rodi, D. J., 3
Rodriguez, E., 17
Rodriguez-Larrea, D., 52
Roessner, C. A., 51
Rojas, A. M., 96

Roomans, G. M., 167, 179, 202
Roopenian, D. C., 232, 247
Rosenbaum, D. M., 132
Rosenberg, J. M., 142
Rosenberg, J. S., 79
Rosenberg, R. D., 79–80
Rosendaal, F. R., 80, 98, 100–101
Rosing, J., 81
Roskoski, R. Jr., 131
Rossjohn, J., 226
Rossmann, M. G., 2
Rothenberg, B. E., 234
Roth, G. J., 104
Roubinet, C., 171
Roumier, A., 172
Rousseau, S. M., 80
Routledge, M. N., 105
Roux, B., 142
Rowland-Jones, S., 30
Roy, C., 171
Roy, K., 168
Rubenstein, R. C., 68
Rudd, P. M., 258–259
Ruddy, D. A., 225, 257
Rudolph, M. G., 236, 250–251, 254
Rugg, E. L., 68
Ruggeri, Z. M., 108
Ruigrok, Y. M., 100
Ruskin, J., 10, 14
Russell, R. J., 3
Rutella, S., 101
Ruvkun, G. B., 4
Ryadchenko, E., 14
Ryan, A. W., 102
Ryan, T., 102
Rybarczyk, B. J., 83
Ryckewaert, J. J., 108

# S

Sadasivan, B., 259
Sadoff, J. C., 5
Saito, H., 80
Saiz-Urra, L., 17
Sakata, Y., 93
Sakharov, D. V., 94

Sakorafas, P., 248
Sala, C., 165
Salem, N., 185
Sali, A., 3, 8
Salimi, N., 32
Sal-Man, N., 131, 135
Salter-Cid, L., 260
Salter, R. D., 224, 238–239, 253
Samama, M., 80
Samna Soumana, O., 138
Samnegard, A., 101
Samraoui, B., 224, 226, 230, 232, 235, 238, 240, 246, 248, 255, 259
Sanchez-Gonzalez, A., 17
Sanchez, L. M., 226, 230, 232, 236, 239, 249–250, 257, 260
Sanchez-Rodriguez, A., 17
Sanchez-Ruiz, J. M., 52
Sanders, P. M., 225
Sandset, P. M., 81
Sane, D. C., 86
Sanger, F., 2
Sansom, M. S., 133–134, 144
Santana, L., 9, 16–18
Santos, D., 68
Saper, M. A., 224, 226, 230, 232, 235, 238, 240, 246, 248, 255, 259
Saridakis, E., 3
Sasisekharan, V., 15
Sassa, S., 46
Sato, F., 4
Satta, Y., 227, 231, 240
Sauls, D. L., 103
Savolainen, V., 99
Sawada-Hirai, R., 234
Sawaya, M. R., 81–82
Saxe, D., 81
Schadick, K., 49
Schad, P. E., 86
Scheiner, T., 97
Schellekens, H., 7
Schellman, J. A., 52
Schild, H., 227, 235, 237, 253
Schirmbeck, R., 5
Schlessinger, J., 138, 153, 164, 170
Schmidt, J. J., 135
Schmieder, S., 166, 169, 178

Schneider, D. E., 104, 131–133, 150, 153
Schneider-Yin, X., 54
Schnieders, J., 7
Schobert, B., 153
Schoenfisch, M. H., 83
Schonbrun, J., 138
Schreiber, U., 196
Schubert, H. L., 49–51
Schubert, W. D., 46
Schulga, A. A., 135, 140, 153
Schulten, K., 95, 133
Schwartz, M. L., 84
Schwartz, S. M., 98
Schwertz, H., 108
Sciortino, F., 10
Scott, A. I., 51
Scott-Browne, J. P., 236, 251, 253
Scott, D. J., 105, 107
Scott, E. M., 105
Scott, J. D., 178
Seaman, C., 67
Seder, R. A., 5
Segelke, B. W., 225–226, 230, 232, 234–235, 245–246, 248–249, 253, 256, 259
Segre, G. V., 169–170
Seidler, U., 165
Sellers, E. M., 6
Sengupta, D., 145–146
SenGupta, S. K., 172
Senkevich, T. G., 22
Sentman, C. L., 226, 232, 245
Sere, K. M., 81
Seth, R., 131
Seubert, A., 69
Seubert, S., 69
Sexton, G. J., 101
Shady, A. A., 54, 62
Shafer, J. A., 84
Shainoff, J. R., 100
Shai, Y., 131, 135
Shapiro, A. D., 93
Sharkey, D. J., 169
Sharma, A., 165
Shawar, S. M., 240, 259
Shaw, A. S., 258
Shaw, C., 167
Shaw, P. H., 67

# AUTHOR INDEX

Shaw, S., 172
Shekar, S. C., 137
Shell, M. S., 2
Sheng, M., 165, 183
Shen, H. B., 8
Shen, M.-Y., 8
Shenolikar, S., 165–166, 169, 172
Shental-Bechor, D., 258
Shepherd, D., 230, 234, 237, 246, 253
Shepherd, K., 105
Sheppard, L., 105
Sher, B., 81
Sherman, L. A., 224, 255
Shieh, T., 137
Shields, R. L., 248
Shiel, J. A., 55
Shilton, B. H., 103
Shimizu, A., 88
Shin, C. K., 83
Shi, P. Y., 5
Shoolingin-Jordan, P. M., 57
Short, D. B., 168, 172
Shuman, H., 95
Shwayder, T. A., 67
Sicheritz-Ponten, T., 8
Sidney, J., 7, 32
Sidobre, S., 237
Siebenlist, K. R., 84, 89, 91, 100–101
Silvain, J., 106
Silva, P. J., 48
Silveira, A., 76, 106
Sim, B. C., 237, 252
Simister, N. E., 225–226, 230, 232–233, 239, 247–248, 253, 256, 260
Simon, M. I., 81
Simons, M., 10
Simpson-Haidaris, P. J., 83, 101
Singh, A. K., 168
Singh, N., 226–227, 230, 236, 239, 249, 257
Singh, T. P., 225–226, 236
Siscovick, D. S., 98, 105
Siudak, Z., 103
Sizemore, N., 165, 169
Sizemore, S., 165, 169
Skolnick, L. M., 67
Skubiszak, A., 107
Slayter, H. S., 81

Slooter, A. J., 98, 100
Slusky, J. S., 131, 138, 154
Smith, A. E., 68
Smith, D. A, 96
Smith, K. J., 252, 254
Smith, S. O., 132, 137
Smith, T. F., 8
Sneddon, W. B., 170, 172
Snijder, E. J., 24–25
Snow, P. M., 225–226, 230, 234, 239, 248–249, 254, 256
Sobel, J. H., 90
Sobol, A. G., 135, 140, 148
Sodt, A. J., 138
Sohl, J. L., 52
Somasiri, A., 169
Song, W., 24
Song, Y., 238
Soria, C., 94, 106
Soria, J., 94, 106
Sotomayor, M., 95
Souaille, M., 142
Sousa Da Silva, A. W., 3
Southwood, S., 7
Soyez, S., 108
Spaan, W. J., 24–25
Sparks, E. A., 95–96
Spectre, G., 108
Speir, J. A., 252
Spies, T., 226–227, 230, 232, 236, 238–239, 250, 254, 257–259, 261
Spitzer, S. G., 80
Spraggon, G., 81, 90
Sridhar, V., 248
Stafford, A. R., 101
Stagljar, I., 169
Stamford, P. J., 48
Stam, N. J., 240
Standeven, K. F., 91, 96–97, 99–100
Stanfield, R. L., 251–252, 254, 258–259
Stanford, N., 104
Stanton, B. A., 168
Staritzbichler, R., 138
Stauffer, F., 46
Stea, B., 166, 170
Stead, N., 80
Steck, S., 5

Stefanova, I., 251
Steinle, A., 238, 254, 257, 261
Steinman, H. M., 90
Steitz, T. A., 15
Stenberg, P., 102
Stenflo, J., 80
Stenkamp, R. E., 84, 88
Stenson, P. D., 54–55
Stephen, P. M., 104
Stephens, H. A., 250
Stepien, E., 103–104, 106–107
Steplock, D., 169, 172
Stern, D. M., 80
Stevens, D. J., 3
Stevens, J., 252
Stickland, M., 86, 99
Stickland, M. H., 86, 97–99
Stirling, Y., 97
Stockner, T., 133–134
Storey, R. F., 104
Story, C. M., 260
Strandberg, B. E., 2
Straub, J. E., 137
Strominger, J. L., 224, 226, 230, 232, 235, 238, 240, 244–246, 248, 255, 259
Strong, D. D., 90
Stronge, V. S., 230, 234, 237, 246, 253
Strong, R. K., 226, 236, 250
Stroynowski, I., 225, 236, 240, 259
Stump, D. C., 93
Stura, E. A., 225–226, 230, 232, 234–235, 244–246, 248–249, 253, 256, 259
Suehiro, K., 108
Sugita, Y., 136–137
Sullivan, J. H., 105
Sullivan, L. C., 226
Sumen, C., 258
Sundback, J., 226, 232, 245
Sun, J., 237
Sun, R., 251
Superfine, R., 95
Suzuki, K., 80
Svensson, J., 104
Svensson, P. J., 80
Svergun, D. I., 180, 182, 202–203, 205
Swendsen, R. H., 142

Sydor, W. J., 104
Syme, C. A., 170, 172
Syrjala, M., 99
Syrtcev, N. P., 133
Szczeklik, A., 103–104
Szuldrzynski, K., 104, 106

## T

Tabaczewski, P., 225
Tadeo, X., 52
Taiwo, J. O., 23
Takada, S., 9
Takada, Y., 107
Takahashi, Y., 165, 174, 185
Takahata, N., 227, 231, 240
Takeda, T., 166, 169, 178
Tamura, K., 19
Tan, A. K., 94
Tang, H., 102
Tang, J. X., 95–96
Tang, K., 102
Tangri, S., 7
Tanguy, M. L., 106
Tans, G., 81
Tan, T. W., 238
Taouil, K., 168
Tarasova, N. I., 131
Tarasov, S. G., 131
Taubenberger, J. K., 4
Tavares, A., 108
Tavella, M. H., 99
Tawfik, D. S., 22
Teague, S. J., 8
Teller, D. C., 84
Tenchini, M. L., 97
Teramura, G., 98
Terawaki, S.-I., 185
Testa, J. R., 171
Teyton, L., 230, 234–235, 246, 250, 253
Thelin, W. R., 165
Thiagarajan, P., 101
Thian, F. S., 132
Thirumalai, D., 137
Thomas, A., 136

Thomas, N. S., 55
Thomas, W., 225, 257
Thompson, E. O., 2
Thompson, G. G., 69
Thornton, J. M., 4
Thorsen, S., 92
Thun, M. J., 105
Thurston, G. D., 105
Tieleman, D. P., 133–134
Tietze, L. F., 48
Timmons, S., 107
Tisdale, M. J., 225
Tjandra, N., 83
Tkach, E. N., 133, 135, 140, 148
Todeschini, R., 12
Toit, L. D., 103
Tokuhira, M., 107
Tokura, S., 84
Tokuriki, N., 22
Tolley, N. D., 108
Tomasello, E., 250
Tomatsu, S., 257
Tonegawa, S., 235, 237
Toomre, D., 170
Tormo, J., 225, 230, 238, 253–254
Tornvall, P., 76, 106
Totterman, K., 99
Toumaniantz, G., 168
Townsend, A., 234
Townsend-Merino, W., 3
Tracz, W., 103–104, 106
Traebert, M., 169
Tranqui, L., 108
Trentinaglia, I., 105
Treutlein, H. R., 134
Trigueros, T., 52
Trotter, K. W., 168, 172
Trumbo, T. A., 99
Tse, C. M., 165
Tsuchihashi, Z., 225, 257
Tsukita, S., 171
Tsurupa, G., 83, 108
Tuley, E. A., 81
Tuppy, H., 2
Tynan, F. E., 245, 252
Tysoe-Calnon, V. A., 239, 255

## U

Ubarretxena-Belandia, I., 131
Ubeira, F. M., 9, 16–18
Uitte de Willige, S., 100–101
Ulloa-Aguirre, A., 69
Ullrich, A., 152
Ulmer, J. B., 5
Ulrich, E. L., 3
Umetsu, K., 102
Undas, A., 103–104, 106–107
Unterreitmeier, S., 131
Upmanyu, M., 95
Uriarte, E., 9, 16–18, 20
Utz, U., 252, 254
Uversky, V. N., 22

## V

Vahtera, E., 99
Valentine, J. L., 104
Valenti, P., 171
Valnickova, Z., 94
van Bleek, G. M., 251
Vandenbroucke, J. P., 80
van der Graaf, Y., 98
van der Hoek, L., 25
van der Westhuizen, F. H., 103
Van de Water, L., 84
Vandlen, R., 170
van Erck, M. G., 109
van Hinsbergh, V. W., 109
van Hylckama Vlieg, A., 100
van Kaer, L., 235, 237
van Solinge, W. W., 98
van Wijk, R., 98
Varona-Santos, J., 16, 18
Varon, D., 108
Vaughn, D. E., 225–226, 230, 234, 239, 247–249, 254, 256, 261
Vazquez-Prieto, S., 9, 18
Vecchio, L. D., 7
Veerapandian, L., 82
Vegh, Z., 251
Vehar, G. A., 80

Veklich, Y. I., 83–84, 89, 108
Velankar, S., 3
Veltkamp, J. J., 80
Vendruscolo, M., 189, 211
Vereshaga, Y. A., 136, 144
Vergoten, G., 136, 139
Verkleij, C. J., 103
Verkman, A. S., 68, 168
Vermeulen-Oost, W., 25
Verpy, E., 185
Vicente, V., 108
Vida, L., 104
Vijaykrishna, D., 23
Vilaire, G., 131, 138, 154
Vilar, S., 17–18
Vilas, R., 9, 18
Villanueva, G. B., 104
Villmow, B., 250
Vita, R., 32
Vitiello, A., 224, 255
Vivier, E., 250
Voelz, V. A., 2–3
Volkmann, N., 81–82
Volkmer, T., 132
Voltz, J. W., 165–166
Volynsky, P. E., 129, 133, 135–136, 140, 144, 148, 153
von Heijne, G., 131
von Klot, S., 105
von Zastrow, M., 166
Vos, H. L., 101
Vracko, M., 11, 13, 17
Vyas, J. M., 240, 259

## W

Waheed, A., 225–226, 236, 257
Walker, F. J., 80
Wallen, H. N., 104
Wallin, E., 131
Wall, J. S., 89, 102
Walpole, N. G., 226
Walshaw, J., 137
Walters, R. F. S., 132
Walters, R. S., 131, 138, 154
Walton, D. J., 103
Wang, C., 185
Wang, F., 137, 172
Wang, H. L., 95, 104
Wang, J., 12, 15, 18–19
Wang, P., 24
Wang, Q., 69
Wang, T. M., 12, 16–20, 28
Wang, W., 15, 94
Wang, X., 15, 17, 26, 56
Wang, Y., 172
Wang, Z., 17, 19, 22
Wanner, V., 225
Wan, X., 172
Ward, E. S., 234, 247
Warner, C. A., 56, 61
Warren, L. A., 80
Warren, M. E., 103
Wartiovaara, U., 99
Watala, C., 104
Waterman, M. S., 8
Watson, C. J., 67
Watt, K. W., 90
Watts, S., 259
Webb, B., 8
Webb, G. C., 84
Webster, G., 183
Weikl, T. R., 2–3
Weiner, D. B., 5
Weinman, E. J., 165–166, 168–169, 172
Weisel, J. W., 83–86, 89–91, 94–97, 99–100, 105, 108
Weissbach, L., 101
Weiss, M. S., 85
Weisz, O., 2
Weitz, J. I., 101
Welch, W. J., 68
Welsh, M. J., 68, 166–167
Weniger, M. A., 69
Wen, Q., 95–96
Werling, R. W., 80
Wesley, P. K., 224, 238–239, 253
Wessling-Resnick, M., 249
West, A. P. Jr., 248–250, 260
Westrick, L. G., 86
West, R. M., 107
Wheeler, C., 259
Whisstock, J. C., 253

Whitaker, A. N., 90
Whitby, F. G., 49
White, T. C., 101
Wichmann, H. E., 105
Wideman, C., 80
Wiek, I., 103
Wiesner, I., 14
Wiesnerova, D., 14
Wigley, D. B., 247
Wijmenga, C., 100
Wild, J., 5
Wiley, D. C., 226, 232, 235–236, 240, 244–245, 251, 258
Wilhelm, S. E., 103
Willcox, B. E., 225, 230, 236, 251, 254, 257, 261
Will, G., 2
Willie, S. T., 226–227, 230, 232, 239, 250, 257–258, 261
Wilson, I. A., 226–227, 230, 232, 234–237, 244–246, 251–254, 256, 259
Wilson, I. J., 99
Windisch, B., 212
Wingren, C., 226–227, 232, 235, 237, 246, 253, 259
Winter, C. C., 253
Witzmann, F. A., 18
Wolan, D. W., 250
Wolberg, A. S., 103–104
Wolfenstein-Todel, C., 100
Wolford, C., 69
Wolf-Watz, M., 186
Wolf, Y. I., 22
Wolthers, K. C., 25
Wolynes, P. G., 211
Wong, R. T., 23
Wong, W., 172
Woo, E. R., 6
Woolfson, D. N., 137
Wormald, M. R., 258–259
Wu, J., 238
Wunsch, C. D., 8
Wun, T. C., 80
Wust, T., 136
Wu, W. L., 24
Wyckoff, H. W., 2, 179
Wyer, J. R., 230, 236, 238, 251, 253–254

## X

Xing, L., 15, 17, 26
Xu, A. S., 104
Xu, W., 61
Xu, Y., 136

## Y

Yadav, S., 225–227, 230, 236, 239, 249, 257
Yakovlev, S., 81, 86, 101
Yamashita, R., 3
Yamazaki, K., 102
Yang, D., 135
Yang, L., 17, 19, 22
Yang, X., 8
Yang, Z., 81–82, 88
Yao, S., 209
Yap, A. S., 165
Yasuda, M., 62
Yau, S. S., 12, 18
Yawman, A., 5
Yeager, M., 138
Yee, V. C., 84
Yin, H., 131, 138, 154
Yip, C. W., 23
Yonemura, S., 171
Yoo, H. W., 54, 56, 61
Yorifuji, H., 84
Yuasa, I., 102
Yun, C. C., 169–170
Yu, S., 81
Yu, X., 8, 17, 19, 22
Yu, Y., 172

## Z

Zaccaï, G., 187, 196
Zaccai, N. R., 246
Zacharski, L. R., 80
Zachos, N. C., 165
Zaitsev, S., 107
Zajonc, D. M., 230, 234–235, 246, 253
Zalewski, J., 104, 106
Zamah, A. M., 170

Zarebski, L., 32
Zawilska, K., 107
Zeff, R. A., 227, 231–232, 240
Zeitlin, P. L., 68
Zekert, S. L., 76
Zemskov, E., 108
Zeng, L., 47, 49, 51, 58
Zeng,W., 166
Zeng, Z., 225–226, 230, 232, 234–235, 245–246, 248–249, 253, 256, 259
Zerva, L., 252
Zhang, J. Z., 102, 134, 136, 141–142, 146
Zhang, M., 185
Zhang, Q., 3
Zhang, S., 236
Zhan, S., 102
Zheng, X., 17, 19, 22
Zhou, J., 234
Zhou, M. M., 47, 49, 51, 58
Zhou, X. Y., 257
Zhu, B., 9
Zhu, H. C., 23
Zhu, W., 14, 16–17, 25, 28–29
Zimmerman, T. S., 80
Zimmer, R., 8
Zito, F. A., 97, 166, 170
Zmudka, K., 103
Zoonens, M., 131
Zuiderweg, E. R., 186
Zupan, J., 9, 11, 15–17
Zwaenepoel, I., 185
Zwick, E., 152

# SUBJECT INDEX

Note: The letters '$f$' and '$t$' following the locators refer to figures and tables respectively.

## A

Activated protein C (APC), 80
Akcasu–Gurol (AG) approach
    NSE data, 195
    rigid-bodies, 193–194
Allele specific alternative splicing (asAS), 102
Allosteric signal dynamic propagation
    domain motion, four-point model
        deuterated complex, 210–211
        multidomain proteins, 211
        NHERF1-FERM complex, 207–208
        NSE data, 209
        PDZ1 domain, 209
        PDZ1–PDZ2 oscillatory motion, 210
    nanoscale protein dynamics
        AG approach, 193–195
        causes, 189, 191
        coupling, 201–202
        diffusion constant, 197$f$
        mobility tensors, 194$f$, 198–201
        native-like topology, 196, 198
        neutron scattering, 191–192
        neutron-scattering techniques, 195
        NSE spectroscopy, 192–193, 195
        quasielastic-scattering intensity, 196
        rotational diffusion, 199–200, 202
        rotational mobility tensor, 198
        Rouse and Zimm model, 196
    nanoscale protein motion
        NHERF1 $^d$FERM and NHERF1 $^h$FERM, 206–207
        NHERF1-FERM complex, 204–205
        rigid-body calculation, NHERF1, 202, 203$f$
        SANS/SAXS data, 205–206
        unbound NHERF1, 203

NSE
    genomic sequence, 212
    nanoscales, protein motion, 213
    NHERF1, 212
    protein motion, 211
    protein dynamics, physical concepts
        creeping movements and underdamped oscillations, 188–189, 190$t$
        Young's moduli, 187
    transmission, 186–187
Atomistic models
    Bnip3 and ErbB2 free energy estimation, 142–144
    PMF, 142
    TM peptides, 144
Avian and swine flus, 24

## B

*Beta-globin* gene, 14
Bone marrow transplantation (BMT), 67–68

## C

Cell signaling
    allosteric signals, nanoscale protein motion
        biophysical techniques, 191–192
        causes, 189, 190
        dynamic propagation, 202–207
        four-point model, 207–211
        NSE, 211–213
        physical concepts, 187–189
        theoretical techniques, 192–202
        transmission, 186–187

Cell signaling (continued)
  autoinhibition and long-range allostery, NHERF1
    description, 178
    disease associated mutation, PDZ domains, 185–186
    domain–domain interactions, 181–183
    PDZ2, C-Terminal domain, 185
    PDZ domains, 183–185
    shape, 179–181
    structure and dynamic studies, 178–179
  proteins move
    CFTR, 164–165
    NHERF (see Na$^+$/H$^+$ exchange regulatory factor)
  transduction, allosteric scaffolding protein interactions
    allosteric modulation, NHERF1, 173–174
    CFTR, 172
    ezrin and ERM proteins, 171–172
    negative cooperativity and feedback loop, 174–178
    NHERF1, 171
CEP. See Congenital erythropoietic porphyria
CFTR. See Cystic fibrosis transmembrane conductance regulator
CG model. See Coarse-grained model
Coarse-grained (CG) model
  advantages and difficulties, 146
  bilayer lipid composition, 145–146
  CG MD simulations, 145
  monomeric and dimeric states, 144–145
Congenital erythropoietic porphyria (CEP)
  C73R UROIIIS, hotspot mutation
    expression, in vivo, 64–66
    structural characterization, in vitro, 62–64
  deficient function, 70
  molecular basis
    mutation, URO-synthase gene, 54–57
    pathogenic mutation, UROIIIS catalytic activity, 57–59
    UROIIIS destabilization, pathogenic mutants, 59–62
  treatment
    marrow suppression, BMT and gene therapy, 67–68
    proteasome molecular chaperones and inhibitors, 68–69
    UROIIIS mutants activity value, 58$t$
  C73R UROIIIS mutation
    expression, in vivo
      intracellular degradation, 66
      M1 cell lines, 64–65
      qRT-PCR, 66
    structural characterization
      A69G and L43V, 64
      far-UV region, 63$f$
      model, 63–64
      tryptophan fluorescence, 62
Cystic fibrosis transmembrane conductance regulator (CFTR)
  C-CFTR, 174
  NHERF and ezrin, 172
  and NHERF1 interaction, 167–168
  PDZ1 domain, 174–177

**D**

Dimerization motifs, TM helices
  "glycophorin-like", 134–135
  sequence-based prediction, 135
DNA sequences
  2D representation, 11–13
  numerical characterization
    2D graph, 13
    3D graph, 13–14
  and RNA, 10

**F**

Fibrin clot structure, thrombosis and vascular disease
  cells interaction and wound healing
    angiogenesis, 108–109
    αIIbβIII integrin, 107–108
    αVβ3 integrin, 108
  clot formation
    cross-linking by factor XIIIa, 89–91
    fibrinopeptides cleavage, 86–88

molecule polymerization into fibers,
    88–89
coagulation system
  antithrombin, 79–80
  APC, 80
  extrinsic and intrinsic pathway, 77, 78f
  FX and prothrombin activation
      factors, 77
  natural inhibitors, 79f
  TFPI, 80–81
  thrombin generation, 78–79
dysfibrinogenemia, 106
elastic properties
  branch-points, 95–96
  factors, 96
  protofibril and strain-hardening, 95
  refolding fibrin domains, 94–95
environmental factors
  air pollution, 105
  glycation, 103
  homocysteine, 102–103
  lysine residue, 104
  oxidative stress, 104
factor XIII (FXIII)
  A- and B-subunits, 84
  activation peptide cleavage, 84–85
  proteins cross-linkage, 85–86
fibrinogen
  biantennary carbohydrate clusters, 83
  calcium, role, 82–83
  αC domains, 83–84
  chromosome 4q23-32, 81
  crystal, 81–82
  molecule, 82f
fibrinolysis (see Fibrinolysis)
heterogeneity in coagulation
  asAS, 102
  dysfibrinogenemias, 96–97
  fibrinogen 420, 101–102
  fibrinogen γ', 100–101
  FXIIIB phenotypes, 102
  polymorphism (see Fibrinogen
      polymorphism)
thrombosis, 105–107
Fibrinogen polymorphism
  BclI variation, 97, 98t

*FXIIIA* gene, 99–100
*FXIIIB gene*, 100
–455G/A variation, 97–99
Fibrinolysis
  alpha 2-AP, 93
  PAI-1 and PAI-2, 93
  plasmin, tPA and uPA, 91–93
  SERPINS, 91
  TAFI, 93–94
*FXIIIA* gene polymorphism, 99–100
*FXIIIB gene* polymorphism, 100

# G

Graphical representation and numerical
    characterization (GRANCH)
  avian and swine flu viruses, 24
  bioinformatics, proteins, 7–9
  coronavirus and flu viruses
    phylogenetic studies, 28–29
    similarity/dissimilarity analyses,
        29–30
  2D method, 25–26
  20D method, 26–28
  DNA sequence
    applications, 14–15
    graphical representation, 10–13
    H5N1 neuraminidase RNA, 12f
    human beta-globin (HUMHBB)
        CGR, 11f
    numerical characterization, 13–14
  dot-matrix graph, 10f
  drugs and proteins, 5–7
  peptide stretch identification, 30–32
  protein basics
    biological function, 3–4
    description, 2
    proteome, 2–3
  protein sequences
    graphical representation, 15–17
    numerical characterization, 17–19
    similarities/dissimilarities and
        phylogeny, 19–21
  SARS coronavirus, 24–25
  viral protein features, 21–23

## H

Helix packing
  conformational search, 136–137
  surface properties complementarity
    3D structure predictions, 140t
    "knob-to-holes", 137
    PREDDIMER algorithm, 139–140
    scoring function, 137–138
    standard prediction approaches, 140–141
    SVM algorithm, 138–139
Heme group-derivatized cofactors, 45
Hereditary hemochromatosis protein (HFE), 225, 248
HFE. See Hereditary hemochromatosis protein
HLA. See Human leukocyte antigen
H5N1 avian flu, 24
H1N1 swine flu, 24
Human heme group biosynthetic pathway, 45f
Human leukocyte antigen (HLA), 224–225

## I

Ischemic stroke, 106–107

## K

Kappa-opioid receptor, 170

## L

Ligand-binding groove
  antigen-binding groove, 244
  CD1, 245–246
  cell-surface receptor, 250
  FcRn, 247–248
  γδ-TCR ligands T10 and T22, 246–247
  HFE and MHC-I similarities, 248
  molecular surface comparison, 243, 246f
  ZAG, 249–250

Ligand-binding residues
  CD1, 234–235
  contact zones, 233–234
  FcRn, 232–233
  immune function, 234
  MIC -A and -B, 236
  nonclassical molecules, T10 and T22, 235
  types, 232
  ZAG, 236
Light chain binding
  residues
    α3 domain and β⊂2M, 238
    human ZAG, 239
  structural basis
    CD1 molecule, 255–256
    class I molecules vs. FcRn, 256
    MHC-I proteins, 255
    MIC-A, 257–258
    quaternary arrangement, 256
    ZAG α3 quaternary arrangement, 257

## M

Major histocompatibility complex I (MHC-I)
  Class I and Class I-related proteins
    features, 225, 226t
    in Protein Data Bank, 227–230t
  description, 224
  glycosyalation
    cellular immune system, 258
    murine CD1 sequence, 259
    N-linked sites, 260–261
    protein sequence, human HFE, 260
    structure and biochemical analysis, 259
  HFE, 225
  HLA, 224–225
  sequence analysis
    description, 228, 230–231
    ligand-binding residues, 232–236
    light chain-binding residues, 238–239
    multiple sequence alignment, 231f
    TCR-binding residues, 236–238
  structure analysis

α1 and α2 domains, 240, 242f
classical and nonclassical
determination, 239–240
comparison by RMSD, 240, 241t
crystallographic studies, 240
ligand-binding groove, 244–250
light chain Binding, 255–258
proteins and αβ-TCR anatomy,
233, 240f
TCR binding, 251–254
TCR, 224
Membrane proteins (MPs) oligomerization,
130–131
MHC class I chain related (MIC-A) proteins
crystal structure, 250
and ZAG, 227
MHC-I. See Major histocompatibility
complex I
Molecular hydrophobicity potential (MHP)
approach, 139
Monte Carlo (MC) search, 136

# N

$Na^+/H^+$ exchange regulatory factor 1
(NHERF1)
allosteric modulation, membrane
complexes
C-CFTR binding, 175f
EB, 174
F-actin filament, 173
FERM binding, 176f
molecular mechanism, ezrin, 173–174
autoinhibition, C-terminal domain, 185
CFTR
fluorescence photobleaching recovery
and particle tracking, 168
polarised expression, 167–168
control, domain–domain interactions
PDZ2 and CT, 181–183
PDZ1 and PDZ2, 183
SANS, 181, 182t
disease associated mutation, PDZ
domains, 185–186
feature, 172–173

function, 166–167
human, domain structure, 166t
long range interdomain allosteric
communications
CT domain, 180–181
SAXS, 179–180
NaPiT2a
mutations, 169
phosphate ions reabsorption, kidney,
168–169
negative cooperativity and feedback loop
allosteric regulation, 177–178
C-CFTR, 174, 177
domain-domain allosteric, 177
PDZ domains, 177
PDZ domains
C-terminal helical extension, 184–185
PDZ1 and PDZ2, 183
PDZ2CT, 183–184
PDZ2 structure, 184t
podocalyxin complexes, 169
tyrosine kinase receptor complexes, 170
$Na^+/H^+$ exchange regulatory factor
(NHERF)
description, 165
NHERF1
CFTR, 167–168
feature, 172–173
function, 166–167
human, domain structure, 166t
NaPiT2a, 168–169
PDZ domains, 165–166
podocalyxin complexes, 169
tyrosine kinase receptor complexes, 170
Neutron spin echo (NSE) spectroscopy
AG formula, 195
and dummy atom rigid-body
calculations, 205
genomic sequence, 212
intermediate scattering function, 192
multicomponent contrast-matching
experiment, 191–192
NHERF1, 212
PDZ1 and PDZ2, 204f
protein motion, 211
SANS, 205

Neutron spin echo (NSE) spectroscopy (*continued*)
  unbound NHERF1, 202–203
NHERF1. *See* Na$^+$/H$^+$ exchange regulatory factor 1
NMR model. *See* Nuclear magnetic resonance model
Nuclear magnetic resonance (NMR) model
  Bnip3 dimer, 148
  comparison, 150–151
  GpA structure prediction, 138

## P

Parathyroid hormone receptor (PTH1R), 170
PBG. *See* Porphobilinogen
PBGD. *See* Porphobilinogen deaminase
Peptide-based drugs, 6
Porphobilinogen (PBG)
  cyclization, 47
  synthase mechanism, 45–46
Porphobilinogen deaminase (PBGD)
  hydroxymethylbilane (HMB), 46
  UROIIIS
    metabolon, 51–52
    preuroporphyrinogen, 57
Proteins
  bioinformatics
    characterization and *in silico*, 8–9
    complexities, 7
    genomic and sequence data, 7–8
    GRANCH, 9
  pharmaceutical, 5–6
  sequences
    amino acids, 19–20
    2D lattice graphs, 16–17
    DNA and RNA, 15–16
    magic circle, 16$f$
    numerical characterization, 17–19
    phylogenetic trees, 21$f$
    zig-zag curve, 17$f$
  similarities/dissimilarities and phylogeny
    amino acids, 19–20
    3D graph, 20$f$, 22
    evolutionary relationships, 19
    viral (*see* Viral proteins)
Proteins data bases (PDB), 3
PTH1R. *See* Parathyroid hormone receptor

## R

ROSETTA energy function, 138
Rouse model, 196
R153Q and E225K mutation, 185–186

## S

SARS coronavirus, 24–25
Site-directed mutagenesis techniques, 4
Support vector machine (SVM) algorithm, 138–139

## T

T cell receptor (TCR) binding
  residues
    antigen presentation, 236
    CD1 isotypes, 237
    classical function, 238
    "functional hot spot", 237
    T22$^b$ *vs.* MHC-Ia, 237–238
  structural basis
    CDR loops, 251–252
    MIC-A and -B, 254
    pMHC complex, 251
    TfR binding, 254
TfR. *See* Transferrin receptor
Tissue factor pathway inhibitor (TFPI), 80–81
TM helices. *See* Transmembrane helices
TM peptides-interceptors, 131
ToxR assay, 148
Transferrin receptor (TfR)
  binding, 254
  residues, 234

Transmembrane (TM) helices
　computational methods, limitation
　　conformational space, 149
　　protein interactions, 149–150
　　self-assembling, 148–149
　helix-helix association
　　atomistic model, 142–144
　　CG model, 144–146
　　implicit membrane, 141–142
　HH dimer structure prediction
　　CG MD simulation, 133–134
　　dimerization motifs, 134–135
　　packing (*see* Helix packing)
　modeling and experimental techniques
　　NMR, 146–148
　　ToxR assay, 148
　NMR model comparison, 150–151
　pharmacological targets, 152
　polytopic MPs 3$^D$ structure, 152–153
　structure
　　prediction, 151
　　and thermodynamics, function and design, 153–154

## U

UROIIIS. *See* Uroporphyrinogen III synthase
Uroporphyrinogen III synthase (UROIIIS)
　catalytic mechanism
　　abiotic condensation, 47–48
　　cyclization, 48*f*
　　human, three-dimentional structure, 48–50
　　preuroporphyrinogen enzymatic conversion, 49*f*
　　structural data, 50–51
　CEP-producing mutation
　　C73R, 55–56
　　genotype/phenotype analysis, 56*f*
　　genotyping, 54
　　recessive characteristic, 56–57
　　single point, 54–55
　destabilization, pathogenic mutants
　　genotype/phenotype relationship, 60–62

　　rESA values, 59–60
　　unfolding rate, 61*f*
　　wild type, 60, 61
　human, ribbon representation, 50*f*
　interdomain flexibility, catalytic reaction
　　PBGD, 51–52
　　"venus flytrap mechanism", 51
　main function, 69–70
　pathogenic mutation, catalytic activity
　　"expressed enzymatic activity", 59
　　HPLC separation, 59*f*
　　preuroporphyrinogen, 57–58
　　residues, 58–59
　　SDS gel electrophoresis, 57
　thermodynamic stability
　　Arrhenius model, 53–54
　　circular dichroism, 52
　　kinetically stable proteins, 52–53

## V

"Venus flytrap mechanism", 51
Viral proteins
　avian and swine flus, 24
　coronavirus and flu viruses
　　peptide stretch identification, 30–32
　　phylogenetic tree, 28–29
　　similarity/dissimilarity analyses, 29–30
　2D method, 25–26
　20D method
　　amino acids axis assignment, 27*t*
　　DNA and protein sequences, 26–28
　　numerical characterization, 28
　features
　　disease prevention and treatment, 23
　　properties, 21–22
　SARS coronavirus, 24–25

## W

Weighted histogram analysis method (WHAM), 142
Western blot analysis, 65*f*

## Z

ZAG. *See* Zinc α2-glycoprotein
Zimm model, 196
Zinc α2-glycoprotein (ZAG)
    ligand binding
        sequence analysis, 236
        structure analysis, 249–250
    light chain binding
        human, 239
        quaternary arrangement, 257

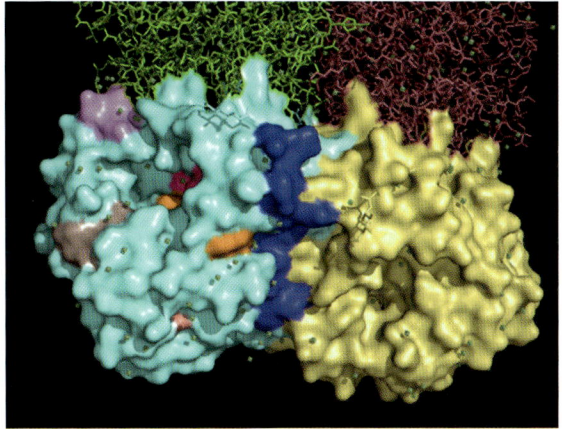

AMBARNIL GHOSH AND ASHESH NANDY, CHAPTER 1, FIG. 10. Conserved surface exposed regions are shown in different colors in the cyan colored monomer of neuraminidase (other monomers are colored in magenta, green, and yellow). Here the six conserved regions are shown in six different colors. The conserved C-terminal portion is shown in blue.

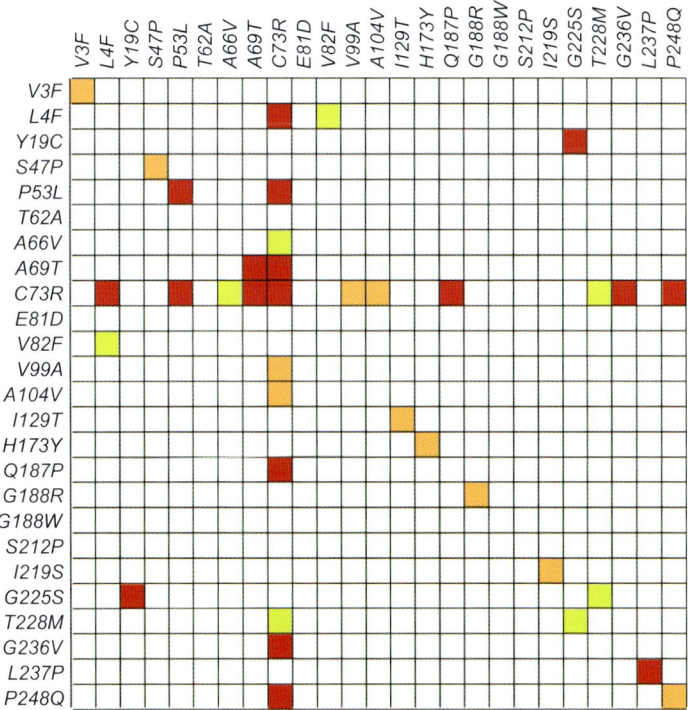

AROLA FORTIAN ET AL., CHAPTER 2, FIG. 7. Genotype/phenotype analysis of the studied cases of congenital erythropoietic porphyria. Mild, moderate, and severe phenotypes are colored in yellow, orange, and red, respectively. Patients with different phenotypes but sharing genotype are represented by triangles.

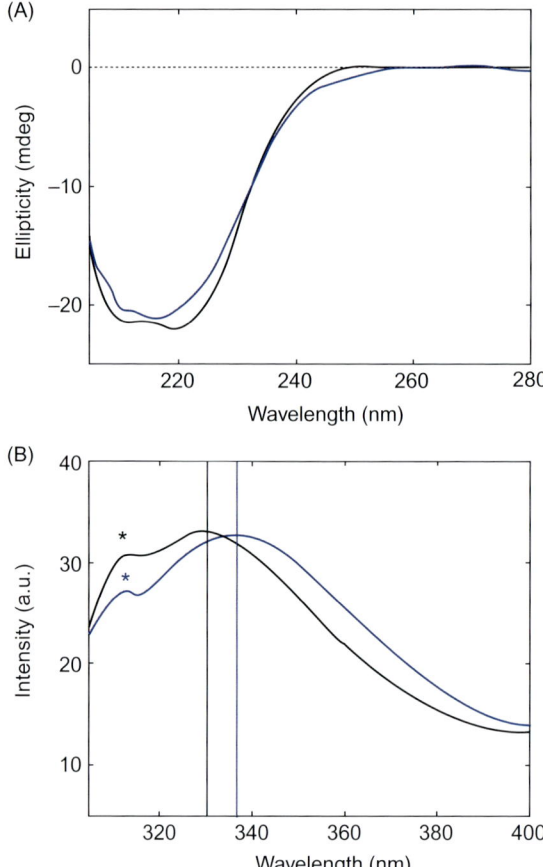

AROLA FORTIAN ET AL., CHAPTER 2, FIG. 10. (A) Far-UV region of the circular dichroism (CD) spectra for wild type (black line) or C73R (blue line) UROIIIS. (B) Tryptophan emission fluorescence spectra for wild type (black line) or C73R (blue line) UROIIIS. The position of the maximum in each spectrum is highlighted by the vertical lines. The peaks labeled with an asterisk correspond to the Raman spectrum of water.

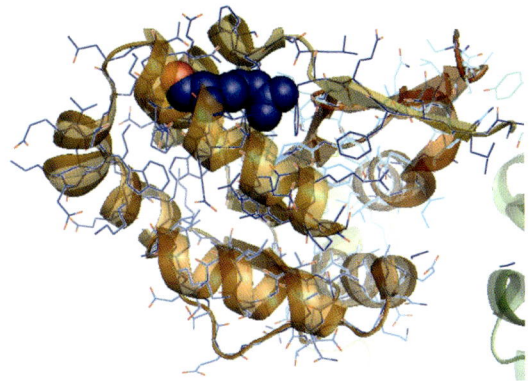

AROLA FORTIAN ET AL., CHAPTER 2, FIG. 11. Overlay of wild-type UROIIIS (1jr2, light brown) and the modeled structure for C73R (dark brown). The side chains for C73 and R73 are represented by red and blue spheres, respectively.

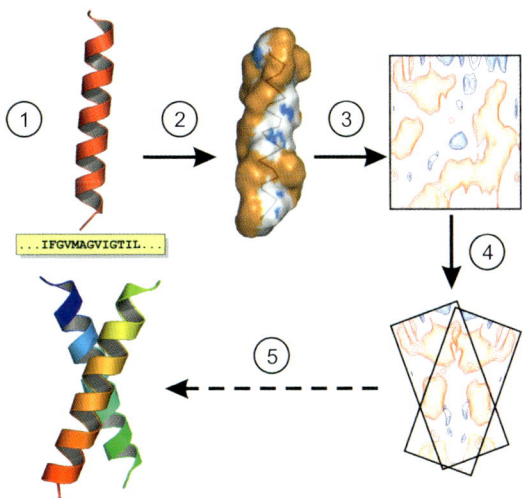

ANTON A. POLYANSKY ET AL., CHAPTER 4, FIG. 1. PREDDIMER—a computational algorithm for prediction of 3D models of transmembrane (TM) helical dimers. The method is based on alignment of the peptides' surfaces in order to achieve the best complementarity of hydrophobic (molecular hydrophobicity potential, MHP) and landscape properties (expressed in terms of 2D maps) of the monomers. Key steps are numbered: 1, building ideal helices from their TM-sequences; 2, calculation and mapping of helices' surfaces (MHP+landscape); 3, projection of surface properties onto the cylinder and building of 2D maps; 4, mutual superimposition of the maps by varying rotation and tilt angles of the helices, and their pairwise comparison using scoring functions; 5, reconstruction of 3D structure of the dimer. Solvent accessible surface of a helix as well as 2D maps are colored according to MHP values. Orange, white, and blue regions correspond to hydrophobic, neutral, and hydrophilic surface patches, respectively.

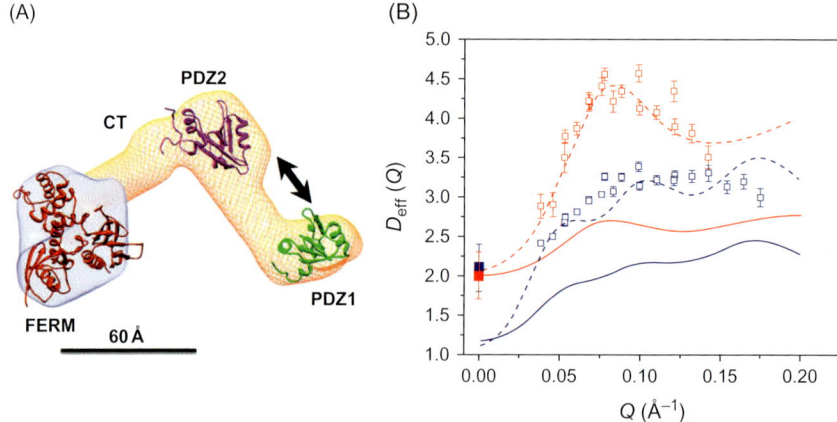

ZIMEI BU AND DAVID J. E. CALLAWAY, CHAPTER 5, FIG. 11. (A) A model representing domain motion between PDZ1 and PDZ2 in the complex. The 3D shape of the complex is reconstructed from SANS (Li et al., 2009). The known high-resolution structure fragments of PDZ1, PDZ2, and ezrin FERM domain (PDB code: 1NI2) are docked into the envelope using UCSF chimera (Pettersen et al., 2004). The arrows represent translational motion between PDZ1 and PDZ2. A length scale bar of 60 Å is shown. (B) Comparing experimental $D_{\text{eff}}(Q)$ of NHERF1·FERM with calculations incorporating interdomain motion between PDZ1 and PDZ2. The calculations were performed using the coordinates of the docked high-resolution structures of different domains. The calculations were performed with only one adjustable parameter $H^T$ for the domain motion model of both the deuterated and the hydrogenated complexes. Open blue square is the NSE data from hydrogenated complex. Open red square is NSE data from hydrogenated NHERF1 in complex with deuterated FERM domain. Solid blue square and solid red square are the self-diffusion constants $D_0$ of the hydrogenated and the deuterated complexes, respectively, obtained from PFG NMR. The calculated curves are shown for NHERF1·$^d$FERM (solid red line for rigid-body model and dashed red line for model incorporating domain motion) and NHERF1·$^h$FERM (solid blue line for rigid-body model and dashed blue line for incorporating domain motion). The comparisons suggest that deuteration of the FERM domain amplifies the effects of protein internal motions detected by NSE.

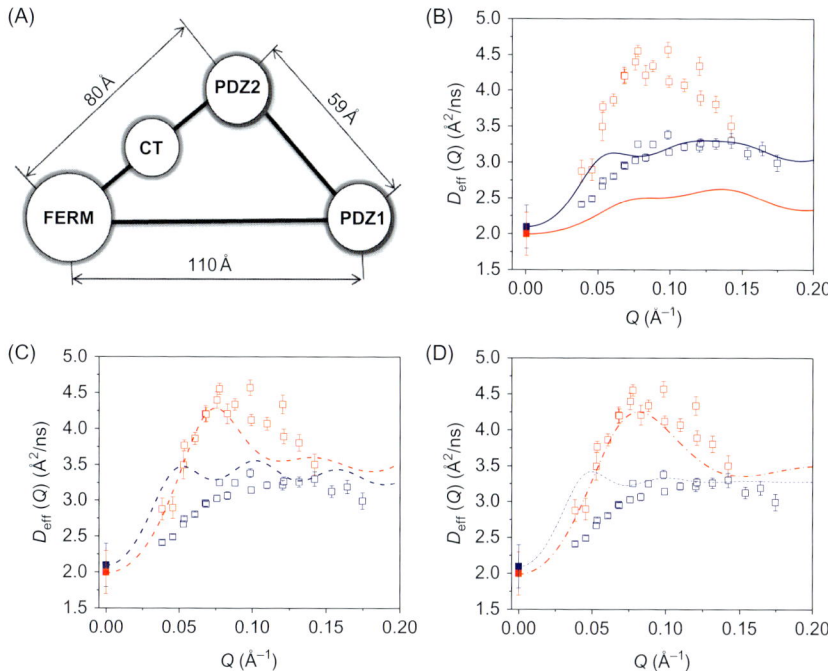

ZIMEI BU AND DAVID J. E. CALLAWAY, CHAPTER 5, FIG. 12. A simple four-point model can well describe domain motion in the complex. (A) The four-point model represents the NHERF1·FERM complex, with the centers of PDZ1, PDZ2, CT, and FERM domain taken from Fig. 3A. (B) Comparing the experimental NSE data with the four-point rigid-body calculations for NHERF1·$^h$FERM (blue open squares are the experimental data and blue solid line is the calculated data) and for NHERF1·$^d$FERM (red open squares are experimental data and red solid line is the calculated data). $D_0$ of NHERF1·$^d$FERM (solid red squares) and NHERF1·$^d$FERM (solid blue squares) from PFG NMR are shown. (C) Comparing the experimental data with calculations assuming interdomain motion between PDZ1 and PDZ2 in NHERF1·$^d$FERM (red dash line) and NHERF1·$^h$FERM (blue dash line). The experimental symbols are the same as in (B). (D) Comparing the experimental data with calculations incorporating interdomain motion between PDZ1 and PDZ2, as well as assuming finite size form factor of spheres of 20 Å radius for the FERM domain and for both PDZ domains in NHERF1·$^d$FERM (red dash dot line) and in NHERF1·$^h$FERM (blue dash dot line).

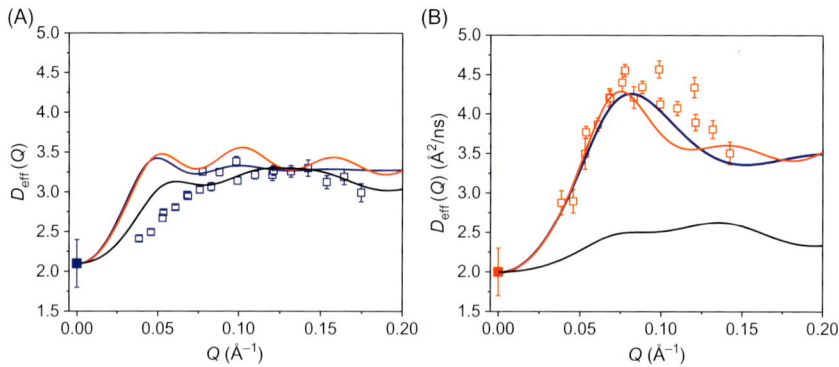

ZIMEI BU AND DAVID J. E. CALLAWAY, CHAPTER 5, FIG. 13. For the hydrogenated NHERF1·$^h$FERM complex, the difference in $D_{eff}(Q)$ between the rigid-body model and domain-motion models is very small, but is significantly increased in the deuterated complex. (A) Comparing the rigid-body calculation with the domain-motion calculation in the four-point model in the hydrogenated NHERF1·$^h$FERM complex. NSE data from the NHERF1·$^h$FERM (blue open squares), the four-point rigid-body model (black line), four-point model incorporating domain motion between PDZ1 and PDZ2 (red line), four-point model incorporating domain motion between PDZ1 and PDZ2 and finite size form factor of 20 Å radius for the FERM domain, PDZ1 and PDZ2 (blue line). $D_0$ at $Q=0$ Å$^{-1}$ as measured from PFG NMR is shown in blue solid square. (B) Comparing the rigid-body calculation with the domain-motion calculation in the four-point model in the deuterated NHERF1·$^d$FERM complex. NSE data from the NHERF1·$^d$FERM (red open squares), the four-point rigid-body model (black line), four-point model incorporating domain motion between PDZ1 and PDZ2 (red line), four-point model incorporating domain motion between PDZ1 and PDZ2 and finite size form factor of 20 Å radius for the FERM domain, PDZ1 and PDZ2 (blue line). $D_0$ at $Q=0$ Å$^{-1}$ as measured from PFG NMR is shown in red solid square.

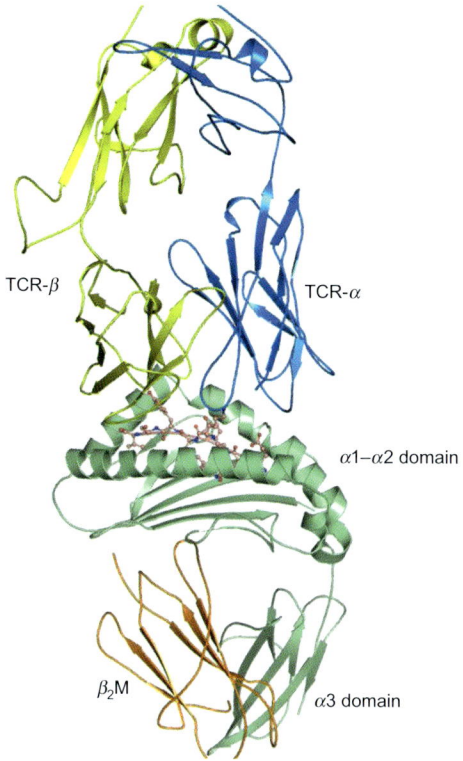

Md. Imtaiyaz Hassan and Faizan Ahmad, Chapter 6, Fig. 2. Architecture of MHC-like molecules showing anatomy of complex formed between MHC class I proteins and αβ-TCR. Class I molecules consist of a heavy chain (light green) and a light β$_2$M chain (orange). The peptide-binding site is formed exclusively by elements of the heavy chain formed by α1α2 platform, whereas the α3 domain is mainly responsible for β$_2$M binding. A view from the TCR binding onto the class I peptide-binding groove where peptide is shown in ball and stick model (reddish). The α (sky blue) and β (yellow) chains of TCR are shown in ribbon diagram. The structure was drawn in PyMOL using the atomic coordinates of pMHC–TCR complex with PDB code 2CKB (Garcia et al., 1998).

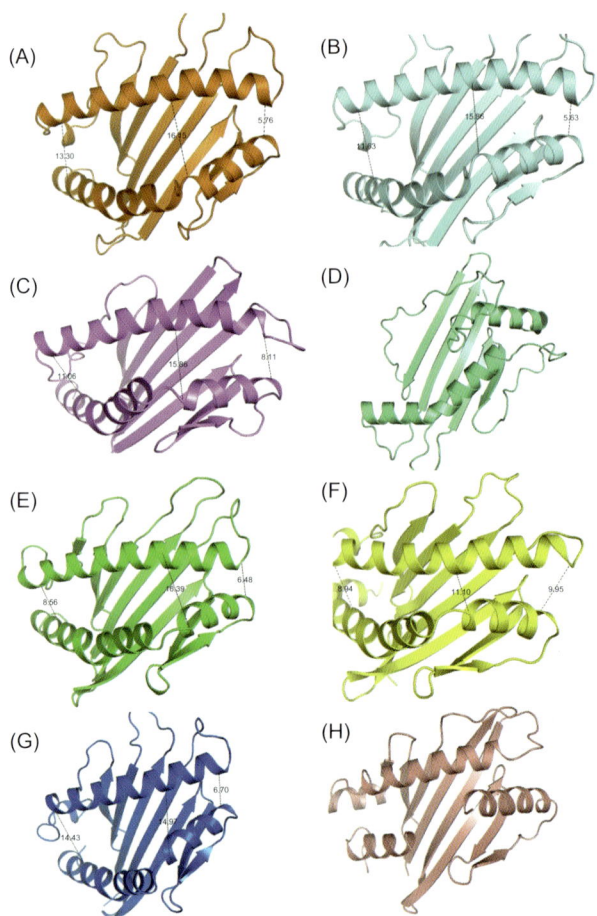

Md. Imtaiyaz Hassan and Faizan Ahmad, Chapter 6, Fig. 3. Cartoon diagram showing top view of the ligand binding α1 and α2 domains of various MHC-like molecules. (A) HLA-A2 is involved in the presentation of peptides (orange); (B) HLA-B27 also binds to peptides (cyan); (C) a hydrophobic binding groove of the CD1 (pink); (D) the nonclassical MHC molecule T22, which is a γδ-T cell ligand (light green); (E) Fc transported molecules FcRn (green); (F) TfR-binding protein HFE (yellow); (G) a view of the α1 and α2 domains of ZAG showing the proposed site for fatty acid binding (sky blue); and (H) the α1 and α2 domains of nonclassical MHC-like molecule MICA showing that its groove is structurally analogous to a class I molecule. The distance between the helices was calculated in PyMol. The structure was drawn using atomic coordinates of PDB code, 2CLR (HLA_A2), 3HLA (HLA_B27), 1C16 (T22), 1A6Z (HFE), 1ZAG (ZAG), 3FRU (FcRn), 1B3J (MIC_A), and 1CD1 (CD1).

Armstrong

A3

—————
v.83